RECENT ADVANCES IN FRESH-WATER BIOLOGY

VOL. II

Prof. K. S. Rao
School of Studies in Zoology
Vikram University
Ujjain (M.P.)

ANMOL PUBLICATIONS PVT. LTD.
NEW DELHI-110 002 (INDIA)

ANMOL PUBLICATIONS PVT. LTD.
4374/4B, Ansari Road, Daryaganj
New Delhi-110 002

Recent Advances in Fresh-Water Biology

First Edition 1997

ISBN 81-7041-753-8 (Set)

PRINTED IN INDIA

Published by J. L. Kumar for Anmol Publications Pvt. Ltd., New Delhi-110 002 and Printed at Mehra Offset Press, Delhi.

INTRODUCTION

The available fresh-water on the biosphere is least when compared to that of the oceans. Despite the small amount of available fresh-water this is of utmost importance to man and the fresh water biota as a whole. Little attention has been given to the complex ecological relations of the fresh-water organisms. An attempt has been made in this book to present our knowledge of fresh-water organisms so far ever compiled. A thorough treatment of this group has been attempted involving the published information in the form of various research papers from different area of the biosphere.

The present title is designed to approach the morphology, anatomy and physiology of fresh-water organisms. It intends to enable the student through his own natural appetency to correlate form and function so that he possesses for subsequent work in either anatomy of physiology or foundation for a real edifice of knowledge.

I express my thanks to my friends and colleagues whose continuous inspiration have initiated me in bringing out this book.

It is a pleasure to acknowledge the unstinting co-operation of the publishers during the publication of this book.

Of course, any shortcomings and errors are entirely my own. I hope that the readers of this book will be kind enough and send me their comments and criticisms.

School of Studies in Zoology
Vikram University
Ujjain (M.P.)

Dr. K.S. Rao

CONTENTS

1

LIMNOLOGICAL STUDIES ON KUNJWANI POND, JAMMU I. PHYSICO-CHEMICAL FEATURES AND ITS SCOPE TO FISH CULTURE

S. Kumar

Department of Bio-Sciences,
University of Jammu, Jammu, 180004, India.

Abstract—A detailed study of the physico-chemical characteristics of a freshwater pond at Kunjwani in Jammu were conducted at four study sites during April 1989 to March 1990. A direct relationship of pH with carbonate and an inverse relationship of pH, dissolved oxygen and carbonate with free carbon-dioxide as recorded during the course of present work. In order to investigate the relative importance and degree of significant relationship among physico-chemical variables the coefficient of correlation were derived.

INTRODUCTION

Jammu and Kashmir is an extensive hilly area, bestowed with lotic and lentic waters. In Jammu province of J and K state are present large number of seasonal and perennial ponds. These varied from cold water, winter frozen, high altitude to warm water ponds of low altitude in the plains. Hitherto, most of these lentic water bodies have been used mainly for drinking, bathing, washing, etc., and a few of these ponds are used for fish culture. In view of the lack of scientific information on the ponds of Jammu province, present limnological investigations of a pond at Kunjwani (32° 42′ 21″ N, 74° 52′ 45″ E at 312 m. a MSL) in Jammu were undertaken to record seasonal fluctuations in physico-chemical parameters.

The Kunjwani pond is a shallow, rainfed and perennial natural pond. Additionally, it also receives tap water and overflow from the water pumping station. Bottom of the pond is soft, rich in macrophytic detritus. There are scattered depressions at its bottom on the sides due to mud removal by local inhabitants. Fish survey has indicated the presence of *Channa punctatus* only. Dabchick (*Podiceps ruficollis*), common Kingfisher (*Haleycycon symrnensis*) and little egret (*Egretta gazetta*) also visit the pond during winter.

MATERIALS AND METHODS

Water samples for the present investigation were collected in plastic canes from five spots of each study sites. At the time of water sampling, air and water temperature (mercury bulb °C thermometer), depth (graduated meter-rod) and transparency (Secchi disc) were also recorded. For dissolved oxygen, however, water samples were collected in separate BOD bottles and fixed at spot. Chemical analysis of water was done according to the Standard Methods for the Examination of water and waste water (APHA, 1975).

RESULTS AND DISCUSSION

Monthly mean variations of certain physico-chemical characteristics are shown in the Fig. 1.1. Annual average, standard deviation and range of these variables are given in Table I. Coefficient of correlation in the form of correlation matrix are presented in Tables II-V. Significant values are indicated by asterisk. The results are discussed as the mean of four study sites.

Temperature

Temperature is one of the most important physical factor which regulates natural process within the environment. Air and surface water temperature varied between 14.13 °C—39.05 °C and 13.93 °C—34.5 °C respectively. Due to shallowness of experimental sites, the water temperature has shown a tendency to follow closely the atmospheric temperature. This type of observation for shallow water bodies is in conformity with the earlier findings of Reid and Wood (1976), Singhal *et. al.* (1985) and Malhotra *et al.* (1986-87). Increased solar radiation due to comparatively longer day length may explain gradual

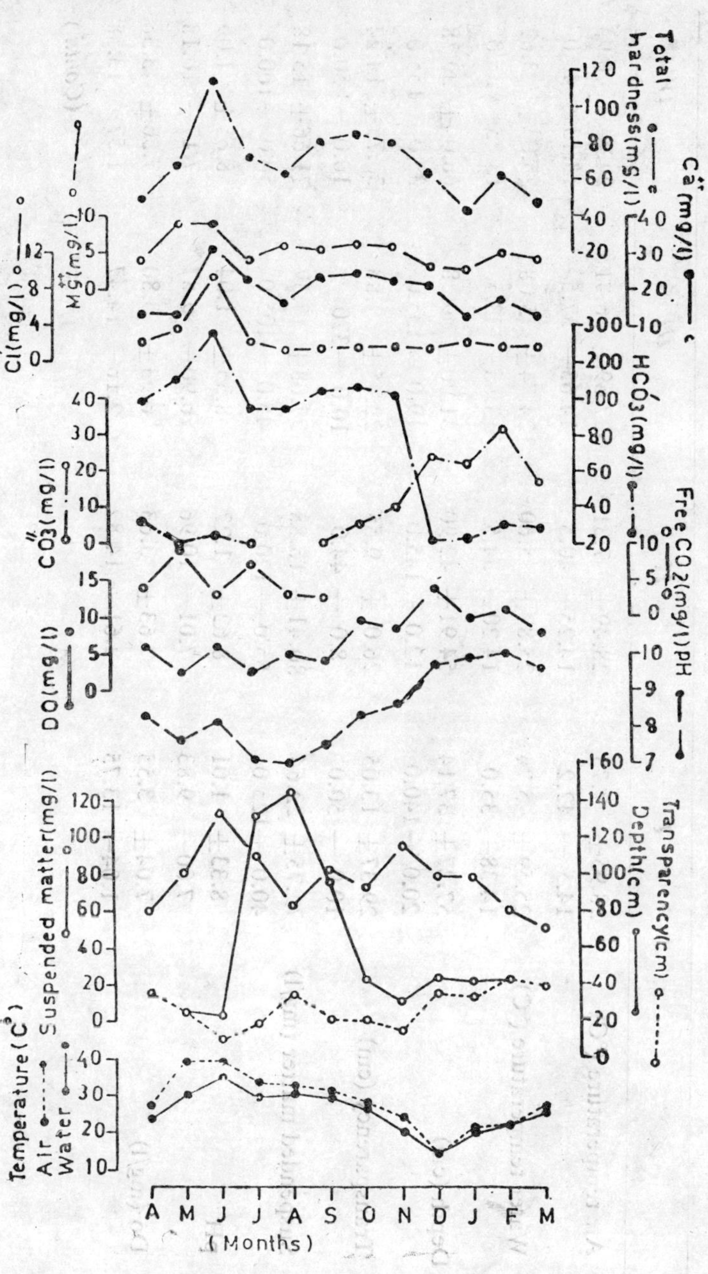

Fig. 1.1. Monthly mean variations in physico-chemical parameters in Kunjwani Pond, Jammu (April 1989 to March 1990).

Table I. Annual average, standard deviation and range of certain physico-chemical characteristics at different Study sites.

	I	*II*	*III*	*IV*
Air temperature (°C)	28.95± 7.75	28.49± 7.31	27.69± 6.41	27.24± 6.95
	14.5 — 42.2	14.25— 40.5	14.63— 37.5	13.13— 40.0
Water temperature (°C)	25.59± 5.74	25.82± 5.60	24.74± 5.08	25.0 ± 3.62
	14.38— 35.0	14.20— 34.5	13.88— 33.5	13.25— 35.0
Depth (cm)	57.27± 37.14	54.91± 42.90	51.20± 39.28	66.0 ± 39.78
	20.0 —140.0	12.0 —145.0	19.0 —135.0	35.0 —155.0
Transparency (cm)	29.37± 13.05	26.0 ± 9.55	22.45± 7.88	30.75± 14.24
	10.0 — 50.0	8.0 — 44.0	10.0 —370	10.0 — 50.0
Suspended matter (mg/l)	79.75± 22.64	80.41± 16.88	77.08± 17.37	71.66± 15.18
	40.0 —125.0	55.0 —120.0	45.0 —105.0	50.0 —100.0
pH	8.33± 1.01	8.62± 1.27	8.35± 1.04	8.65± 1.09
	7.00— 9.83	7.01— 10.96	6.90— 9.87	7.17— 10.18
Do (mg/l)	7.04± 3.55	7.63± 3.63	6.93± 3.80	7.56± 3.36
	1.04— 13.75	2.61— 14.82	2.16— 14.47	1.57— 13.58

(*Contd.*)

	I	*II*	*III*	*IV*
FCO_2 (mg/l)	3.29± 6.09	2.06± 2.82	2.58± 3.09	1.45± 1.80
	0.0 — 21.75	0.0 — 10.07	0.0 — 9.66	0.0 — 4.82
CO_3^- (mg/l)	10.22± 10.76	10.61± 13.75	8.26± 10.79	10.25± 11.57
	0.0 — 25.52	0.0 — 38.03	0.0 — 32.70	0.0 — 33.36
HCO_3^- (mg/l)	88.78± 61.46	103.35± 77.98	104.44± 76.9[illegible]	93.97± 67.91
	21.82—246.78	10.85—296.62	21.35—291.66	14.54—277.17
Cl^- (mg/l)	2.59± 2.47	2.36± 1.72	2.73± 2.53	2.53± 2.25
	1.25— 10.56	1.43— 7.61	1.28— 10.74	1.20— 9.70
Ca^{++} (mg/l)	18.59± 7.68	18.92± 5.25	20.81± 4.87	18.37± 5.97
	7.68— 36.50	12.10— 28.18	12.57— 29.57	10.86— 30.95
Mg^{++} (mg/l)	4.81± 1.92	6.33± 3.12	4.93± 2.87	4.79± 2.11
	1.46— 7.28	2.41— 14.69	0.67— 9.91	2.17— 9.10
Total (mg/l) hardness	61.28± 24.15	70.25± 22.06	60.02± 23.96	61.34± 24.12
	18.66— 92.17	34.70—108.30	22.72—111.18	25.64—111.18

Table II. Correlation matrix for Kunjwani Pond, Jammu at Study Side I.

	AT	*WT*	*D*	*TR*	*SM*	*pH*	*DO*	FCO_2	CO_3^{--}	HCO_3^-	*Cl–*	Ca^{++}	Mg^{++}	*TH*
AT	1.00	0.950	0.171	—0.45	0.036	—0.80	—0.84	0.618	—0.715	0.786	—0.570	0.34	0.29	0.414
WT		1.00	0.31	—0.34	0.367	—0.795	—0.812	0.449	0.736	0.759	0.552	0.362	0.296	0.42
D			1.00	—0.06	0.10	0.598	—0.398	0.056	—0.492	—0.026	—0.295	0.040	0.113	0.210
TR				1.00	—0.873	0.43	0.273	—0.051	0.289	—0.754	—0.478	—0.799	—0.637	—0.735
SM					1.00	—0.721	—0.330	0.175	—0.094	0.648	0.643	0.628	0.425	0.430
pH						1.00	0.708	—0.576	0.842	—0.597	—0.121	—0.204	—0.251	—0.361
DO							1.00	—0.632	0.753	—0.468	—0.169	—0.022	—0.120	—0.247
FCO_2								1.00	—	0.158	0.03	—0.36	—0.219	—0.247
CO_3^{--}									1.00	—0.523	—0.083	—0.082	0.061	0.027
HCO_3^-										1.00	0.784	0.756	0.519	0.621
Cl–											1.00	0.647	0.371	0.358
Ca^{++}												1.00	0.687	0.823
Mg^{++}													1.00	0.896
TH														1.00

Asteric indiate significant coorelation—

$n=12$

$p<0.05$

$p<0.01$

$p<0.001$

Table III. Correlation matrix for Kunjwani Pond, Jammu at study site-II

	AT	*WT*	*D*	*TR*	*SM*	*pH*	*DO*	*FCO_2*	*CO_3^{--}*	*HCO_3^-*	*Cl^-*	*Ca^{++}*	*Mg^{++}*	*TH*
AT	1.00	0.964	0.135	—0.643	0.363	—0.746	—0.752	0.539	—0.744	0.782	0.589	0.326	0.641	0.800
WT		1.00	0.231	—0.521	0.282	—0.710	—0.748	0.503	—0.517	0.736	0.574	0.245	0.537	0.734
D			1.00	0.332	—0.307	—0.524	—0.443	0.660	—0.269	—0.213	—0.357	—0.090	—0.281	—0.077
TR				1.00	—3.738	0.512	0.428	—0.148	0.740	—0.889	—0.638	—0.537	—0.711	—0.865
SM					1.00	—0.276	—0.355	0.209	—0.490	0.755	0.742	0.710	0.357	0.657
pH						1.00	0.839	—0.710	0.916	—0.613	—0.198	—0.549	—0.451	—0.674
DO							1.00	—0.725	0.795	—0.490	—0.196	—0.166	—0.312	—0.546
FCO_2								1.00	—	0.257	0.189	0.292	0.09	0.326
CO_3^{--}									1.00	0.782	—0.316	—0.427	—0.586	—0.625
HCO_3^-										1.00	0.811	0.561	0.712	0.883
Cl^-											1.00	0.876	0.496	0.626
Ca^{++}												1.00	0.026	0.656
Mg^{++}													1.00	0.688
TH														1.00

Asteric indicate significant correlation.

$n=12$

$p<0.05$

$p<0.01$

$p\leqslant 0.001$

Table IV. Correlation matrix for Kunjwani Pond, Jammu at study site-III.

	AT	*WT*	*D*	*TR*	*SM*	*pH*	*DO*	*FCO_2*	*CO_3*	*HCO_3*	*Cl*	*Ca*	*Mg*	*TH*
AT	1.00	0.933	0.366	−0.198	0.994	0.819	−0.896	0.574	−0.602	0.701	0.549	0.026	0.522	0.805
WT		1.00	0.231	−0.190	0.030	−0.702	−0.740	0.392	−0.488	0.642	0.576	0.032	0.292	0.890
D			1.00	0.056	0.014	−0.380	−0.257	0.038	−0.256	−0.171	−0 224	−0.026	−0.144	0.116
TR				1.00	0.038	0.155	0.155	0.165	0.064	−0.234	−0.126	−0.028	0.555	0.046
SM					1.00	−0.014	0.023	0.003	−0.027	0.098	0.000	0.030	−0.022	0.135
pH						1.00	0.887	−0.569	0.721	−0.596	−0.251	0.007	−0.172	0.869
DO							1.00	−0.631	0.670	−0.594	−0.392	0.045	−0.261	0.056
FCO_2								1.00	—	0.141	0.353	−0.003	0.750	0.206
CO_3									1.00	−0.537	−0.244	−0.141	0.259	−0.042
HCO_3										1.00	0.771	0.124	0.313	0.457
Cl											1.00	0.104	0.264	0.346
Ca												1.00	0.018	0.172
Mg													1.00	0.255
TH														1.00

Asteric indicate significant correlation—

$n=12$

$p<0.05$

$p<0.01$

$p<0.001$

Table V. Correlation matrix for Kunjwani Pond, Jammu at study sites-IV.

	AT	*WT*	*D*	*TR*	*SM*	*pH*	*DO*	FCO_2	CO_3^{--}	HCO_3^-	*Cl–*	*Ca++*	*Mg++*	*TH*
AT	1.00	0.975	0.458	–0.330	0.311	0.010	–0.664	0.790	–0.725	0.885	0.648	0.268	0.772	0.731
WT		1.00	0.347	–0.349	0.265	–0 742	–0.598	0,797	–0.698	0.770	0.580	0.340	0.715	0.762
D			1.00	–0.052	–0 013	–0.608	–0 414	0.486	–0.383	–0.049	–0.205	0 072	–0.022	0.114
TR				1 00	0 828	0.576	0.716	–0 504	–0.283	–0 700	–0 418	–0.821	–0.735	0 801
SM					1.00	–0 341	–0.359	–0 522	–0.462	0 592	0 525	0.671	0.513	0.577
pH						1.00	0.775	–0 794	0.927	–0 600	–0 237	–0 361	–0.205	–0 680
DO							1.00	–0.730	0.741	–0.417	–0 246	–0 170	–0 436	–0 406
FCO_2								1.00	—	0.717	0 657	0 958	0 167	0.639
CO_3^{--}									1.00	–0.810	–0.336	–0 432	0.799	–0 734
HCO_3^-										1 00	0.812	0 718	0 922	0.924
Cl											1 00	0 556	0 608	0.615
Ca++												1.00	0 646	0 775
Mg++													1.00	0.887
TH														1.00

Asteric indicate significant correlation—

$n=12$

$p<0\ 05$

$p<0.01$

$p<0.001$

increase in both air and water temperature from December to May/June. Maximum record to water temperature in June may also be attributed to low macrophytic production and highest load of suspended matter (112.5 mg/l). Suspended matter is known to absorb more heat from insolation (Reed, 1962 ; Reid and Wood, 1976 ; Dutta, 1978). Timms (1970) reported that turbid waters are warmer than the clear ones under the same circumstances. Similarly, a gradual reduction in solar illumination may explain fall in both air and water temperature observed from July to December. A direct relationship of water temperature with bright sunshine and its duration is also worked out by Munawar (1970) and Harshey *et al.* (1982).

Depth

Depth varied between 21.75 cm —143.75 cm showed bimodal peaks in Kunjwani pond. The first annual increase was recorded from July to September with highest value in August, followed by a decline upto November. Second increase in water level was measured from December to February, with second annual peak in December. Monsoon and inflow of water from watershed may explain first annual increase in water spread. Similarly, the effect of winter rains is attributed to slight rise in water level recorded during December—February. Rise in water level during monsoon and winter rains has also earlier been reported by Dutta (1978) and Malhotra *et al.*(1986-87).

Fall in water level from March to June was associated with summer evaporation of water at high temperature. Lowest water level was recorded in June with highest water temperature (34.5 °C).

Transparency

The transparency values ranged between 9.5 cm (June)—42.25 cm (February), remained low during May-November and comparatively high from December to April. Decline and decomposition of submerged macrophytes. agitation by biotic interferences and increased production of plankton may explain low values recorded in June. Increase in suspended matter during macrophytic decomposition is earlier reported by Winberg *et al.* (1972). An inverse relationship of turbidity with vegetation has also earlier been reported by Robel (1961). Floating macrophytic decline and their decomposition may explain low values of transparency recorded during October and November. The higher values of transparency noticed from December to April (32.75—42.25 cm) may be attributed to the presence of mats of submerged macrophytes minimising the shore erosion and resuspension of soft bottom sediments through water agitation.

Suspended matter

The suspended matter varied between 50.00 mg/l in March and 112.5 mg/l in June. The higher values of suspended matter recorded in June, July and November with highest June peak. This June peak in suspended matter is attributed to the resuspension of soft sediments by biotic interferences. Comparatively low values of suspended matter from December to March related with the presence of thick mats of submerged macrophytes.

Hydrogen-ion concentration pH

The Kunjwani pond waters are alkaline and ranged from 7.02 (August) to 9.98 (February). Present annual range of pH is higher than the earlier reports of Verma (1964) for Dalsagar tank (8.2—8.8) ; Mathew (1975) for Govindgarh tank (8.0—8.2) ; Vasisht and Sharma (1975) for an Urban pond, Ambala (7.0—7.3) ; Singh and Sahai (1978) for Jalwania pond (7.1—8.4) ; Harshey *et al.* (1982) for a fish tank, Jabalpur (6.2—7.2) ; Kaul (1983) for Lalpari talab Rajkot (8.1—8.4) ; Kant and Raina (1985) for two ponds, Jammu (6.0—8.1 and 6.5—9.3) ; Patil *et al.* (1985) for a fish tank, Jabalpur (7.0—8.0) ; Singhal *et al.* (1985) for Jyotisar fish farm at Kurukshetra (5.7—8.8) ; Singhal *et al.* (1986) for the ponds in CIFRI, Karnal (7.4—8.2) ; Shaify *et al.* (1986) for Motia pond Bhopal (7.3—8.7) and Singh and Rai (1988) for Badatalab M.P. (6.5—7.9).

Present wide range of annual pH is due to the presence or absence of free carbondioxide and carbonate, and macrophytic diversity during one or other month at one or other study site (Welch, 1952) ; Hutchinson, 1957 ; Cole, 1975 ; Wetzel, 1975 ; Reid and Wood, 1976 ; Jhingran, 1982 and Goldman and Horne, 1983).

Complete absence of carbonate and presence of free carbon-dioxide (2.56—9.19 mg/l) may explain low value of pH recorded during May to September (Fig. 1.1). Similarly, complete absence of free carbon-dioxide and presence of carbonate (4.65—32.36 mg/l) may explain the higher value of pH noticed from October to March. Direct relationship of pH with CO_3'' and an inverse with free carbon-dioxide is well documented by Gonzalves and Joshi (1946), Zafar (1964), Swarup and Singh (1979) and Jhingran (1982).

Among biotic factors, higher photosynthetic activity due to increased production of submerged macrophytes (*Potamogeton orispus* and *Nitella mucronata*) may support an increase in pH observed from October to March. This type of observation also reported by Das and Srivastava (1956), Moitra and Bhattacharya (1965), Banerjee and Raychoudhary (1966) and Dutta and Malhotra (1987). Similarly, summer fall in photosynthesis due to increased production of floating macrophytes (*Lemna minor*, *Wolffia arrhiza* and *Azolla pinnata*), increase in water level and suspended matter fall in transparency and low DO may explain low values of pH recorded during May to September.

Dissolved Oxygen

Dissolved oxygen in Kunjwani pond varied between 2.59 mg/l (July)—14.15 mg/l (December). DO concentration remained low during April to September and higher from October to February. Winter rise in DO has also earlier been worked out by Singh *et al.* (1980) and Rao (1986) and it appears to be due to its greater solubility, reduced microbial decomposition of dead organic matter and low organismal respiratory demand at low temperature and *increased prolific* growth of submerged macrophytes. DO enrichment with increased production of submerged macrophytes is also reported by Sreenivasan (1966), Sahai and Sinha (1969) and Sahai and Srivastava (1976). Absence of free carbon-dioxide from October to March appears as an additional factor explaining present increase in DO. Inverse relationship of DO and free CO_2 is well known (Welch, 1952 ; Hutchinson, 1957 and Kadlac, 1962). In Kunjwani pond, April to September low values of DO inspite of increased photoperiod may be due to its utilization in decomposition of macrophytes which starts decline after April and completely disappeared in July, and allogenetic dead organic matter brought into pond from the catchment area during early monsoon rains. Low DO retaining capacity of water due to increased organismal respiratory demand at high temperature may also support this low values of dissolved oxygen (Rao, 1986).

Free carbon-dioxide

Free carbondioxide was completely absent from October to March. The value ranged from 0.0–9.19 mg/l. April to September free carbon-dioxide presence in this pond appears to be the effect of gradual rise in temperature accelerating the decomposition of auto-genetic organic matter and its influx through rain water. Rise in free carbon-dioxide during rains earlier also reported by Dutta (1978) and Jhingran (1982).

Carbonate

The carbonate values showed a range from 0.0–32.36 mg/l. The present study showed an inverse relationship of carbonate with free carbon-dioxide. Similar type of observations also reported by Reid and Wood (1976), Goldman and Horne (1983) and Patil *et al.* (1985).

Bio-carbonate

The bicarbonate value varied between 20.92 mg/l (December)–278.06 mg/l (June) with bimodal peaks. Its first annual increase was observed during April to June with highest

June peak and second increase from September to November with October peak. December to March low values of bicarbonate is attributed to low rate of recycling of nutrients caused by reduced activity of microbes during winter at low temperature and increased utilization of bicarbonate in absence of free CO_2 by submerged macrophytes which recorded their luxuriant growth during this period.

According to Welch (1952). Pal *et al.* (1960) and Goldman and Horne (1983) some submerged macrophytes develop encrustation of marl upon the leaves, petioles, stems and other exposed parts. Increase in bicarbonate values from April to June (97.67—149.83 mg/l) may be due to fall in water level (Fig. 1.1). Bicarbonates increase with fall in water level also reported by Cook and Powers (1958) and Singhal *et al.* (1986). Death and decomposition, and conversion of marls produced by submerged macrophytes into soluble form by free CO_2 may also explain the increase in bicarbonate from April to June. Ruttner (1963) and Swarup and Singh (1979) stated that the accumulation of lesser quantities of carbonate during summer to the liberation of free CO_2 in the process of decomposition of bottom deposits, which possibly resulted into conversion of insoluble $CaCO_3$ into soluble Ca $(HCO_3)_2$.

In Kunjwani pond, there is sudden fall in bicarbonate recorded in July and December to March. Dilution due to monsoon and winter rains may explain present low values of bicarbonate. Fall in bicarbonate concentration as a consequence of dilution is a well-known limnological phenomenon and has also reported by Bamforth (1958), Das and Akhtar (1970) and Singhal *et al.* (1986).

Second annual increase in bicarbonate concentration recorded from August to November may be attributed to the fall in photosynthesis due to greater production of floating macrophytes, and water refilling and release of salts from mud flats left dried during summer season.

Chloride

The chloride values ranged from 1.27 mg/l (September) to 9.65 mg/l (June). Present **chloride content** is low in comparison with the earlier findings of Singh (1965) for a small lake near Delhi (30 mg/l) ; Harshey *et al.* (1982) for a tank in Jabalpur (36—130 mg/l) ; Kant and Rama (1985) for two ponds (2.63—12.0 mg/l ; 2.64—220 mg/l) at New University Campus, Jammu ; Kant and Rampal (1987) for the non-polluted (2.6—22.0 mg/l) and polluted (13.5—67.0 mg/l) ponds, Jammu, Patil *et al.* (1985) for a fish farm at Jabalpur (5.2—12.5 mg/l) ; Singhal *et al.* (1985) for unregulated rural pond Jyotisar (8.5—37.1 mg/l) and Singhal *et. al.* (1986) for a pond at CIFRI (8.6—38.0 mg/l), Karnal ; Adoni and Joshi (1987) for Sagar lake (32—95 mg/l) and Singh and Rai (1988) for Bada talab, M.P. (35.51 mg/l). This low value of chloride in this pond as compared with the observations of earlier workers (op. cit.), is because of insignificant sources of animal origin (Thresh *et al.* 1944 ; Zafar, 1964 and Adoni and Joshi, 1987).

A look at the Fig. 1.1 reveals high values of chloride from April to June, with annual highest value in the month of June. Present summer increase in Cl^- is in conformity with the earlier observations of Munawar (1970) and Harshey *et al.* (1982). An accelerated activity of decomposers, decomposing aquatic macrophytes (submerged forms) which died during this period and water evaporation at summer high temperature may explain higher values of chloride from April to June. Chloride enrichment as a consequence of decomposition of macrophytes is also reported by Kundangar and Zutshi (1985), Sarwar (1987) and Untoo (1989).

Calcium, Magnesium and total hardness

The present study recorded the annual mean variation of calcium, magnesium and total hardness as 12.49 mg/l (June) 31 30 mg/l (January), 2.64 mg/l (June)—8.68 (January) and 42.00 mg/l (January)—113.80 mg/l (June) respectively. Seasonally calcium, magnesium and total hardness showed bimodal increase. Their first annual increase was recorded from February to June followed by a decline during July, August. (Fig. 1.1). Again during September and October there is an increase in these three parameters, followed by a decline upto January.

High concentration of Ca^{++}, Mg^{++} and total hardness recorded from April to June may be attributed to the decline in submerged macrophytes (*Potamogeton* sp. and *Nitella* sp.) and conversion of marls produced by these plants into soluble form by free carbon-dioxide. Singhal *et al.* (1986) also reported higher concentrations of Ca^{++} and Mg^{++} during summer season.

On the basis of present findings, it is concluded that Kunjwani pond is highly suitable for fish culture. Among various chemical parameters, annual variation of pH (7.02–9.98), DO (2.68–14.15 mg/l), free CO_2 (0.00–9.19 mg/l), Carbonate (0.06–32.36 mg/l) and bicarbonate (20.92–278.06 mg/l) appears as favourable for carp production. This range for one or other chemical parameters has been considered as optimum for fish culture by Ellis *et al.* (1946), Dubey and Verma (1966), Banerjea (1967), Muddana *et al.* (1970, 71), Parmeswaran *et al.* (1971) and Jhingran (1982).

In order to exploit this pond for fishery development, it is suggested that fish larvae (in advanced stage) of *Gyprinus carpio*, *Hypophthalmichthyes* sp. *Catla catla*, *Cirrhinus mrigala* etc. may be introduced in the month of July (at beginning of monsoon) and allowed to grow upto April/May, before harvesting. During this period of ten to 11 months, most of the culturable fishes are known to grow successfully to table size (Hora and Pilley, 1962 ; Alikunhi, 1966 and Lakshmanan *et al*, 1971).

ACKNOWLEDGEMENT

The Author is grateful to Head, Department of Bio Sciences, University of Jammu for laboratory facilities and Dr. S.P.S. Dutta for kind help rendered for the successful completion of the work.

REFERENCES

Adoni A.D. and Joshi, G. (1987). Physico-chemical regime of three fresh water bodies in and around Sagar (M.P.). *Geobios new Reports*, 6 (1) : 43-46.

Alikunhi, K.H. (1966). Synopsis of biological data on common carp *Cyprinus carpio* Linnaeus, 1758 (Asia and the Far east). FAO Fish. Synop. 31, 1.

A.P.H.A. (1975). *Standard methods for the examination of water and waste water*, 12th Amer. Publ. Hlth, Assoc. Inc., New York.

Bamforth, S.S. (1958). Ecological studies on the planktonic protozoa of a small artificial pond. *Limnol. and Oceanogr.*, *3*: 1-12.

Banerjea, S.M. (1967). Water quality and soil condition of a fish pond in some state of India in relation to fish production. *Ind. J. Fish.*, *14* (1-2) : 115-144.

Banerjee, A.C. and Raychoudhury, N.C. (1966). Observations on some physico-chemical features of the Chilka lake, *Indian J. Fish*, *13* : 395-429.

Cole, G.A. (1974). *A text book of Limnology.* The G.V. Mosby Company, Saint Louis.

Cook, A.H. and Powers. C.F. (1958). Early biochemical changes in the soils and water of artificially created marshes in New York. *New York Fish and Game.* J., *5* : 9-65.

Das, S.M. and Akhtar, S. (1970). A report on fresh water plankton from the Dal lake, Kashmir. *Kashmir Science, 7* (1-2) : 133-137.

Das, S.M. and Srivastava, V.K. (1956). Quantitative studies on fresh water plankton pt. II, Correlation between plankton and hydrological factors. *Proc. Nat. Acad. Sci., India, 26* B (4) : 243-253.

Dubey, G.P. and Verma, M.N. (1966). Hydrobiological studies on the Budhwasi tank Seoni, Madhya Pradesh with special reference to fish production. *Ichthyologica*, 7 : 59-66.

Dutta, S.P.S. (1978). *Limnology of Gadigarh stream (Miran Sahib, Jammu) with special reference to Consumers Inhabiting the stream.* Ph.D. Thesis, Univ. of Jammu.

Dutta, S.P.S. and Malhotra, Y.R. (1987). Diel variations in some hydrological parameters of Gadigarh stream, Jammu. *Hydrobiol*, *3* (6) 43-50.

Ellis, M.M., Westfall, B.A. and Marion, D. Ellis (1946). Defemination of water quality. *U.S. Fish and Wildl. Serv. Res. Pep.*, *9*: 1-122.

Goldman, C.R. and Horne, A.J. (1983). *Limnology.* Mc Graw Hill International Book Company, New Delhi.

Gonzalves, E.V. and Joshi, V. (1946). Fresh water algae near Bombay T. The seasonal succession of algae in a tank of Bandra. *J, Bom. Nat. Hist. Soc.*, *46* : 154-176.

Harshey, D.K., Patil, S.G. and Singh, D.F. (1982). Limnological studies on a tropical fresh water fish tank of Jabalpur, India. 1. The abiotic factors. *Geobios new Reports, 1* (2) : 98-102.

Hora, S.L. and Pillay, T.V.R. (1962). Handbook on fish culture in the Indo-Pacific Fisheries region, *FAO Fish Biology, Tech. Pap., 14* : 203.

Hutchinson, G.E. (1957). *A treatise on Limnology.* Vol. I. John Wiley and Sons, New York, London, : 1015 pp.

Jhingran, V.G. (1982). *Fish and Fisheries of India.* Hindustan Publishing Corporation, India : 954 pp.

Kadlec, J.A. (1962). Effects of a draw down on a water fowl impoundment. *Ecology, 43* (2) : 267-281.

Kant, S. and Raina, A.K. (1985). Limnological studies of two ponds in Jammu I. Qualitative and quantitative distribution of phytoplankton, *Zoologica Orientalis, 2* (1-2) : 89-92.

Kant, S. and Rampal, R.K. (1987). Limnology of polluted and unpolluted ponds in Jammu. Proc. Nat. Symp on Environmental pollution and Pesticide Toxicology and 8th Annual Session of Academy of Environmental Biology, India. ed. Shashi Kant : 47-57.

Kaul, S (1983). Limnological studies of an aquatic body Larpari talab at Rajkot. *Ind. J. Ecol., 10* (1) : 124-128.

Kundangar, M.R. and Zutshi, D.P. (1985). Environmental features and plankton communities of two Himalayan rural lakes. Proc. nat. Symp. Pure and Applied Limnology edited by A.D. Adoni, *Bull. Bot. Soc. Sagar, 32* : 40-47.

Lakshmanan, M.A.V., Sukumaran, K.K., Murty, D.S., Chakraborty, D.P. and Philipose, M.T. (1971). Preliminary observations on the intensive fish farming in fresh water ponds by composite culture of Indian and exotic species. *J. Inland Fisheries Society*, India, *III* : 1-21.

Malhotra, Y.R., Dutta, S.P.S. and Suri, S.N. (1986-87). Limnology of a fish pond in Regional Research Laboratory, Jammu. *Matsya, 12-13* : 174-177.

Mathew, P.M. (1975). Limnology and Productivity of Govindgarh Lake, Rewa, (M.P.). *J. Inland Fish. Soc.*, India, *7* : 16-24.

Moitra, S.K. and Bhattacharya, B.K. (1965). Some limnological factors affecting plankton production in a fish pond in Kalyani, West Bengal (India). *Ichthyologica, 4* : 8-12.

Muddana, V., Halappa, G. and Rajagopal, K.V. (1970). Preliminary observations on Paddy cum fish culture of Hebbal. *Ind. J. Fish., 17* (1-2) : 105-110.

Muddana, V., Rajagopal, K.V. and Chandrasekhariah, H.W. (1971). The scope of utilizing paddy field as fish nurseries, Mysore, *J. Agri Sci., 5* (4) : 447-465.

Munawar, M. (1970). Limnological studies on fresh water ponds of Hyderabad I. Biotope. *Hydrobio ogia, 35* : 127-162.

Pal, B.P., Kundu, B.C., Sundaralingram, V.S. and Venkataraman, G.S. (1960). *Charophyta.* Indian Council of Agricultural Reseaıch, New Delhi.

Parmeswaran, S., Radhakrishan, S., Selvaraj, C. and Bhuyan, B.R. (1971). Fish yield in Assam ponds kept under different ecological conditions. *Ind J. Fish., 18* : 67-83.

Patil, S.G., Harshey, D.K. and Singh, D.F. (1985). Limnological studies of a tropical freshwater fish tank of Jabalpur, M.P., *Geobios new Reports., 4* : 143-148.

Rao, K.J. (1986). Studies on the seasonal and diel variations in some physico-chemical conditions of a freshwater pond under prawn culture. *Proc. nat. Symp. Fish and Env.*, 96-102.

Reed, E.P. (1962). *Limnology and fisheries of the Saskatchewan river in Saskatchawan.* Fisheries report No. 6. Fisheries Branch Department of Natural Resources, Province of Saskatchewan : 1-48.

Reid, G K. and Wood, R.D. (1976). *Ecology of Inland waters and estuaries.* D van Norstrand Company, New York.

Robel, R J. (1961). Water depth and turbidity in relation to growth of Sago pond weed. *J. Wild Life Management, 25* (4) : 436-438.

Ruttner, F. (1963). Fundamentals of Limnology 3rd ed. (Translated from German by D.G. Frey and F.E.J. Fry). University of Toronto Press, Toronto, Ontario, Canada, : 295 pp.

Sahai, R. and Sinha, A.B. (1969). Investigations on bioecology of inland waters of Gorakhpur (U.P.) India. 1. Limnology of Ramgarh lake. *Hydrobiologia, 34* : 433-447.

Sahai, R. and Srivastava, V.C. (1976). The physicochemical complexes and its relationship with the macrophytes of Chilwa lake. *Geobios. 3* : 15-19.

Saify, T., Chaghtal, S.A., Parveen, A. and Durrani, I.A. (1986). Hydrology and periodicity of Phytoplankton in the sewage fed Motia pond. Bhopal (India). *Geobios, 13* (5) : 199-203.

Sarwar, S.G. (1987). Species composition and seasonal variations of periphyton on *Ceratophyllum demersum* Linn. in Waskur lake, Kashmir *Geobios New Reports. 6* : 114-118.

Singh, M. (1965). Phytoplankton periodicity in a lake near Delhi I. Seasonal fluctuations of the physico-chemical characteristics of water, *Phykos, 4* : 61-68.

Singh, D F. and Rai, M.K (1988). Studies on the limnology of Bada Talab, Chhindwara, M P. *J. Environ Biol*, *9* (1) : 69-71.

Singh, S B. and Sahai, R. (1978). Seasonal fluctuations of zooplankton population in relation to certain physico-chemical factors of a pond. *Geobios, 5* : 228 230

Singh, R.K , Srivastava, N.P. and Desai, V.R. (1980). Seasonal and diurnal variations in physico-chemical conditions of water and plankton in lotic sector of Rihand reservoir (U.P.) *J. Inland. Fish India, 12* (1) : 100-111.

Singhal, R.N., Swarn, J. and Davis, R.W. (1985). The relationship among physical, chemical and plankton characteristics of unregulated rural ponds in Haryana, India. *Trop. Ecol , 26* : 43-53.

Singhal, R.N., Swarn, J. and Davis, R.W. (1986). The physico-chemical environment and the plankton of managed ponds in Haryana, India. *Proc Indian. Acad. Sc.*, (Anim,-Sci.) *95* (3) : 353-363.

Srinivasan, A. (1966). Limnology of tropical impoundments. Upland impoundments of Nilgiris (Madras). *Phykos, 8* : 10-16.

Swarup, K. and Singh, S.R. (1979). Limnology of Suraha lake (Ballia) I. Variations in the water quality. *J. Inland. Fish Soc., India, 11* (1) : 22-33.

Thresh, J.C., Suckling, E.V. and Beak, J.F. (1944). *The Examination of water Suppliee* (ed.) E.W. Taylor.

Timms, B.V. (1970) Aspects of the limnology of five small reservoirs in New South Wales. Proc. Linnean. Soc., New South Wales, *95* (1) ; 46-59.

Untoo, G.M. (1989). *Ecological studies on Hygam Wetland with special emphasis on winterfood of migratory Dabbling ducks*, Ph.D. Thesis, Univ. of Jammu, Jammu, India.

Vasisht H.S. and Sharma, B.K. (1975). Ecology of a tropical urban pond in Ambala city of the Haryana State. *Ind. J. Ecol. 2* (1) : 79-86.

Verma, M.N. (1964). Hydrobiological survey of Dal Sagar tank, Seoni (M P.) with special reference to the effect of *Microcystis aeruginosa* (Kurtz) bloom in fish production for suggestion and for improvement of the tank fishery *Ichthyologica, 3* : 37-46.

Welch, P.S. (1952). *Limnology.* McGraw Hill Book Co., New York, : 539 pp.

Wetzel, R.G. (1975). *Limnology,* W.B. Saunders Co., Philadelphia.

Winberg, G.G., Bobitsky, V.A., Gaurilov, S I , Gladky, G.V. Zakharenkov, I., Kovalevskaya, R Z., Mikheeva, T.M , Nevyadomskaya, P.S Ostapenya, A P., Petrovich, P G., Pateniko, J S and Yakushko, O.P. (1972). Biological productivity of different types of lakes in : *Productivity problems of freshwater Kaximierg Dolny Poland.* Edited by Z. Kajak and A.K. Ilkowska, : 383-404.

Zafar, A.R. (1964). On the ecology of algae in certain fish ponds of Hyderabad, India II. Distribution of unicellular and colonial forms. *Hydrobiologia, 24* (4) : 556-566.

2

OBSERVATIONS ON COMMUNITY ANALYSIS OF ZOOPLANKTON FROM GANDHISAGAR RESERVOIR MANDSAUR M.P. INDIA

Usha Choubey

School of Studies in Zoology
Vikram University, Ujjain

Abstract—Studies on community analysis of Zooplankton in Gandhisagar reservoir were conducted at six study sites *viz* Kampura, Dhabla, Kharawda, Sanjeet, Chandwasa and Dam site. 55 Zooplanktonic species were recorded, out of which the protozoa was represented by 12 species, Rotifera by 15, Cladocera by 19, ostracoda by **3**, Cyclopoida by 3 and calanoida by 2. Copepod formed the most dominant plankters of the GSR. Highest density recorded in the month of Oct. (672 org/l). Cladocerans were the second dominant group of Zooplankton, Rotifers and Protozoans comes next in the order of abundance.

INTRODUCTION

Zooplankton form an important intermediary step in the grazing food chain in aquatic biolop and any ecosystem. Analysis at the specific level can not entirely disclose the functional mechanism of the aquatic ecosystem, unless the dynamics of zooplankton community is adequately known. The distribution pattern of zooplankton has been well documented by several works. Das and Shrivastava (1956), Vasisht and Dhir (1970), Vasisht and Sharma (1975), Awtramani (1980) Sharma (1983), Vasisht and Jindal (1980) and Saxena (1982). Gandhisagar reservoir has not been accorded much attention for detailed scientific study in this regard, but for the works of Rao *et. al.* (1987*a*, 1987*b*, 1988*a* and 1988*b*) ; Rao and Shrivastava (1989), Dubey and Chatterjee (1976), Kartha (1987) and Rao) 1987*e*, 1990) on the morphometric, biological and fisheries aspects of this reservoir. Present study reported community analysis of Zooplanktonic organisms in Gandhisagar reservoir.

MATERIALS AND METHODS

Data for the present study collected from six selected study sites in the reservoir. *viz.*, (1) Rampura (2) Dhabla (3) Kharawda (4) Sanjeet (5) Chandwasa and (6) Dam site, for the sack of brevity, data recorded from Rampura study site (1) was only given, but comparision with the data recorded from other study sites were made, wherever felt necessary. The methodology used for analysis of Zooplanktonic communities and for species categorisation is based on the works of Edmondson (1959), Pennak (1978), Sehgal (1983), Tonapi (1980), Michal and Sharma (1986), Wetzel (1974), APHA (1985). The diversity indices were derived based on the formaly-provided by Shannon Weaver (1963).

RESULTS AND DISCUSSION

1. Zooplanktonic groups and their species composition

The qualitative samples wherein the materials from all five study sites have been observed. Show an annual species diversity of 57 species, out of which 12 species, to protozoa, 15 Rotifera, 19 Cladocera, 5 copepoda, 3 Ostraloda and 2 Coelentrata (Table 1). The bulk of Zooplanktonic assemblages of this reservoir is contributed primarily by cladocerans followed by Rotifers, Protozoans, Copepods and ostracods. Groupwise relative percentages of the zooplanktonic groups is as per details given Cladocera 33.33%, Rotefera 26.31%, Protozoa 21.05% Copepoda 8.77%, Ostracoda 5.26% and Coelenterates 1.3%.

2. Temporal Variation

Temporal variations of the zooplanktonic populations of this reservoir include the monthly variations in the species diversity and density at all the five study sites, but in description data from study site 1 (Rampura) only was considered. The annual average of the zooplanktonic populations at surface had a species diversity 28.83, while at the bottom the species diversity was 30.25. Maximum zooplanktonic populations at the surface (32) and at the bottom (30) occurred in the month of September and October, while the minimum species diversity at the surface (11) and the bottom (10) occurred in the month of July.

Table 1. Qualitative list of Zooplankton at GSR during 1988.

Protozoa

1. *Difflugia* sps.
2. *Arcella* sps.
3. *Nebela* sps.
4. *Actinophrys* sps.
5. *Stentor* sps.
6. *Vorticella* sps.
7. *Colpidium* sps.
8. *Opercularia* sps.
9. *Metopus* sps.
10. *Parmulina* sps.
11. *Heleopera* sps.
12. *Euglypha* sps.

Rotifera

1. *Keratella tropica*
2. *Keratella quadrata*
3. *Polyarthra vulgaris*
4. *Ptygura* sps.
5. *Trichotria* sps.
6. *Philodina* sps.
7. *Filinia longiseta*
8. *Monostyla* sps.
9. *Asplanchna brightwelli*
10. *Floscularia* sps.
11. *Collotheca* sps.
12. *Beauchampia* sps.
13. *Brachionus forficula*
14. *Brachionus calyciflorus*
15. *Brachionus falcatus*

Cladocera

1. *Daphnia carinata*
2. *Daphnia pulex*
3. *Daphnia magna*
4. *Daphnia lumholtzi*
5. *Daphnia dubia*
6. *Ceriodaphnia regaudi*
7. *Moina* sps.
8. *Moina macrocopa*
9. *Leydiqia quadrangularis*
10. *Diaphanosoma brachyurum*
11. *D. sarsii*
12. *Simocephalus serrulatus*
13. *Simocephalus* sps.
14. *Polyphemus* sps.
15. *Alonella dentifera*
16. *Alona quadrata*
17. *Alona quttata*
18. *Alona quadrangularis*
19. *Alona rectangula*

Ostracoda

1. *Cypris obensa*
2. *Cyprinotus* sps.
3. *Stenocypris* sps.

Copepoda

Cyclopoida

1. *Thermocyclops crassus*
2. *Mesocyclops thermocyclopoides*
3. *Ergasilus* sps.

Calanoida

1. *Heliodiaptomus viduus*
2. *Eodiaptomus lumholtzi*
3. *Nauplius larvae*

Coelenterates

1. *Limnocnida indica*
2. *Craspedacusta sowerbyi*

Insecta

1. *Chaoborus larvae*

As for the zooplanktonic density the highest density occurred in the month of May (1207 org/l) at surface but the (1306 org/l) maximum zooplanktonic density in the bottom occurred in the month of June 1360 org/l. In general higher zooplanktonic densities occurred in summer, while least density occurred in the rainy season. Within three seasons the highest total zooplankton density of 4554 org/l at surface and 5094 org/l at bottom was recorded in summer (March, April, May and June). The lowest 3145 org/l at surface and 3490 org/l at bottom was recorded in monsoon season (July, August, September and October).

Copepods formed the most dominant plankters of the GSR ranging from 11.91% to 75.0% at the surface and 31.57% to 82.75 at the bottom, of the total zooplankton. They showed constantly high density in almost all the months of the year except for August and September. A bimodal distributional pattern can be observed in this group with the first peak in October (672 org/l) and second in May and June (548 org/l). Three species of Cyclopoid and two species and Calanoids are recorded (Table 2, Fig. 1*a*).

Cladocerans were the second dominant group of zooplankton of GSR. The density of cladocera ranged from 9 to 348.2 org l at surface in the month of December and June respectively. Their density in the bottom ranged upto 33.7 to 326.3 org/l in the months of December to May respectively. In this group *Daphnia lumholtzi* formed 27.14% and *Diaphanosoma brachyurum* formed 16.02% of the total Cladoceran plankton (Table 2). The kite drawings (Fig. 1*b*) showing the annual fluctuations of the Cladocera and the community shows a bimodal pattern of occurrence showing two peaks, one in the month of September and the second in June.

The Rotifera comes next in the order of abundance after Cladocera. This also exhibited bimodal pattern of occurrence showing a smaller peak in August (260.2 org/l) and larger peak in December (347.1 org/l). Minimum densities (87.3 org/l) of this species occurred in the month of July (Fig. 1*c*). The species diversity of Rotifera ranged upto 12 *Keratella tropica* was the most dominant plankter in Rotifera. Next in order come *Brachionus falcatus* and *Filinia longiseta*.

The protozoan density percentage varied from 2.14% in July to 32.41% in September at surface. Their respective densities ranged upto 12.8 org/l in July to 256.9 org/l in September. They also exhibit a bimodal annual distribution showing the largest peak in September and a smaller peak in May. The protozoans of GSR show as pecial diversity of 11 and constitute dominant population of *Difflugia* occurring throughout the year and *Arcella* occurring from August to January and April to June (Table 2 Fig. 1*d*).

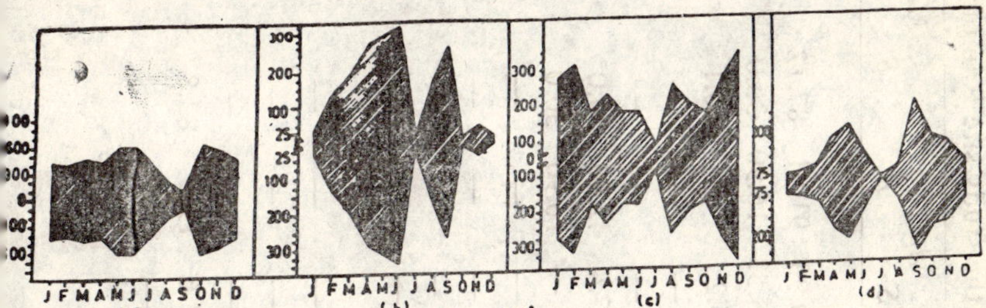

FIG1

The Ostracoda comes last in the order of abundance. They exhibit a bimodal pattern showing two peaks, one in November (160 org/l) and the other in February (133 org/l). This group shows a species diversity of 3, with dominating population of *Stenocypris* sps.

Bathymetric distribution of major zooplanktonic groups

Vertical distribution of protozoans, shows its occurrence from its surface upto 84 feet. Protozoans show comparatively lesser density at 84 feet depth. The population of protozoans were recorded upto 50 to 60 feet depths after which their numbers gradually decline. The maximum populations were recorded in the month of May and June at 15 and 9 feet depth respectively and minimum population in the month of July at 24 feet depth. The vertical distribution of rotifers ranging from the surface upto 84 feet depth. Rotifers shows comparatively lesser density upto 84 feet depth. Its population gradually decline with increase in depth. Maximum populations of Rotifera were recorded in the month of November and January at surface waters and minimum in the month of July at 24 feet

Table 2. Details of density of Zooplankton (org./l) at the study site Rampura.

Name of the Species	*Jan.*		*Feb.*		*Mar.*		*Apr.*		*May*		*June*	
	S	*B*	*S*	*B*	*S*	*B*	*S*	*B*	*S*	*B*	*S*	*B*
PROTOZOA												
1. *Difflugia* sps.	23.2	45.9	41.4	64.5	62.0	75.9	70.95	91.91	87.12	104.6	88.0	130.0
2. *Arcella* sps.	18.4	—	—	—	—	—	35.53	16.90	36.0	14.8	18.0	—
3. *Actinophirys* sps.	18.4	—	22.0	—	22.0	—	35.2	—	30.1	—	18.1	—
4. *Stentor* sps.	12.0	—	—	—	—	—	—	—	—	—	—	—
5. *Vorticella* sps.	12.0	—	—	—	—	—	—	—	—	—	—	—
6. *Euglyphs* sps.	—	8.01	22.0	30.0	33.0	53.9	37.4	42.90	30.0	28.0	—	—
7. *Heleopera* sps.	—	2.43	—	—	—	—	—	16.90	30.0	18.2	—	—
8. *Sphenoderia* sps.	—	—	—	—	—	—	—	—	—	—	—	—
9. *Colpidium* sps.	—	—	—	—	—	—	—	—	—	—	—	—
10. *Opercularia* sps.	11.6	8.01	—	—	—	—	—	—	—	—	—	—
11. *Metopus* sps.	—	7.0	—	—	—	—	—	—	—	—	—	—
ROTIFERA												
1. *Trichotria* sps.	3.6	—	—	—	—	—	—	—	—	—	—	—
2. *Ptygura* sps.	35.2	12.6	—	—	—	—	—	—	—	—	—	—
3. *Beauchampia* sps.	—	12.6	—	—	—	—	—	—	—	—	—	—
4. *Keratella tropica*	37.6	25.2	46.8	18.0	58.0	18.7	72.6	19.5	82.8	14.8	86.4	19.5
5. *Polyarthra vulgoris*	23.2	2.9	—	—	—	—	—	—	—	—	—	—

(*Contd.*)

Name of the Species	*Jul.*		*Aug.*		*Sep.*		*Oct.*		*Nov.*		*Dec.*	
	S	***B***	***S***	***B***	***S***	***B***	***S***	***B***	***S***	***B***	***S***	***B***
PROTOZOA												
1. *Difflugia* sps.	12.5	—	20.3	24.8	22.0	45.0	44.0	58.6	73.7	72.6	51.3	67.0
2. *Arcella* sps.	—	—	14.7	—	20.1	23.7	25.0	—	33.0	—	28.4	—
3. *Actinophirys* sps.	—	—	20.3	—	16.0	—	15.0	—	3.96	—	3.24	—
4. *Stentor* sps.	—	—	—	—	16.16	18.0	5.3	4.9	—	—	—	—
5. *Vorticella* sps.	—	—	—	—	13.6	5.4	8.0	7.7	—	—	—	—
6. *Euglypha* sps.	—	—	—	—	13.6	16.2	22.0	36.9	—	—	—	—
7. *Heleopera* sps.	—	—	—	—	43.2	31.5	16.0	17.6	7.7	—	—	—
8. *Sphenoderia* sps.	—	—	20.3	8.0	41.6	14.4	19.4	9.9	13.2	8.8	—	—
9. *Colpidium* sps.	—	—	—	—	44.0	7.2	3.0	2.3	3.3	8.8	—	—
10. *Opercularia* sps.	—	—	—	—	—	9.6	16.0	7.48	20.5	8.8	18.0	7.0
11. *Metopus* sps.	—	—	20.3	4.0	44.0	7.2	3.0	7.4	15.0	8 8	13.0	11.0
ROTIFERA												
1. *Trichotria* sps.	—	—	—	—	—	—	2.5	—	12.1	—	—	—
2. *Ptygura* sps.	—	—	—	—	43.2	14.4	10.0	10.7	8.1	—	—	—
3. *Beauchampia* sps.	—	—	—	—	12.0	11.7	8.0	9.9	—	—	—	—
4. *Keratella tropica*	42.5	2.5	81.9	4.0	46.1	4.5	76.0	28.0	68.2	17.6	54.0	12.5
5. *Polyarthra vulgaris*	—	—	—	—	—	—	—	—	15.4	12.4	23.4	—
6. *Asplanchna* sps.	—	—	—	—	—	—	10.0	—	53.9	20.0	56.8	12.0

Table 2. (*Contd.*)

Name of the Species	Jan. S	Jan. B	Feb. S	Feb. B	Mar. S	Mar. B	Apr. S	Apr. B	May S	May B	June S	June B
6. *Asplanchna* sps.	22.4	—	34.2	12.0	—	—	—	—	—	—	—	—
7. *Brachionus falcatus*	71.4	37.8	78.3	45.0	88.0	39.6	71.6	14.3	33.6	31.2	31.2	31.0
8. *B.calyciflorus*	6.4	13.0	79.2	—	88.0	—	—	—	—	—	—	—
9. *B. forficula*	15.2	—	22.5	—	—	—	—	—	—	—	—	—
10. *Philodina* sps.	—	—	—	—	—	—	—	—	—	—	—	—
11. *Filinia longiseta*	—	—	19.9	—	66.0	—	94.6	—	61.5	—	61.5	—
12. *Floscularia* sps.	—	—	—	—	—	—	—	—	—	—	—	—
13. *Collotheca* sps.	—	—	34.2	—	—	—	—	—	—	—	—	—
OSTRACODA												
1. *Cypris obensa*	—	28.8	—	40.0	—	21.1	—	16.9	—	16.9	—	16.8
2. *Cyprinotus* sps.	—	19.0	—	45.0	—	34.1	—	39.0	—	31.2	—	30.2
3. *Stenocypris* sps.	—	30.6	—	48.0	—	38.5	—	40.3	—	33.8	—	32.6
CLADOCERA												
1. *Daphnia carinata*	—	—	39.6	73.0	66.0	70.9	72.2	53.3	—	28.6	28.8	26.6
2. *D. pulex*	—	—	—	—	—	11.0	109.7	88.4	91.2	88.4	88.4	88.4
3. *D. lumholtzi*	—	—	59.4	93.0	86.0	93.0	70.0	106.6	91.2	97.5	95.5	97.5
4. *Moina macrocops*	—	—	—	—	—	—	—	—	—	—	—	—
5. *Moina* sps.	—	—	—	—	—	—	—	—	—	—	—	—
6. *Ceriodaphnia regaudi*	6.4	2.7	—	—	—	—	—	—	60.5	44.2	61.5	44.2

(*Contd.*)

Name of the Species	July S	July B	Aug. S	Aug. B	Sep. S	Sep. B	Oct. S	Oct. B	Nov. S	Nov. B	Dec. S	Dec. B
7. *Brachionus falcatus*	22.4	2.9	61.6	25.6	15.0	—	19.0	41.8	16.7	18.8	77.4	—
8. *B. calyciflorus*	22.4	10.4	—	—	—	—	8.1	7.9	2.5	15.8	77.0	—
9. *B. forficula*	—	—	14.5	—	—	—	18.0	19.8	27.1	17.7	41.4	22.5
10. *Philodina* sps.	—	—	—	—	15.0	9.5	5.3	2.7	4.8	2.5	17.1	22.5
11. *Filinia longiseta*	—	—	61.6	12.8	41.6	9.9	3.1	6.6	4.5	—	—	—
12. *Floscularia* sps.	—	—	—	—	—	9.9	13.0	8.4	22.0	—	—	—
13. *Collotheca* sps.	—	—	—	—	29.6	7.5	3.1	2.3	2.9	—	—	—
OSTRACODA												
1. *Cypris obensa*	—	—	—	—	—	20.9	—	25.5	—	45.1	—	44.0
2. *Cyprinotus* sps.	—	—	—	—	—	26.3	—	42.9	—	29.7	—	39.4
3. *Stenocypris* sps.	—	—	—	—	—	23.7	—	43.7	—	41.5	—	31.8
CLADOCERA												
1. *Daphnia carinata*	10.0	2.0	20.3	21.6	27.2	29.4	5.0	44.0	—	—	—	—
2. *D. pulex*	—	—	—	10.4	10.4	9.5	—	—	5.6	—	—	—
3. *D. lumholtzi*	24.0	10.0	76.3	86.4	16.0	51.3	—	—	—	—	—	—
4. *Moina macrocopa*	—	—	20.3	43.2	20.0	30.4	—	—	—	—	—	—
5. *Moins* sps.	—	—	16.16	9.5	16.16	9.5	—	—	—	—	—	—
6. *Ceriodaphnia reqaudi*	—	—	20.3	43.2	62.4	48.16	—	—	—	—	—	—

Table 2. (*Contd.*)

Name of the Species	Jan. S	Jan. B	Feb. S	Feb. B	Mar. S	Mar. B	Apr. S	Apr. B	May S	May B	June S	June B
7. *Diaphanosoma brachyurum*	20.0	34.2	19.8	48.0	22.0	11.0	35.0	35.1	42.0	23.4	42.0	23.4
8. *D. sarsil*	—	—	—	—	—	—	—	—	—	—	—	—
9. *Alona guttata*	—	—	—	—	—	—	—	—	—	—	—	—
10. *Alonella dentifera*	—	—	—	—	—	—	—	—	—	—	—	—
11. *Sinocephalus serrulatus*	4.0	9.0	19.8	36.2	22.0	23.6	44.0	35.1	30.0	44.2	32.0	44.2
CYCLOPOIDA COPEPODA												
1. *Thermocyslops crassus*	19.6	56.97	19.8	24.0	22.2	27.6	22.0	16.9	30.0	29.6	29.6	30.2
2. *Mesocyclops thermocyclopoides*	19.1	29.85	29.8	22.0	22.2	23.6	35.2	53.3	32.0	34.2	30.0	48.8
3. *Ergasilus* sps.	11.04	8.0	12.6	8.0	22.2	11.0	35.2	16.9	30.0	14.8	30.8	14.8
4. *Copepodites (Cyclopoid)*	155.1	120.1	73.6	109.0	128.8	227.9	107.8	114.4	32.4	58.4	32.8	79.8
CLNOIDA												
1. *Heliodiapomus viduus*	54.0	95.58	47.7	62.00	75.2	102.85	89.2	146.1	91.3	224.5	95.0	241.2
2. *Eediaptomus lumholtzi*	48.40	139.11	58.43	96.0	80.4	115.85	86.9	103.7	144.0	165.7	129.4	175.7
3. *Copepodite (Calanoid)*	22.08	59.4	19.8	48.00	22.20	35.2	35.2	66.1	30.0	36.4	36.4	50.2
4. *Nauplius larvae*	73.44	93.6	159.2	87.0	158.8	91.5	105.4	208.8	111.2	115.5	165.1	134.8

Name of the Species	July		Aug.		Sep.		Oct.		Nov.		Dec.	
	S	B	S	B	S	B	S	B	S	B	S	B
7. *Diaphanosoma brachyurum*	15.5	30.0	20.3	32.4	45.6	72.2	5.9	6.40	18.0	15.34	9.0	11.2
8. *D. sarsil*	—	—	—	—	20.0	4.5	5.0	5.06	—	6.30	—	—
9. *Alona guttata*	—	—	—	—	16.16	2.40	—	—	14.2	15.4	—	22.5
10. *Alonella dentifera*	—	—	—	—	16.16	—	—	—	—	—	—	—
11. *Sinocephalus serrulatus*	6.0	32.4	7.0	32.4	24.0	58.0	—	—	—	—	—	—
CYCLOPOIDA COPEPODA												
1. *Thermocyslops crassus*	10.08	30.48	19.6	30.48	8.0	12.35	23.9	26.2	13.5	25.18	19.6	21.2
2. *Mesocyclops thermocyclopoides*	10.08	32.40	19.6	32.40	8.0	17.57	42.8	47.0	55.3	78.40	19.6	21.2
3. *Ergasilus* sps.	—	—	—	—	16.00	7.5	22.0	31.9	14.1	31.5	13.5	22.5
4. *Copepodites (Cyclopoid)*	67.26	75.96	—	51.94	—	41.8	183.4	200.0	168.7	321.20	171.2	257.6
CLANOIDA												
1. *Heliodiaptemus viduus*	90.0	131.0	26.0	127.0	—	108.6	145.0	71.3	101.8	121.1	49.5	89.5
2. *Eodiaptomus lumholtzi*	121.7	75.0	61.60	72.6	16.0	23.8	103.4	75.13	99.3	139.2	25.0	56.10
3. *Copepodite (Calanoid)*	60.0	44.4	11.0	40.4	—	14.2	5.9	16.5	8.5	18.8	44.1	56.2
4. *Nauplius larvae*	79.6	110.6	48.6	144.8	57.0	69.9	143.0	195.0	159.1	87.53	173.90	192.4

S=Surface B=Bottom

depth. The vertical distribution of cladocera shows its occurrence from surface upto 84 feet depth. Cladocera also shows lesser density upto 84 feet depth. Its populations gradually decline at greater depth. Maximum population of this group were recorded in the month of May and June 15 and 9 feet depth respectively and minimum in the month of October, November and December at 78 feet and 84 feet depth. The vertical distribution of Copepoda population shows its occurrence from surface upto 84 feet depth. The maximum populations were recorded in the month of June or in summer season at 9 and 18 feet depth and 80 feet depth in the month of October, and the minimum populations were recorded in the month of September at 60 feet depth.

Zooplankton species diversity Indices

In the present study the zooplanktonic diversity indices were derived by the application of Shannon Weaver Index (1963) for the Rampura study site (Table 3*ab*).

Species diversity derived by this method of zooplankton populations show the highest "I" values in the month of November and April (3.34) at the surface waters and (3.54) at the bottom in the month of October. Minimum species diversity occurred in the month of July 2.42 and 2.46 at surface and bottom respectively. The "I" values were above 3 from October to May showing this seven months period as the most optimum for zooplanktonic population diversity. The same index was used for deriving the species diversity of different zooplanktonic groups. It was noticed that for Protozoa the "I" values in the surface ranged between 0.26 to 2.14 showing the maximum and minimum values in the month of August and September respectively. In the bottom these values ranged between a maximum of 0.69 and minimum of 0.11 in the same months. For Rotifera the "I" values were maximum in the month of December (1.0) and minimum of June (0.43) at the surface. In the bottom maximum values were recorded in the month of October (0.6) and minimum in the month of March (0.18). Cladocera "I" values ranged between 0.10 in the month of December and 1.09 in the month of September at the surface. In the bottom these values ranged between 0.10 to 0.92 in the month of November and September respectively. These values for copepoda recorded in the month of October (1.8) and minimum in the month of September (0.43) at the surface. Maximum values were recorded in the bottom in the months of April (2.28) and minimum in the month of September (0.6).

Seasonality and periodicity of Zooplankton population

Table 2 show the seasonality and periodicity of the various zoolpanktonic populations at all the five study sites in GSR. Among protozoans *Difflugia* sps, occurred throughout the year. Maximum population of this genus occurred in summer. Least populations occcurred in the rest of the period of the year (Table 2). *Arcella* sps. *Actinophrys* sps. *Vorticella* sps, *Stentor* sps., *Heleopera* sps., *Sphenoderia* sps., *Colpidium* sps., *Opercularia* sps., and *Metopus* sps., populations occurred in monsoon and early winter (September-November) with fast rate of multiplication in September. *Euglypha* sps., however could be recorded from September, October and January to May. Protozoans occurred least in the premonsoon and winter season. Among Rotifera the dominant *Keratella tropica* and *Brachionus falcatus* populations show bimodal fluctuations, maxima with summer and least populations occurred in winter. *Brachionus calyciflorus, B. forficula* occurred from October to March only. *Filinia longiseta* occurred from February to November. *Philodina* sps., *Floscularia* sps. occurred from September to November. *Polyarthra vulgaris* occurred only in winter *Ptygura* sps. are comparatively sparsely represented, ranging in their occurrence from September to January. *Trichotria* sps. was also comparatively sparsely represented in October and November. All these species flourished in the monsoon season. Amongst Ostracoda the species *Cypris obensa, Cyprinotus* sps. and *Stenocypris* sps. are represented from September to June. Among the cladocerans populations *Diaphanosoma brachyurum* occurred throughout the year, with the maximum population in summer. *Daphnia carinata, D. lumholtzi* populations occurred from February to October. *Simocephalus serrulatus, Daphnia pulex*, show a single summer maxima in the month of April. *Moina* sps. and *M. macrocopa* occurred only in the months of August and September. *Alona guttata* and *Alonella dentifera* show a restricted occurrence in the months of September, November and December and only in September respectively. *Ceriodaphnia regaudi* occurred from May to September (Table 2).

Among the Cyclopoid population *Mesocyclops thermocyclopoides, Thermocyclops crassus* and *Ergasilus* occurred throughout the year with the maximum population in summer and winter. Among the calanoids the species *Heliodiaptomus viduus* and *Eodiaptomus lumholtzi* occurred throughout the year with the maximum population in summer.

Spatial distribution

Under this the horizontal column and bottom distribution pattern may be considered. Amongst the protozoa *Heleopera, Sphenoderia, Colpidium, Opercularia* and *Metopus* sps. may be established as predominantly Epilimnic species. A number of the Thecamoebae like *Difflugia, Arcella* etc. may be considered as (Table 2) the profundal forms. The ciliates like *Stentor, Vorticella*, etc. seem to have a restricted distribution in this reservoir in the rainy season and represent predominantly spilimnic forms. The group Rotifera represented by the dominant rotifers by *K. tropica, B. falcatus* and *Filinia longiseta* have higher densities in the surface water and may be classified as spilimnic species. The Rotifers species *Trichotria, Polyarthra. Asplanchna, B. calyciflorus, Philodina* could be collected from this reservoir in the rainy reason only. These forms represent one of the earliest community to flourish in an aquatic ecosystem and predominantly epilimnic. For Rotifer species like *Ptygura* sps. *Beauchampia* sps. *Floscularia* sps. and *Collotheca* sps., represent the profundal or Benthic forms ranging in the profundal area of the Metalimnion. The cladocera in general are surface organisms. However *Simocephalus serrulatus*, in this reservoir have shown greater densities in the bottom in all the months.

During the present study a bimodal distribution of zooplankton was observed. Two peaks were noticed first in summer

(1207 org/l) and second peak in winter (1138 org/l) months. The summer peak constituted the primary peak of the year. The bimodal periodicity of zooplankton over an yearly cycle has been well documented by several workers (Das and Shrivastava 1956, Vasisht and Dhir 1970, Vasisht and Sharma 1975, Awtramani 1980, Sharma 1983, Bohra 1976, Vasisht and Jindal 1980, Saxena 1982, and Pahwa and Mehrotra 1966). But Tandon and Singh (1972) from Nangal observed three peaks of zooplankton from Lakotia lake, while Sumitra (1969) reported only one peak in October from ponds at Madurai. The zooplanktonic peaks mostly coincided with Phytoplankton peaks there by showing direct correlation. This relationship was reported by several workers Shetty *et. al.* (1961), Pahwa and Mehrotra (1966), Tandon and Singh (1972), Sharma (1983), Bhatnagar (1982). Similar observations were made in present study (Table $r=0.90$, $P < 0.01$). But on the other hand, according to grazing theory, the zooplankton and phytoplankton should show an inverse correlation, (Anderson *et. al.* 1955). But it has been found that the zooplankton consume only a certain portion of the total phytoplanktonic community and therefore the above relationship is restricted primarily to the nanoplankton (Porter 1973, 1977) because of easy grazing on small size plankton Kajak (1980) has reported that zooplankton are unable to control the blooms of bigger algae as the feed on smaller ones upto 30 mm. size. Saunders (1969). Ruttner and Kolisko (1974). Moore (1978, 1980) reported dependence of zooplankton on detritus and associated bacterial flora.

The maximum plankton yield is governed mainly by favourable balance between various ecological conditions. Numerous workers like Singh (1966), Vasisht (1968), Mathew (1972) and Vasisht and Sharma (1975) found that low temperature is conducive to plankton production. In the present study optimum temperature for zooplankton development was found to be 22°C to 28°C, the temperature was neither too high nor too low. At higher temperature their number was reduced. Moore (1981) has stated that temperature was limiting to the development of all major zooplanktonic groups. But in the present study the annual observation could not derive any significant correlation between temperature and total zooplankton population. These observations thus appear to be in conformity with the values given by earlier workers like Prasad (1956), Rao (1987, Gupta (1989), Sharma (1983) and Joshi (1987).

Ruttner and Kolisko (1974), Davis (1954) and Gupta (1989) stated that a number of abiotic and biotic environmental circumstances act simultaneously. Das (1978) reported that pH values between 7.3 to 8.4 are considered to be favourable for the growth of planktonic organisms. Joshi (1987) reported highly alkaline sagar lake which has maximum zooplanktonic populations.

Zooplankton density and temporal variations

Among the total zooplanktonic organisms copepods and developing copepodite stages come first in order of abundance in GSR. They exhibited a bimodal pattern with a first peak (548 org/l) in may and June and second peak (672 org/l) in October. Out of this this the cyclopoids population had a density of 92 org/l and the calanoid population shows 235 org/l. The rest of copepod density is due to the' arge number of copepodite and Nauplius larvae recorded in the month of May. The smaller second peak which occurred in the month of October has a general copepod density of 672 org/l, out of which 88.7 org/l of cyclopod and 251 org/l of calanoids and the rest of the copepod density due to the large number of copepodite and Nauplius larvae. The present study thus established copepoda as a largest group in the zooplanktonic organisms in GSR. The calanoid populations were almost double the number of cyclopoid. In general the copepod populations are greater in the bottom of the reservoir (763 org/l in June) when compared to surface cyclopoid population (600 org/l in June). The present study recorded only two dominant species of cyclopoida and two species of calanoida, while reporting the preliminary observations of the copepod fauna of GSR Babar and Choubey (1987) reported 7 species of cyclopoids and 3 species of calanoids, but this study needs further systematic exploration and hence deserves further systematic evaluations.

Second most dominant copepod populations in total zooplankton were reported by Bhattacharya and Saha (perscomm) from Gunti reservoir. In the Shallow zone of Tungabhadra reservoir, Govind (1969) reported four peaks of zooplankton populations out of which one was dominated by rotifers and the others by copepods. Sreenivasan (1970) reported dominant population of cyclopoid copepods in total zooplankton from Aliyar reservoir. Jha *et. al.* (1981) reported copepods population abundant throughout the year from Gobindsagar reservoir, Ayyappan and Gupta (1980) reported calanoid population ranging from 810 to 4790 units/l and cyclopoid population 340 to 1510 from Ramasamundra tank. Dad (1981) reported highest population in the month of June from clear water area of Chambal river. Joshi (1987) reported second most dominant population of copepods throughout the year from Sagar lake. Pahwa and Mehrotra (1966) reported that copepoda was second most dominant in total zooplanktonic population throughout the year wherein they constituted 294 and 646 units/l during the peak month of June from Ganga river. In this regard it resembles to GSR copepods peaks were recorded in June. Peaks of copepod population in summer were reported by Kofoid (1908) from Illinios river and George (1986) in the upstream of Tawa river. Gupta (1989) reported copepod population throughout the year in Gulabsagar and Gangloan, wherein they constituted 21.27% and 16.21% respectively. This could be due to the highly polluted nature of these two aquatic bodies of Jodhpur. The present study reports population of copepods in GSR, wherein constitute 11.91% (September) to 75.00% (October) of copepods out of total zooplanktonic population. Rao (1987) reported a range of 11.67 to 84.58% of copepods in total zooplankton, from Rangasagar lake of Udaipur.

Among the total zooplanktonic population, cladocera come second in order of abundance in GSR. They exhibited a bimodal pattern with a major peak in June (348 org/l) and minor peak (274 org/l) in September. In the Shallow zone of Tungabhadra reservoir Govind (1969) reported second dominant cladoceran peak in March to May. Ganapathi and Pathak (1969) reported 30% zooplankton with second most dominant

Table 3a. Details of monthly species diversity indicies of zooplankton and phtoplankton as study site 1 (Rampura) during 1988

	Jan.		Feb.		Mar.		Apr.		May		June	
	S	B	S	B	S	B	S	B	S	B	S	B
1. Zooplankton Values as per Shannon Weaver Index (1963)	3.31	2.75	3.01	3.16	2.95	2.95	3.34	3.13	3.03	2.79	2.79	2.84
2. Phytoplanktonic D. Values by Margalef's Index (1967)	1.04	1.06	1.12	.99	1.15	1.09	1.10	1.07	1.06	.939	1.04	.912

	Jul.		Aug.		Sep.		Oct.		Nov.		Dec.	
	S	B	S	B	S	B	S	B	S	B	S	B
1. Zooplankton Values as per Shannon Weaver Index (1963)	2.42	2.46	2.87	2.76	3 35	3.18	2.68	3.54	3.34	3.33	2 79	2 93
2. Phytoplanktonic D. Values by Margalef's Index (1967)	.527	.097	.631	.31	.80	.50	.916	.764	.77	.74	1.02	.854

S=Surface B=Bottom

Table 3b. Shannon Weaver Index between species (zooplankton) with group diversity

	Jan.		*Feb.*		*Mar.*		*Apr.*		*May.*		*June*	
	S	*B*	*S*	*B*	*S*	*B*	*S*	*B*	*S*	*B*	*S*	*B*
(1) Protozoa	.458	.283	.30	.27	.36	.32	.51	.46	.56	.39	.31	.22
(2) Rotifera	.99	.41	1.16	.26	.65	.18	.55	—	.44	—	.43	—
(3) Cladocera	.13	.14	.45	.67	.54	.54	.75	.71	.74	.75	.84	.77
(4) Copepoda	1.59	2.14	1.29	1.25	1.30	2.20	1.25	2.28	1 37	1.42	1.37	2.00

	Jul.		*Aug.*		*Sep.*		*Oct.*		*Nov.*		*Dec.*	
	S	*B*	*S*	*B*	*S*	*B*	*S*	*B*	*S*	*B*	*S*	*B*
(1) Protozoa	—	—	.26	.11	1.14	69	.74	.62	.53	.30	.40	.30
(2) Rotifera	—	—	.91	—	.94	.30	.80	6.0	9.8	.50	1.0	—
(3) Cladocera	.12	.34	.68	.92	1.09	1.01	.27	.17	.10	.10	—	—
(4) Copepoda	1 80	2.20	.67	1.40	.43	.60	1.8	1.4	1.2	1.8	1.2	2.0

S=Surface B=Bottom

cladoceran population from Sayaji sarovar. June or July constituted the peak months cladoceran population according to Ganapathi (1964), who reported 27.3% cladoceran population from Almati reservoir. Abraham (1980) reported that cladoceran population were observed very rarely in Bhavanisagar reservoir. Jha *et. at.* (1981) reported that abundance of cladoceran populations during study period from Gobindsagar reservoir. Ayyappan and Gupta (1980) reported dominant cladoceran population in the rainy season at Ramasamundra tank. George (1986) reported cladoceran populations were present in the upstream and absent in down stream of Tawa river. Kar *et. al.* (1987) reported cladoceran population was the second dominant population from river Ib. George *et. al* (1980) reported cladoceran population as the second most dominating population (26.3%) during study period (Monsoon season) from Ujjain area. Rao (1987) reported cladoceran population third in order of abundance from lake Rangasagar Udaipur.

Rao *et. al.* (1987*e*) reported seasonal polymorphism in the cladoceran populations from three fresh water bodies of Ujjain. Gupta (1989) reported dominant population of cladoceran (31.11%) in Gulabsagar and restricted population of 3.98% at Gangloan. Pahwa and Mehrotra (1966) reported that calodceran occurred mainly during May, July and December in moderate number 14 units/l), while during the rest of the months they were either all together absent or present only in very negligible numbers from river Ganga. Gautam (1988) reported that cladoceran populations showed definite periodicity and registered maxima during September, January and February in upper lake Bhopal. Das and Shrivastava (1956 *a*) showed peak periods of cladocerans in December and January from Fish tank in Lucknow. Vasisht (1968) reported their maxima during May and June from Sukhana lake (Chandigarh). In the present study excepting *Diaphanosoma* and *Daphnia*, no cladocerans could be recorded in the winter season as reported by several workers for which the reason may be low temperatures and other physico-chemical regimes in these aquatic bodies. In the present study a highest peak of cladoceran population recorded in summer It can be said that the summer peak of this population prefer higher temperature, conductivity, calcium, total alkalinity and pH. This group shows positive relationship between temperature and conductivity ($r=0.70$ $P<0.01$ and $r=0.73$ $P<0.01$) respectively. Contrary reports by some authors, Rao (1984) reported the cladoceran distribution in the lake Jaisamand to be influensed by the bottom anoxic layer. In the present study peak of cladoceran populations were recorded in the month of September. It can be said this population also prefer high nutrients content, and this season was reproductive period of this group. The present study established *Daphnia, carinata, D. lumholtzi, Diaphanosoma brachyurum* and *Simocephalus serrulatus* as the most dominant cladoceran populations of GSR.

Among the total zooplanktonic orgnisms rotifers come third in the order of abundance in GSR. They exhibited a bimodal pattern with a major peak (347 org/l) in December and a minor peak (260 org/l) in August rotifer populations were reported as a dominant group by several workers. Govind (1969) reported a rotifer peak in February (24.7%) out of the total zooplankton from the shallow zone of Tungbhadra reservoir. Pahwa and Mehrotra (1966) reported rotifer population from Ganga river. Where in they constituted 61.5 to 94.4% exceeding 2000 unit/l, in June and a smaller peak in December. Ganapathi and Pathak (1969) reported a most dominant rotifer population from Sayaji sarovar. Ayyappan and Gupta (1980) reported rotifers as dominant population from Ramasamudra tank. It range of 0.14 to 69.97% of the total in October and August respectively. George *et. al.* (1980) reported from Ujjain aquatic bodies that rotifers were the largest single group (57.9%) represent throughout the period of the study (monsoon months). Chaourasia and Adoni (1986) reported most dominant rotifer populations in Sagar lake. The dominance of rotifers over other planktonic fauna was also reported by Moitra and Bhoumick (1968). Vasisht and Sharma (1975) and Kar *et. al.* (1987) from river Ib. However in the reservoirs rotifers usually show a bimodal pattern as recorded in the present study. There seems to be a general confirmity in this regard. George (1966) recorded maximum number of rotifer species during summer. In the present study, the population dynamics of *Keratella tropica* and *Brachionus falcatus* show the same trend, but other species show a different trend, these populations are absent in summer and show a peak occurrence in winter, contrary to this Gupta (1989) from two ponds near Jodhpur reported a major rotifer peak in August and in February. Thus the present study shows an intermediary position in which 3 species of rotifers shows summer peaks and 9 species absent in summer and show monsoon and winter peaks.

The present study reported preference of higher temperature range by *Keratella tropica*, *Brachionus falcatus* and *Filinia longiseta* and preference of a lower temperature range by the other nine rotifer species. Contrary reports exist in this regard, where as preference of higher temperature range was reported by Pahwa and Mehrotra (1966), Govind (1969). Das and Shrivastava (1956), Gulati (1964) and Vasisht (1968) observed summer maxima in tropical waters. In the present study also rotifer showed positive relationship with water temperature ($r=0.40$ $P<0.05$), D.O. ($r=0.63$ $P<0.01$), transparency ($r=0.85$ $P<0.01$) and conductivity ($r=0.33$ $P<0.05$), while Green (1960) and Shetty *et. al.* (1961) reported a lower temperature range favoured the population build up of rotifera. Vasisht (1968) pointed out the correlation between pH and rotifers fauna.

Among the total zooplanktonic organisms protozoans come fourth in the order of abundance in GSR. They exhibited a bimodal pattern with a major peak (156 org/l) in September and a minor peak (213.0 org/l) in May and least population occurred in August. Abraham (1980), Joshi (1987), Jha *et. al.* (1981) and Dad (1981) reported protozoans throughout the year in Bhavanisagar reservoir, Sagar lake, Gobindsagar and Chambal river respectively. Ayyappan and Gupta (1980) reported protozoans in Februrry from Ramasamudra tank. Pahwa and Mehrotra (1966) reported protozoans in very small number from Ganga river. Rao (1987) reported protozoans population abundance throughout the year but least population occurred in July and August months, resembling the pattern in GSR reported above. Rao (1987) reported direct relationship of protozoans incidence with pH, CO_2 and D.O. Least occurrence in monsoon was relegated to a reduction of these values from Rangasagar lake. The present author supports this view on the basis of the recorded data from GSR.

Ostracoda contribution in GSR come last in order of abundance ranging from 6.11% to 13.7%. They exhibited bimodal pattern in their distribution showing two peaks, one in November (116 org/l) and second in February (133 org/l). Bhattacharya and Saha (pers. comm.) reported 15% Ostracod population of the total zooplankton from the Gumti reservoir representing the third dominant planktonic group. Jha *et. al.* (1981) reported abundant Ostracod population from Gobindsagar reservoir. Ayyappan and Gupta (1980) reported Ostracod only in the month of March from Ramasamudra tank. Gupta (1989) reported Ostracoda in the month of July, October, November, May and June from Gulabsagar. Bhattacharya and Saha (1988) reported Ostracoda as the second most dominant group in total zooplankton from Gumti river. Mahadik (1983) reported Ostracod population in 9 months period, from aquatic bodies of Ujjain area.

Zooplanktonic species diversity index

In the present study species diversity calculated by Shannon Weaver index I values for Protozoa as ranging between 0.26 in August and 1.14 in September, for Rotifera the same values ranged between 0.44 in May to 1.0 in December, for Cladocera the same values ranged between 0.10 in November to 1.09 in September and for copepoda these values ranged between 0.43 in September to 1.8 in October. Such information were not available on other Indian reservoirs. The total zooplanktonic diversity indices in this reservoir ranged from a minimum of 2.42 in July and a Maximum of 3.34 in November and March. Sreenivasan (1964) reported lesser values from Bhavanisagar reservoir and David *et. al.* (1969) reported from Tungabhadra reservoir. Values greater than the GSR, I values given above were described from Gobindsagar reservoir by Jha *et. al.* (1981). It may hence be suggested that with regard to zooplanktonic species diversity, GSR may be assigned an intermediary status.

ACKNOWLEDGEMENTS

The author expresses her sincere gratitude to Dr. K.S. Rao for his kind help in preparing the manuscript. The author is also thankful to Prof. and Head of the Department, School of Studies in Zoology, Vikram University, Ujjain for providing the necessary facilities to complete this work.

REFERENCES

Abraham, M. (1980) : Studies on the plankton of Bhavanisagar Reservoir with special reference to climatological conditions inflow and water level, *J. Inland. fish. Soc. India Vol.* 12 No. 2 pp. 25 to 38.

Anderson, G.C., Comita, G.W. and Engstron, H.V. (1955) : A note on the phytoplankton Zooplankton relationship in two lakes in Washington. *Ecology* 36 : 756-759.

APHA, AWWA and WPCF (1985) : *Standard methods for the examination of water and waste water.* 16th edition, APHA Inc. New York, 1268.

Awatramani, G M. (1980) : *Limnological studies of Sagar lake,* Ph. D. Thesis, Univ. of Sagar, Sagar. M.P. 204.

Ayyappan, S. and Gupta, T.R. Chandrashekharan (1980) : Limnology of Ramasamudra Tank. *J. Inland Fish. Soc. India Vol.* 12, (No. 2) : 1-12.

Bhattacharya, T. and Saha, R.K., (Perse. Comm.) : A comparative study of the Physico-chemical conditions of the water and planktons of Gomti reservoir, open channel and river below Hydel Project in Tripura.

Bhattacharya, T. and Saha, R.K. (1988) : A comparative study of the physico-chemical properties and plankton density of two streams and a Tributary of Gomti River in Tripura. *Geobios* 15 : 201-206.

Bhatnagar, G.P. (1982) : *Limmology of lower lake Bhopal with reference to sewage pollution and entrophication.* N A.B Project report 1979-1982. 77.

Chaourasia, S.K. and Adoni, S.K. (1986) : Seasonal variation of Rotifers in an Eutrophic lake *Proc All. India* Sem. on water quality. Around. Urban Ecosystem Ed/K.S. Unni, Govt. P.G. College, Chhindwara, M.P. 107-116.

Dad, N.K. (1981) : *Limnological studies on Chambal river with reference to pollution.* Ph. D. Thesis, Vikram Univ. 227.

Das, S.M. and Shirivastava, V.K. (1956) : Quantitative studies on freshwater plankton, I. Plankton of a fish tank in Lucknow, India. *Proc. Nat. Acad. Sci. India* 36 (2) B : pp. 85-92.

Das, S.M (1978) : High pollution in lake Nainital, U.P. as evidenced by biological indicators. *Sci. and Cult.* 44 (5) : 236-237.

David, A., Ray, P., Govinda, B.V., Rajagopal, K.V. and Banerjee, R.K. (1969) : Limnology and Fisheries of Tungabhadra reservoir. *Bull. Cent Inland. Fish Res. Inst.* 13 : 1.188.

Davis, C.C. (1954) : A preliminary study of the plankton of cleaveland harbour area Ohio. II. The distribution and quantity of the phytoplankton. *Ecol. Monogr.* 23 : 321-347.

Delorme, L.D. (1970a) : Freshwater ostracods of Canada. Part I Sub-family Cypridinae, *Can. J. Zool.* 48 : 153-168.

Delorme, L.D. (1970b) : Freshwater ostracods of Canada. Part II. Sub-family cypridopsinae, Herpetro-cypridinae and family cyclocy-prididae. *Can. J. Zool.* 48 : 253-266.

Delorme, L D. (1970c) : Freshwater ostracods of Canada Part IV. Families Ilyocypridiade, Motodromadidae Darwinuidae, Cytheridae and Entocytheridae. *Can. J. Zool.* 46 : 1251-1259.

Dubey, G.P. and Chatterjee, S.N. (1976) : "Case study of Gandhisagar Reservoir, M.P. India". *IPFC 76 Sym. 27* July.

Edmondson, W.T. (1959) : *Freshwater Biology* : John. Willey and Sons. Inc. New York.

Fernando, C H. (1980) : The freshwater Zooplankton of Sri Lanka, with a discussion of tropical freshwater Zooplankton composition *Int. Revueges. Hydrobiol.*, 65 : 85-125.

Ganapathi, S.V. and Pathak, C.H. (1969) : Primary productivity in the Sayaji Sarovar (a man made lake) at Baroda. *In. Seminar on the Ecology and fisheries of freshwater reservoir.* ICAR at CIFRI, Barrackpore 27-29.

George, J.P., Chaturvedi, S.P., Rao, K.S. and Swarup, H. (1980) : Limnological studies on the lentic habifats in and around Ujjain. II Diurnal variations in a temple tank containing abundant plankton. *Matsya* 6 : 9-18.

George, M.P. (1986) : Effects of Bottom Ash (Sarni Thermal Power Station) on the distribution of Zooplankton in Tawa river. *Proc. All. India. Sem. on water quality Around Urban Ecosystem* (Ed.) K.S. Unni 18-23.

Govind, B.V. (1969) : Bottom fauna and Macrovegetation in the Tungabhadra reservoir and their role in the food chain of fish com-munities. *Proc. Sem Ecol Fish. F.W. Reservoir.* Sponsored by ICAR at CIFRI Barrackpore from Nov. 27-29.

Green, J. (1960) : Zooplankton of the river Sokoto, The rotifera. *Proc. Zool. Soc.*, London, 135, 491-523.

Gulati, R.C. (1964) : *Limnological studies on some North Indian lakes and Reservoir.* Ph. D. Thesis Univ. Delhi.

Gupta, S. (1989) : *Pollution Ecology of some ponds in Urban vicinity of Jodhpur*. Ph. D. Thesis University of Jodhpur. 1-234.

Jha, B.C., Kaushal, D.K. and Rao, Y. Rama. (1931) : Observations on qualitative abundance of plankton and Periphyton in Gobindsagar Reservoir. *J. Inland. Fish. Soc. India*, Vol. 13, No-2. Dec. 126-129.

Joshi, G. (1987) : *Diurnal studies of Physico-chemical properties of Limnological importance*. Ph. D. Thesis. Dr. Harisingh Gaur Vishwa Vidhyalaya, Sagar 204.

Kartha, K.N. (1989) : Studies on experimental Trawl fishing in Gandhisagar Reservoir, Ph. D Thesis. Vikram University, Ujjain.

Kar, G.K., Mishra, P.L., Das, M.C. and Das, R.C. (1987) : Pollution studies in river Ib III : Plankton population and primary productivity. Indian *J. Environ Hlth*-29(4).

Khan, M.A. and Ejike, C. (1984) : Limnology and plankton periodicity of Jos plateau water reservoir, Nigeria, West Africa. Hydrobiologia 114 : 189-199.

Kofoid, C.A. (1903) : Plankton studies IV. The Plankton of Illinois river, 1994-1899 with introductory notes upon the hydrography of the Illinois river and its basin. Part I. Quantitative investigations and general results. *Bull. III Nat. Hist. Sury.* 6(2) : 95-624.

Kofoid C.A. (1908) : Plankton study V. The plankton of the Illinois river 1894-1899 Part II. Constituent organisms and their seasonal distribution. *Bull. III nat. Hist. Sury.* 8 (1 and 2) : 1-361.

Mahadik, P. (1983) : *Studies on some crustacean fauna (Ostracoda and conchosteraca) and their population dynamics in the equatic bodies of Ujjain area.* M. Phil. Thesis Vikram University, Ujjain. 113.

Mathew, P.M. (1972) : *Limnological investigation on the plankton of Govindgarh lake and its correlation with Physico-chemical factors. Primary productivity and fish productivity.* Ph. D. Thesis, Agra Univ. India.

Moitra, S.K. and Bhoumick, M.L. (1968) : Seasonal cycles of rotifers in a freshwater fish pond in Kalyani, West Bengal. *Proc. Symp. Adv. Trop. Ecol.* Part I : 359-367.

Moore, W.J. (1978) : Distribution and abundance of phytoplankton cycles in 153 lakes, rivers and pools in the Northwest Territories *Can. J. Biol.* 56 : 765-1173.

Moore, W.J. (1986) : Zooplankton and related phytoplankton cycle in a eutrophic lake. *Hydrobiologia.* 74 : 99-104.

Moore, W.J. (1981) : Influence of temperature, photoperiod and trophic conditions on the seasonal cycles of phytoplankton and zooplankton in two deep subarcfic lakes on northern Canada. *Int. Rcvue. Ges. Hydrobiol.*, 66(5) : 745-770.

Pahwa, D.V. and Mehrotra, S.N. (1966) : Observations on fluctuations in the abundance of plankton in relation to certain hydrological conditions of river Ganga. *Proc. Nat. Acad. Sci. India*, 36(2) : 157-189

Pennak, R.W. (1978) : *Freshwater invertebrates of the United States*, 2nd edition, John Wiley and Sons Inc , New York, 803.

Porter, K.G (1973) : Selective grazing and differential digestion of algae by Zooplankton. *Nature*, 244 : 179-180.

Porter, K G. (1977) : The plant animal interface in freshwater ecosystem. *Am Sci.* 65 : 159-170

Prasad, R.P. (1956) : Further studies on the plankton of the inshore waters of Mandapam. Indian *J. Fish.*, 3 : 1-42.

Raghunathan, M B. (1986) : Cladocera (crustacea) of certain urban water bodies *Proc. All India Sem. on water quality around urban Ecosystem.* (ed.) K.S. Unni. 5-8

Rao, K S , Shrivastava, S., Gadia, S. and Patidar, D. (1987d) : on the occurrence of *Corvospongilla lapidosa* in the benthic sediments of Gandhisagar reservoir. *Perspectives in Hydrobiology*. Eds/K S. Rao and S. Shrivastava. Sec. IV (37) : 189-191

Rao, K.S Shrivastava, S., Sreenivasan, P. and Choubey U., (1987c) : Studies on the cyclomorphic changes and environmental relations in some cladocerans of Ujjain. "*Perspectives in Hydrobiology*". Eds/ K.S. Rao and S. Shrivastava Sec. IV (36) : 183-186.

Rao, K.S. and Shrivastava, S. (1989a) : Studies on Biological monitoring of water quality in Chambal and Khan river of Central India. *Geobios* 16 : 78-82.

Rao, K.S. and Shrivastava, S. (1989b) : Hydrobiological studies on Gandhisagar reservoir : Observations of water discharge and silt load from Kshipra river region of the Catchman area. *The Vikram* (In Press).

Rao, K.S., Gupta, D.R., Choubey, U. and Sreenivasan, P. (1987b) : Benthic studies in Gandhisagar reservoir with special reference to Macrozoobenthos. *Perspectives in Hydrobiology*. Eds/K.S. Rao and S. Shrivastava, Sec. IV (34) : 173-178.

Rao, K.S., Jain, S., Dubey, R., Shrivastava, S., Choubey, U. and Pandya, S.S. (1987c) : Ecological significance of aquatic insect populations at "Kaliadeh" Kshipra river Ujjain. *J. Hydrobiol. Vol.* 3 No. (6) : 83-92.

Rao, K.S., Kartha, K.M., Gupta, D.R., Pandya, S.S. and Iyer, H.K. (1988a) : Studies on Morphometry and Hydrology of Gandhisagar reservoir with special reference to its fisheries. *J. Fish. Tech. Cochin.* Vol. 25 No. 1 1. 21-28.

Rao, K.S., Shrivastava, S. Choubey, U. (1988b) : Surficial sediments of Gandhisagar reservoir. *Arch. Hydrobiol.* 147-160.

Rao, M.G. (1987) : *Synecology of the lake Rangasagar in relation to Limnology and Eutrophication.* Ph. D. thesis, Sukhadia University, Udaipur, pp. 189.

Ruttner, F. and Kolisko, A. (1974) : Plankton Rotifers. Biology and Taxonomy. *Die Binnen. Gerasser Suppl.*, 26 : 1-146.

Saunders, G.W. (1969) : Some aspects of feeling in Zooplankton. In Eutrophication-causes, consequences, correctives. *Nat. Acad. Sci. Washington*, 556-573.

Saxena, M.M. (1982) : *Limnological studies of freshwater reservoir, Sardarsamand.* Ph. D. Thesis. Univ. of Jodhpur, Jodhpur India, 122.

Sehgal, K.L. (1983) : *Phytoplanktonic copepods of freshwater Ecosystem.* Published by Interprint, Mehta House, Maraina II, New Delhi 169.

Shannon, C.E. and Weaver, W. (1963) : *The mathematical theory of communication* : Urbana University of Illinois Press 177.

Shetty, H.P., Saha, S.B. and Ghosh, B.B. (1961) : Observations on the distribution and fluctuations of plankton in the Hooghly Mattah Estuarine system, with notes on their relation to commercial fish landings. *Indian. J. Fish.* 8(2) : 326-363.

Singh, M. (1966) : Planktonic algae from Najafgarh lake III. *Res. Bull.* (M.S.) *of Punjab Univ.* 17(1-2) : 71-75 and 99-103.

Sreenivasan, A. (1970) : Limnological studies on Parambikulam Aliyar Project I Aliyar reservoir (Madras State), India. *Schewiz. Hydrol. birkhauserverlog Basel* 32. 405-417.

Sumitra, V. (1969) : *Limnological studies of some tropical ponds.* Ph. D. Thesis Madras University.

Tandon, K.K. and Singh, H. (1972) i Effect of certain Physico-chemical factors on the plankton of the Mangal lake *Proc. Ind. Acad. Sci.* 76 (1-8) 15-25.

Tonapi, G.T. (1980) : *Freshwater Animals of India* Oxford and IBH Publishing Co. New-Delhi. 341.

Vasisht. H.S. (1968) : Limnological studies of Sukhna lake, Chandigarh (India) *Proc. Symp. Recent. Adv. Trop. Ecol.* (Part I), 316-325.

Vasisht, H.S. and Dhir, S.C. (1970) : Seasonal distribution of freshwater zooplankton of four tanks, Chandigarh (India) *Ichthyol ogica*, 10 : 44-56.

Vasisht, H.S. and Sharma, B.K. (1975) : Ecology of a typical urban pond in Ambala city of Haryana state (India). *J. Ecol.* 2(1) : 79-86.

Vasisht, H.S. and Jindal, R. (1980) : Rheological Survey of a pucca stream of Patiala (Panjab, India) *Limnologica* (Berlin), 12(1) : 77-83.

Wetzel, R.G. (1975) : "*Limnology*" W.B. Saunders Co. Philadelphia, U.S.A., 743 pp.

3

PRIMARY PRODUCTION BASED TROPHIC CLASSIFICATION—ITS SUITABILITY FOR KASHMIR LAKES—CASE STUDY OF A WARM MONOMICTIC LAKE

K.K. Vass

Central Inland Capture Fisheries
Research Institute (ICAR)
Barrackpore : West Bengal

INTRODUCTION

Natural eutrophication is a slow process of enrichment and is part of aging phenomenon. In nature eutrophication is a beneficial process in that it produces enhanced productivity in the water mass. However, man-made accelerated eutrophication of inland waters has been viewed as an environmental degradation by many workers (Rodhe 1969 ; Vollenweider 1984). This invariably results in deterioration of water quality, which in many cases causes significant economic loss.

Multifactorial approach has been used to evaluate and monitor trophic condition in inland waters. Among other parameters, for rapid trophic assessment, primary production data has been extensively used. Different trophic classifications have been developed based on yearly production estimates and shape of production profiles (Vinberg 1961, Findenegg 1965).

Considerable information about general limnology is available on a number of lakes in the Kashmir Himalayas (Kaul and Vass, 1970 ; Zutshi and Vass 1972 ; Kaul 1967, Vass 1973 ; Zutshi and Vass 1982 ; Vass *et al.* 1988). These workers have classified, Kashmir lakes on various physico-chemical and production features, the cut off concentrations for different parameters have been basically developed on European lakes. However, an attempt has been made here to assess the suitability of this production based trophic classification for Kashmir lakes, since climatically Kashmir is different from the European conditions.

AREA OF STUDY

Apart from drawing comparison with other lakes, investigations were conducted on Naranbagh lake situated about 35 km north of Srinagar. The moderate inland climate of Kashmir gives it relatively high summer temperatures and not too low winter temperature. Lake remains ice-free round the year. Geologically the catchment area consists mainly of altered rocks of various geological age from Purana carboniferous to Eocena. Main morphometric features are tabulated below which give a picture of basis characteristics.

Table 1. Showing the main morphometric features of lakes

Lake surface above sea level (m)	1587
Area of the lake surface (ha)	23.8
Maximum depth (m)	5.0
Mean depth (m)	1.38
Maximum length (m)	780
Maximum width (m)	410
Volume ($m^3 \times 10^3$)	330
Development of volume	0.83

RESULTS

Physico-Chemical Limnology

The data on secchi disc transparency is presented in Fig. 1. In general, it was high in winter and spring while low values were recorded in summer and autumn. Light extinction coefficient varied from 0.34 to 0.8. Thermal behaviour of the lake followed a seasonal trend with no clear-cut thermocline detected. Seasonal pattern of water chemistry is set in Table-2. Concentrations of various parameters indicate that lake water to be moderately enriched.

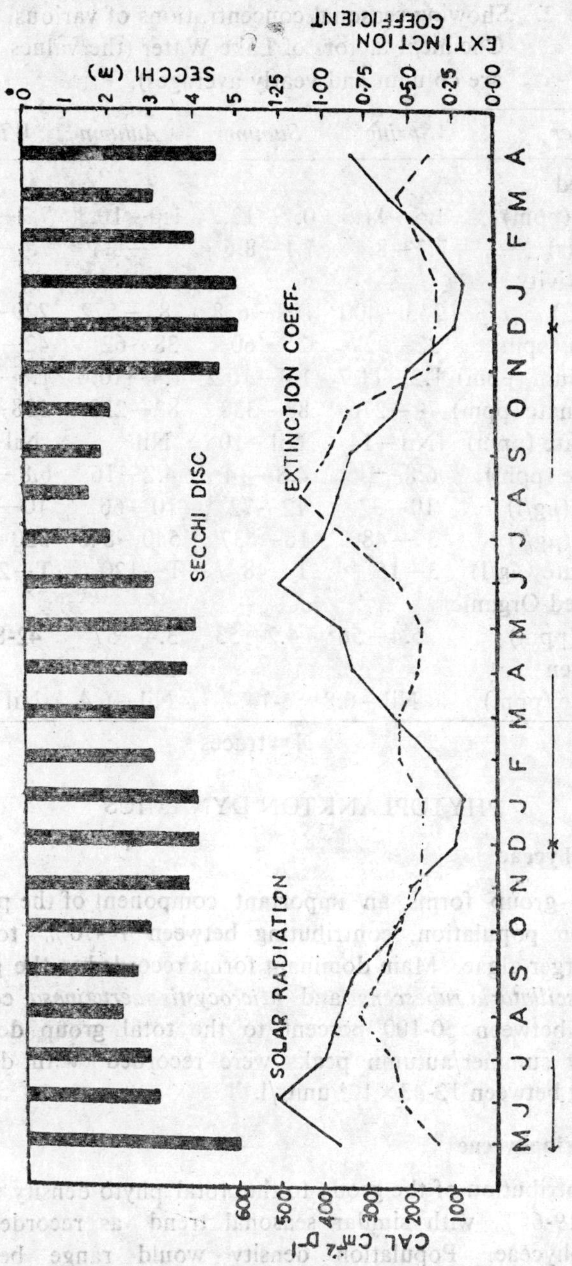

Fig. 1. Showing relationship of Secchi disc, Extinction coefficient and solar radiation.

Table 2. Showing seasonal concentrations of various Chemical factors of Lake Water (the values are column and yearly averages).

Parameter	*Spring*	*Summer*	*Autumn*	*Winter*
Dissolved Oxygen (ppm)	1.5—11.3	0.7—12.2	1.0—10.3	7.7—14.0
pH (units)	7.2—8.4	7.1—8.6	7.2—8.4	7.4—8.2
Conductivity (μS/25°C)	253—400	198—638	283—572	229—419
Calcium (ppm)	32—92	25—60	38—62	42—64
Magnesium (ppm)	1.2—11.7	1.4—10.2	1.4—10.0	1.8—13.4
Bicarbonate (ppm)	79—276	88—330	82—230	108—168
Carbonate (ppm)	Nil—14	Nil —10	Nil	Nil—1.4
Chloride (ppm)	6.8—14	4.8—14	4.2—16	6.8—8.8
Silicate (μg/l)	10—32	12—72	10—66	10—24
Nitrate (μg/l)	32—480	16—637	540—840	320—600
Phosphate (μg/l)	3—15	T—48	T—120	T—24
Dissolved Organic matter (ppm)	5.4—56	4.2—53	3.4—57	42-84
Hydrogen Sulphide (ppm)	Nil—6.8	6-14.4	Nil—6.4	Nil

T=traces

PHYTOPLANKTON DYNAMICS

Cyanophyceae

The group forms an important component of the phytoplankton population, contributing between 10-70% to the total larger algae. Main dominant forms recorded in the group were *Oscillatoria rubescens* and *Microcystis aeruginosa*, contributing between 50-100 percent to the total group density. Usually summer/autumn peaks were recorded with density ranging between 12-43 $\times 10^4$ units/l.

Bacillariophyceae

Contribution of the group to the total phyto-density varied from 19-65% with similar seasonal trend as recorded for cyanophyceae. Population density would range between

5-22 × 10^4 μ/l. Group was mainly represented by *Fragillaria*, *Eunotia* and *Asterionella*.

Chlorophyceae

Mainly represented by *Cosmarium*, *Pediastrum*, *Scenedesmus* and *Ankistrodesmus*. The group recorded mainly summer peaks with density ranging between 0.5-18 × 10^4 μ/l contributing 6-34% to the total phytoplankton density.

Chrysophyceae

Dinobryon was the representative of this group, the species appeared to favour low temperature since its peak was usually recorded during autumn months. It contributed 11-48% to the total phytoplankton density, recording a population range of 5-60 × 10^4 μ/l.

Algal biomass

A comparison of biomasses of the different groups showed that Cyanophyceae maintained high level of fresh-weight. Its maximum contribution of 95% was recorded in September-October and minimum of 72% in February. Bacillariophyceae usually recorded small biomass, with maximum contribution of 15% to the total biomass. Similarly contribution of Chlorophyceae and Chrysophyceae was 14% and 2% respectively to the total algal biomass.

PRIMARY PRODUCTION

In an aquatic system, algal photosynthesis adds chemical energy and organic matter to ecosystem, much of which becomes available to consuming organisms. The range of variation in daily integral photo-assimilation depends on the seasonal changes of environmental conditions. In the present study, the daily rates of phytoplankton production ranged between 93—272 mg cm^{-2} d^{-1} in winter and 1380–2844 mg cm^{-2} d^{-1} in spring/summer (Fig. 2). Fig. 3. shows primary production as a function of depth. Normally, unimodal peak was registered with p-max at 2 m, but at times bimodal trend was noted with secondary peaks at 1 m and 3 m zones.

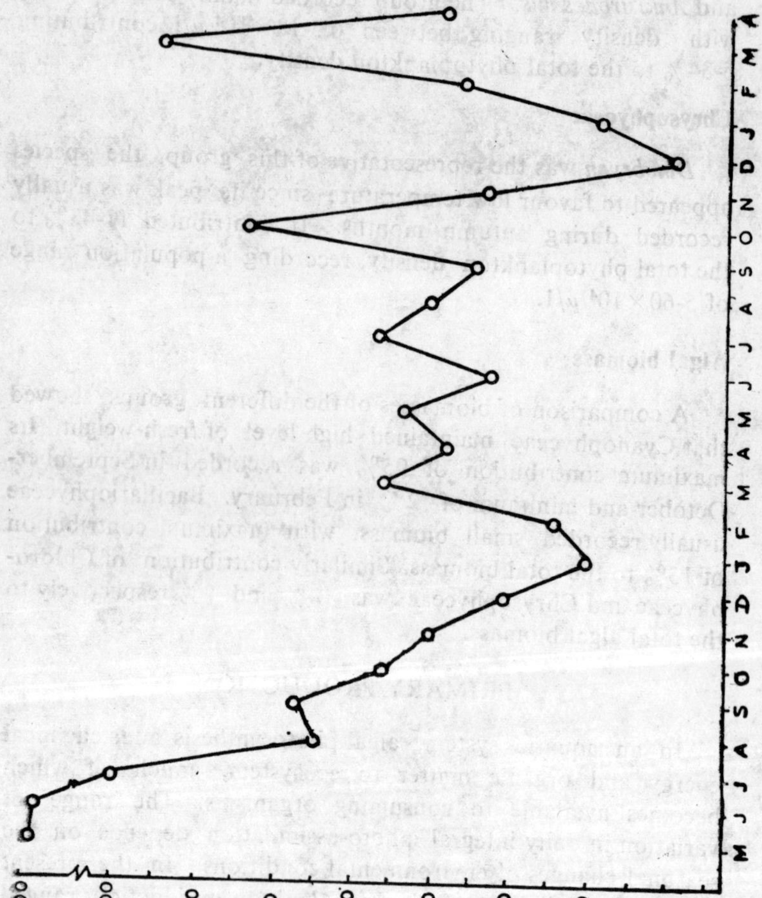

Fig. 2. Seasonal trend of daily phytoplankton Productivity.

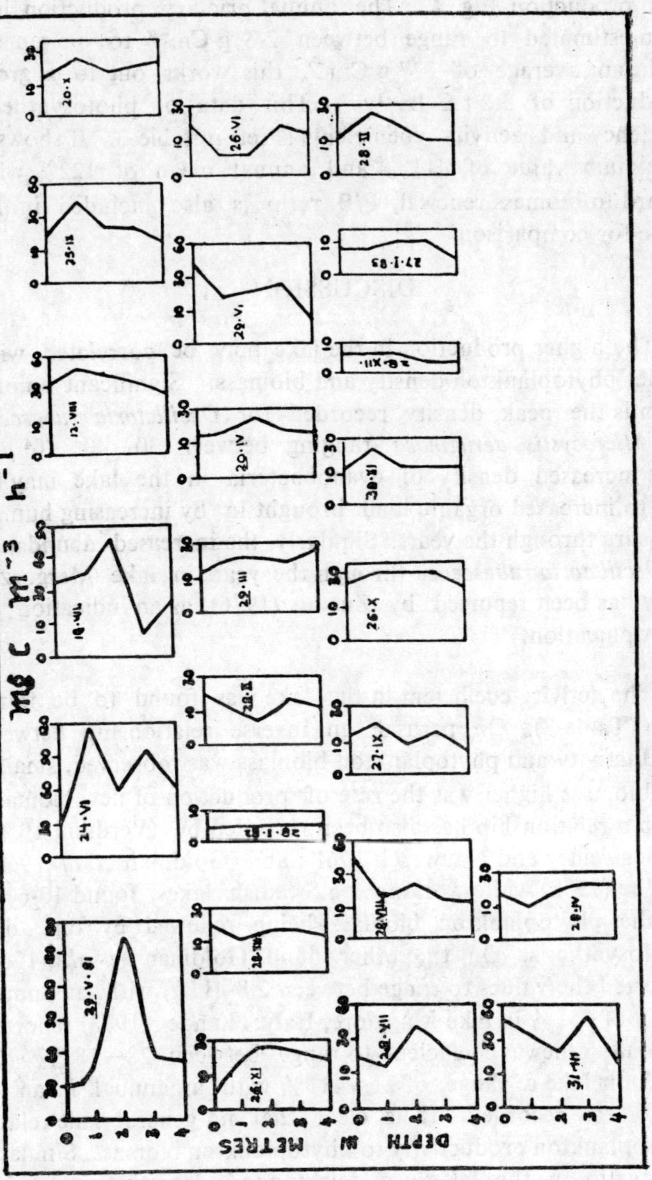

Fig. 3. Primary production as a function of depth on various dates.

The production curve almost follows the biomass fluctuations and similar pattern was evident in case of solar radiation and production Fig. 4. The annual primary production has been estimated to range between 249 g Cm^{-2} to 405 g cm^{-2} with an average of 327 g Cm^{-2}, this works out to a gross production of 3.2 t C $ha^{-1}y^{-1}$. This data on photosynthetic efficiency and activity coefficient is set in table-3. It shows a maximum value of 411% and annual mean of 122% with regard to biomass renewal, P/B ratio is also included in the table for comparison.

DISCUSSION

The higher production in the lake may be correlated with higher phytoplankton density and biomass. Significant among them is the peak density recorded for *Oscillatoria rubescens* and *Microcystis aeruginosa* ranging between $30-40\times10^4$ μ/l. This increased density of cyanobacteria in the lake may be due to increased organic load brought in by increasing human pressure through the years. Similarly, the increased abundance of *Oscillatoria rubescence* through the years in lake Mergozzo. Italy has been reported by Zutshi (1976) as an indication of eutrophication.

The activity coefficient in the lake was found to be fairly high (Table-3). On perusal, an inverse relationship between productivity and phytoplankton biomass was observed, smaller the biomass higher was the rate of production of new biomass. Such a relationship has also been recorded by (Verduin, 1959 ; Vollenweider and Nauwerck 1961 ; and Goldman *et. al.* 1968). Rodhe (1958) while working on Swedish lakes, found 10-60% of the phytoplankton biomass being renewed by one day photosynthesis. On the other hand Goldman *et al.* (1968) reported the values to range between 2.8-411% with an annual mean of 151% in lake Maggiore, Italy. Lande (1973) reported the daily renewal coefficient to range between 28—187%. In Kashmir lake a range of 21—411% with an annual mean of 122% was recorded. It is clear that no general rule relates phytoplankton productivity to phytoplankton biomass. Similarly P/B ratio in the lake was fairly good depicting a range of

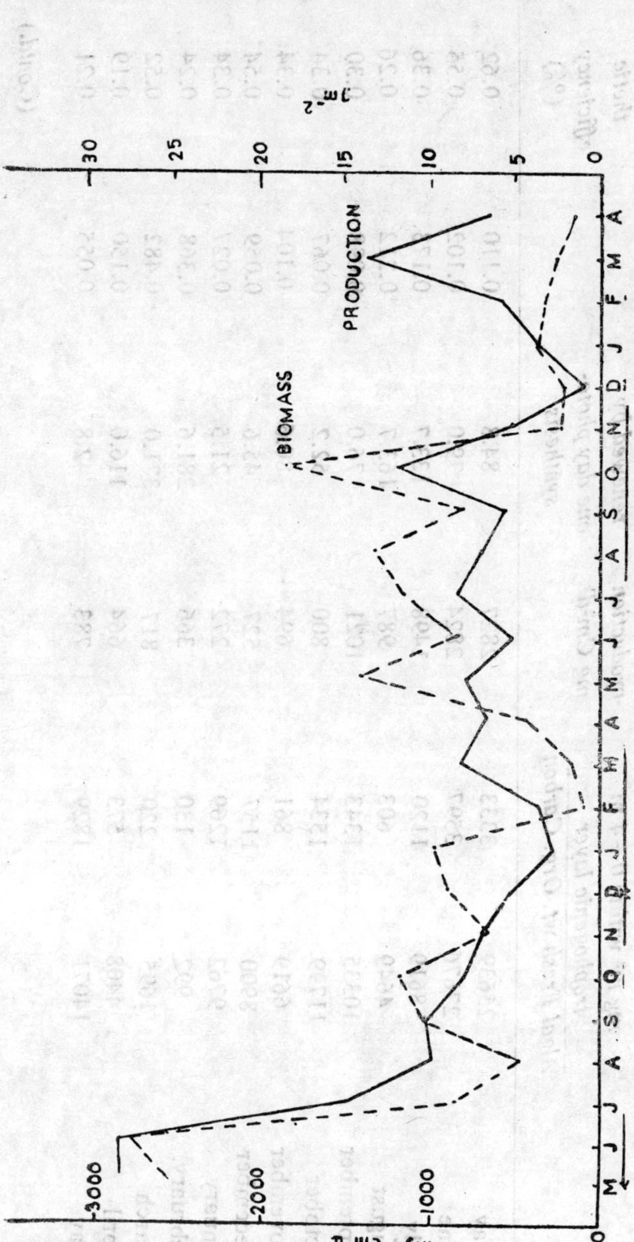

Fig. 4. Relationship between phytoplankton biomass and primary production.

Table 3. Activity Coefficient for Kashmir Lake within 0.4 m layer

Month	*Phytoplankton biomass mg m^2 within 0—4 m trophogenic layer*		*Primary production mg Cm^2d^1*	*% of Biomass renewed by one day photo-synthesis*	*P/B*	*Photosyn-thetic efficiency (%)*
	Algal Fresh wt.	*Org. Carbon*				
May	25639	3333	2827	84.8	0.110	0.62
June	27676	3597	2824	79.0	0.102	0.55
July	8619	1120	1498	133.7	0.173	0.36
August	4649	603	987	163.7	0.212	0.26
September	10335	1343	1021	76.0	0.098	0.30
October	11799	1534	800	52.2	0.067	0.34
November	6619	861	694	80.6	0.104	0.34
December	8900	1157	527	45.6	0.059	0.54
January	9762	1269	272	21.5	0.027	0.34
February	992	130	366	281.6	0.368	0.24
March	1695	220	817	371.0	0.482	0.52
April	4408	573	664	116.0	0.150	0.19
May	14071	1829	783	42.8	0,055	0.21

(*Contd.*)

June	8475	1102	527	47.9	0.062	0.10
July	11649	1514	838	55.4	0.071	0.20
August	13281	1726	713	41.3	0.053	0.19
September	8087	1052	553	52.6	0.068	0.16
October	18750	2438	1196	49.1	0.064	0.51
November	2281	296	519	175.1	0.227	0.21
December	2139	278	93	33.7	0.043	0.096
January	3739	486	357	73.6	0.095	0.45
February	3264	424	584	137.8	0.178	0.38
March	2581	335	1379	411.0	0.534	0.55
April	1566	204	664	326.2	0.424	0.19
Mean	8790.7	1142.7	896.8	122.0	0.1[illegible]9	0.31

0.027 to 0.513 this too does not suggest a positive correlation with production.

TROPHIC CLASSIFICATION

According to Vinberg (1961), Rodhe (1969), and Vollenweider *et al.* (1974), the lakes of higher trophic level attain primary productivity of 75–300 g $Cm^{-2}y^{-1}$. Classification as suggested by these workers is tabulated below :

Table 4. Trophic classification on Annual Carbon fixation

Classification type	*Authority*		
	Vinberg (1961)	*Rodhe (1969)*	*Vollenweider et al. (1974)*
Oligotrophic	7–25	<100	<30
Mesotrophic	25–75	100–200	30–200
Natural Eutrophic	75–250	—	—
Culturally Eutrophic	—	200	—
Highly Eutrophic	—	—	300
Polluted	350–700	—	—

All values as g C $m^{-2}y^{-1}$

While comparing the production values of the Naranbagh lake with above ranges it appears that the lake has exceeded the eutrophic limit by about 100 g Cm^{-2}. As per the comparative table 5 of production ranges for Kashmir biotopes given by Wanganeo (1984), present lake would fall under moderate category.

From the data presented in Tables 4 and 5, it appears that European standards of trophication based on annual carbon fixation may not hold true for Kashmir lakes. It may be emphasized here that due to moderate inland climate in Kashmir Valley compared to Europe, the lakes here tend to show higher production. This is also supported by the observation made by Vollenweider (1978), who emphasized that due to their geographic location, lakes of India, Africa, and Indonesia may tend to show higher production and nutrient status. Therefore, caution may be exercised while applying the production based trophic classification developed on European lakes.

Table 5. Comparative primary production values for different lakes in Kashmir region

Name of lake	*Elevation (m.a.s.l.)*	*Primary production g* $Cm^{-2}y^{-1}$	*Authority*
Surinsar	604	363.2	Zutshi *et al* (1980)
Mansar	666	145.2	do
Dal	1587	215.2	Vass (1982)
Anchar	1587	304.8	Wanganeo (MSS)
Manasbal	1587	195.0	Wanganeo (1984)
Trigam	1587	816.8	Khan (1978)
Nilnag	2180	136.2	Zutshi *et al.* (1980)
Naranbagh	1587	327.2	Present study

ACKNOWLEDGEMENT

The author is grateful to Dr. Arun G. Jhingran, Director, Central Inland Capture Fisheries Research Institute, Barrackpore for his keen interest, guidance and permission to present this paper at a National seminar. Heartfelt thanks are also duc to former colleagues of author at the Cold water Fisheries, Research centre of CIFRI, Harwan-Kashmir for their help in field and lab. work.

REFERENCES

Findenegg, I. (1965). Relationship between standing crop and primary Productivity. *Mem Ist. Ital. Idrobiol,* 18 suppl : 271-289.

Goldman, C,R., Gerletti M., Javornicky P.; Santolini U.M. and Amezaga, E D. (1968). Primary productivity, Bacteria, Phyto-Zooplankton in lake Maggiore correlations and Relationships with Ecological factors. *Mem. Inst. Ital. Idrobiol, 23* : 49-127.

Kaul, V. and Vass, K.K 1970. Studies on the productivity of an emergent aquatic plant, *Typha augustifolia*—Proc. 57 Ind. Sci. Congr. Abs.

Kaul, V. (1977). Limnological survey of Kashmir lakes with reference to trophic status and conservation. *Int. J. Ecol Environ. Sci.* 3 : 29-44.

Lande, A. (1973). Studies on phytoplankton in relation to its production and same physical and chemical factors in lake Svinsjon. *Arch. Hydrobiol 72* (1) : 71-86.

Rodhe, W. (1958). Primary productionn Und Sectypen, *Verh. int. Ver. Limnol., 13* ; 121-141.

Rodhe, W. (1969). Crystallisation of eutrophication concepts in Northern Europe. In Eutrophication, causes, consequences. correctives *Proc. Nat. Acad. Sci,* 50-64, Washington D.C.

Vass, K.K. (1973). *Studies on hydrobiology and primary productivity of some Kashmir* lakes. Ph. D. Thesis Kashmir University pp. 200.

Vass, K K. ; Raina, H.S. : Langer, R.K. and Shyam Sunder (1988). Energy Transfer in a Kashmir Himalayan Sar (Lake). *Int. Revne ges. Hydrobiol. 73* : 547-558.

Verduin, J. (1959). Photosynthesis by aquatic communities in North-western Ohio, *Ecology. 40* : 377-383.

Vinberg, G.G. (1961). Modern conditions and problems in the study of primary production of waters. *Referat. Zhur. Biol. Minsk* 22 (Zh 329) : 11-14.

Vollenweider, R.A. and Nauwerck A.—(1961). Some observations on the 14C method for measuring primary *Verh. Inst. Vern. Limnol 14* : 135-139.

Vollenweider, R.A. ; Munawar, M. and Stadelman—(1974). A comperative review of phytoplankton and primary prodcution in the Laurentian Great lakes. *J. Fish Res. Bd. Canada 31* : 739-762.

Vollenweider, R.A. (1979). Eutrophication. The problem on a world scale. *In Acta convention on Eutrophication,* Rome, Italian National Research Council, AC/2/45-70.

Vollenweider, T.A. (1984). Eutrophication control. *Nature and Resources* Vol. XVI No. 3 : July-Sept.

Wanganeo, A. (1984). Primary production characteristics of a Himalayan lake in Kashmir. *Int. Revneges. Hydrobiol 69*(1).

Zutshi, D.P. (1976). Phytoplankton productivity, Algal dynamics, and Trophic status of Lake Margozzo (Northern Italy). *Mem. Ist. Ital. Idrobiol. 33* : 223-256.

Zutshi, D.P. and Vass, K.K. (1982) Limnological studies on Dal lake III. Biological features. *Proc. Indian Nat. Sci. Acad* B, *48* : 234-241.

Zutshi, D.P. and Vass K.K.—(1972). Trends in eutrophication of Kashmir lakes. *UGC Symp. Hydrobiol Icthy*, Kashmir University, Srinagar 11-12.

4

ALTITUDE RELATED LIMNOLOGICAL STUDIES ON HIMALAYAN LAKES

S. Kumar

Department of Biosciences,
University of Jammu, Jammu.

Abstract—Comparative limnological studies on high altitude Himalayan Lake ecosystem have been given. The lakes are situated between 600 m and 5910 m altitude and are exposed to climatic conditions that vary from cold deserts of Ladakh lake Pangong to sub-tropical lakes Surinsar and Mansar. An attempt has been made to provide information on some important altitude related limnological variations of certain Himalayan lakes.

INTRODUCTION

A large number of natural freshwater lakes exist in Jammu and Kashmir which are of considerable importance. These waterbodies are situated at different altitudes (Table-I). These lakes are of great importance as they are source of fish fodder and many food plants. They also provide many recreational facilities like bird watching, swimming, surf-riding and boating to tourists. Besides, there has been extensive land use for vegetable cultivation (Kaul and Zutshi, 1966) and residential purposes. Limnological studies on high altitude lakes have been undertaken by Zutshi and Vass (1970, 73), Zutshi *et al.* (1972, 1980), Sehgal (1980), Zutshi (1985), Kant (1985), Khan and Zutshi, (1980) and Kumar (Communicated). In the present communication an attempt has been made to compare Limnological features of five lake types distributed within the Himalayan region from Jammu to Ladakh.

Table-I. Range of Altitude for Himalayan Lakes

Series	*Altitude (m)*	*Examples*
I	Above 600	Mansar, Surinsar.
II	1585-1600	Dal, Anchar, Naranbagh, Manasbal, Trigam.
III	2000-2500	Nilnag.
IV	3000-4000	Alipather, Gangabal, Nundkel, Sheshnag, Kounsarnag.
V	4241-5910	Pangong Tso, Tso Morari

DESCRIPTION OF STUDY AREA

Himalayan Lakes are mainly of five different types (Table-I) with respect to their origin, altitude, biota and duration and extent of ice-cover.

(*i*) I Series

Siwalik lakes are oval to sub-oval in shape. They are fault basin, non-drainage types. The main supply of water is by precipitation, the springs present within the basin being of minor importance, Monsoon period in lake region extends from July to September. Seepage losses are minor, therefore, the water residence time and renewal period of lakes is quite high *viz.* Surinsar and Mansar lakes.

(*ii*) Series

The Valley lakes are situated between Pir-Panjal and inner Himalayan region, occuring mainly along the course of river Jhelum. These Valley lakes have been further sub-divided into three types based on the presence or absence of feeding Channels. The sub types are :

(*a*) *Drainage lakes*

Drainage lakes, having the presence of both inflow and outflow channels, *viz.* Wular lake, Hokarsar lake, Dal and Anchar lakes.

(*b*) *Semi-drainage*

Semi-drainage, where there is no inflow channel but the outflow channels are present *e.g.* Mansbal lake, lake Surinsar, lake Mansar (artificial outflow).

(*c*) *Non-drainage*

Na-drainage, having neither any inflow nor any outflow channel. *e.g.* Trigam lake and Bodsar lake (May be Mansar also if artificial inflow is not considered valid).

(*iii*) **Series**

The lakes present in lower fringes of Pir-Panjal range (also on outer Karewas), come into existence due to tectonic activity. They occupy the altitudinal range between 2000 m to 2500 m right in the midst of pine forests. *e.g.* Nilnag lake.

(*iv*) **Series**

These lakes are situated on inner Himalayas partly along the Pir-Panjal ranges between 3000 m – 4000 m altitude. These are glaciated lakes, originated during the third Himalayan glaciation and formed by embankments across the line of drainage. These embankments may have been caused by deposits of available debris from a side slop or by advance of a side glacier with its lateral moraines. There are no human settlements close to lake and surroundings are undisturbed. During dry periods there is a fall in the water level which has been recorded even upto less than 3 meters as in case of Alipather lake, *e.g.* Alipather, Gangabal Nundkol, Sheshnag and Kounsarnag.

(*v*) **Series**

These lakes are present under cold desert conditions of Ladakh and situated between altitudes of 4241 m and 5910 m. These included both fresh and saline waterbodies of different surface areas and depths and either closed or drainage types *viz.* Lake Pangong Tso and Tso Morari.

MATERIAL AND METHODS

The Material for the present study was collected from the Siwalik lakes, Jammu. For collection of plankton, a conical (144 μ mesh size) net was used. Collections were preserved in five present formalin and analysed by using the Sedgwick-Rafter counting cell according to the pıocedure described by Welch (1948). The air and waste temperature were recorded at the time of collection and penetration of light was noted with the help of a Secchi-disc. Water samples were analysed according to the Sandard Methods for Examination of Water and Waste Water (APHA, 1976).

RESULTS AND DISCUSSION

The results of altitude related limnological features on Himalayan lakes are given in the Tables II and III. Siwalik lakes (Mansar and Surinsar) are situated in the Sub-tropical monsoonic region. The annual rainfall is about 150 cm. The Valley lakes collect a rainfall about 75 cm. The rainy periods spreads over winter and spring months (January—April). In Ladakh total precipitation does not exceed 3.05 cm.

Around Siwalik lakes winter is usually and dry atmospheric temperature remains well above the freezing point ; maximum summer temperature varies between 35°C and 41°C. In valley lakes summer temperature ranging from 28°C to 30°C. In winter it falls below 0°C. In high altitude Kashmir lakes winter temperature remains below the freezing point. (Table II). In the mountain lakes (above 3000 m) water has thick-ice cover which lasts upto seven to eight months. The growing period usually starts from June. In case of Nilnag lake this period ranged from April to November and in case of Alipather lake from June to October. But in Ladakh lakes, the Ice-cover period ranged between four to eleven months depending on the salinity of water and altitude. (Table III).

The value of dissolved oxygen content decreases from lower altitude to higher altitude. The maximum Value recorded in Siwalik Varies 13.8—14.2 mg/l (Winter season) and minimum values observed in Alipather lake (6.5 mg/l) (Summer season).

Table II. Physico-chemical Characteristics of Different Lakes

Parameters	*Surinsar*	*Mansar*	*Dal*	*Anchar*	*Naranbagh*	*Manasbal*	*Trigam*	*Nilnag*	*Alipather*	*Pangong Tso*
Altitude (m)	604	710	1584	1583	1583	1584	1583	2180	3200	5910
Area (ha)	29	45	1250	700	24.3	280	11	6.3	8	—
Max. depth (m)	26	37	6	2.75	5	12	2.25	6.5	6.5	—
Temperature (°C)	31	30	30	28	29	29	30	25	15	—
Secchi (m)	1.6-2.25	2-5	0.4-5	0.25-1.75	1.2-4.0	2.5-5.5	0.25-0.6	1.1-2.3	1.2-1.9	—
Stratification type	WM	WM	WM-PM	WM-PM	WM-PM	WM	P	D	D	—
Thermocline depth (m)	5-9	6-9	ND	ND	ND	5-9	ND	2-4	2-4	—
pH	7.2-8.7	7.4-8.9	7.4-9.4	7.2-9.6	7.6-8.7	7-8.2	8.2-9.0	8.1	7.10	7.9
DO (mg/l)	1.1-14.2	6.5-13.8	7.2-11.6	10.9	10.7	6.8-10.5	10.2	5.1-10.0	6.5	—
Alkalinity as $CaCO_3$ (mg/l)	135	116	70-120	51-141	97-186	62-186	250-475	90	18	935
Conductivity (μS)	420	300	100-216	102-275	250-350	110-350	700-1070	182	14	18900
PO_4-P (μg/l)	40	14	5-20	5-30	6 24	5-15	20-47	—	—	—
NO_3-N (μg/l)	—	—	15-30	14-74	30-55	8-37	100-235	—	—	—

WM — Warm Monomictic
PM — Polymictic
D — Dimictic
B — Pond type
ND — Not distinct

Courtesy : Zutshi, D.P. and Khan, M.A. (1978). On lake typology of Kashmir. Environ. Physiol. Ecol. Plant ; 465-472.

Table III. Limnological features of lake types.

Parameters	*Siwalik*	*Valley*	*Forest*	*Mountain*	*Ladakh*
Altitude (m)	above 600	1585-1600	2000-2500	3000-4000	4241-5910
Origin	Tectonic activity	River valley oxbow	Earth movement	Glacial daming mountains	Tethys Sea ?
Topography	Hilly	Flat land	Foot hills		
Period of ice cover (months)	absent	absent or very short	3-4	7-8	4-11
Secchi transparency	high	moderate-high	moderate	low	low
Specific conductivity	high	moderate	moderate low	low	highest (exception)
Macrophytes Qualitative	all-life forms	all-life forms	floating types	absent	absent (*potamogeton* species)
Phytoplankton	high	moderate high	moderate	low	very low/ absent
Zooplankton	high	moderate high	moderate	low	very low/ absent
Fish population	high	moderate high	moderate	low	very low/ absent

In the severe, winter the Valley lakes freeze and a few cm. thick ice cover gets formed. The temperature just below the ice is zero and it increases with increasing depth. This would point out that lakes behave both as monomictic and dimictic The summer surface water temperature of high mountain lakes does not usually exceed 15°C. Which thus cannot be classified as polar lakes of latitudinal gradient (Welch, 1952). A large number of these lakes develop summer stratification with a temperature difference of about 9°C between epi-limnion and hypolimnion. The depth of thermocline is about 4 m in case of Alipather lake classified as dimictic (with inverse stratification during winter). In this part of Himalayan 2000 m altitude seems to be dividing line between the monomictic and dimictic lakes. The Ladakh lakes are dimictic resembling high altitude Kashmir lakes (Zutshi *et al.*, 1980).

In Siwalic lakes the thermocline is present between 3 m and 9 m. Secchi transparency 5.5 m has been recorded for Manasbal Valley lakes. Low transparency were observed in Nilnag and Alipather lakes. It may be due to extraneous matter especially silt/glaciated slit. The low transparency in Ladakh lakes related to low temperature and plankton density or/and to combined impacts of detritus and plankton.

The specific conductivity values of lakes vary widely. In Alipather lake it is very low (18 μ Scm^{-1}) and in Trigam lake very high (1070 μScm^{-1}). The water of other lakes has conductivity values below 400 μ Scm^{-1}. But in Ladakh lakes especially Pangong Tso the value of conductivity goes to 18,900 μ Scm^{-1}. Which would suggest otherwise a high trophic status of this lake, but high conductivity is due to high electrolyte content (Cl^- 3200 mg/l).

Alkalinity values in the lakes follow almost the same pattern (Table II) the contents of electrolytes decreases with the increase in altitude (except Pangong, which is specialized saline lake). Cations and Anions in high altitude lakes are low. Lakes and lower altitude are at a higher level of enrichment then those at high altitudes.

The high altitude lakes are devoid of any macrophytes. Because their bottom is generally rocky and climate unfavourable for macrophytic growth. The poor macrophytic growth may be because of ice-cover on the surface. Macrophytic vegetation of these lakes otherwise does not show any definite relationship with altitude. Generally the lake with higher transparency develop high density of submerged forms and there water is turbid, floating forms dominate. But in the case of Ladakh lakes only *Potamogeton* sp. occurs in shallow lakes.

The total number of phytoplankton qualitatively decreases due to increase in altitude. The high production of phytoplankton in mountain lakes is due to high population of nanoplankton which are able to make optimum utilization of solar energy at high altitude (Zutshi, 1985). Phytoplankton population was scarcely in Ladakh lakes and was represented by few species of diatoms and Chlorophyceae (*Chlamydomonas* only). Zooplankton follow the same pattern as phytoplankton. At high altitude in Nilnag lake the zooplankton dominated by rotifers and in Alipather lake a pigmented copepod *Acanthodiaptomus* sp. And a pigmented *Daphnia* (*D. tibetana*) have been collected from deeper waters during ice free period. Pigmentation being the adaptation to a higher dosage of UV and visible short wave ligat (Loffler, 1969).

The lakes with their altitudes ranging from 600 m to 5910 m in J and K Himalayas showing a progressive decrease in their productivity from lower to higher. The trophic structure becomes more simpler as one goes higher and same holds true for other physico-chemical and biological characteristics.

Managemeat of Himalayan lakes

Study the rate of eutrophication is an important aspect to be attend upon, so as to understand the ageing of lake. Lack of Scientific approach has already caused dystrophy of a high altitude lake in the region *e.g.* Sanasar.

Exessive pumping out of water lead to reduction in water level which along the peripheries of lakes expose the littoral habitat to greater hazards which may ultimately be deleterious to aquatic biota and lake itself. Therefore, it is suggested that withdrawl of water should be allowed only upto the extent which match the inflow of water into the lake. Any mismatch in this regard may enhance the possibility of dystrophy of the lake.

Tourist input at the lakes should be regulated to allow functioning of lakes as a self-sustainable system with ability to absorb the hazards arising out of tourist influx.

Habitat manipulation of the lake in attempt of beautification should be controlled. This should be attempted only with the advice of an experienced limnologist.

Encroachment or construction of houses in the land area along the lake shore either by local villagers or by Tourism Department should be thoughtfully planned at alternate sites.

The impact of cremation ground (located in lake catchment) should be investigated and suitable measures evolved to offset the impact, if any.

The higher aquatic vegetation should be removed from lakes completely. It should not, however, be merely cutting the vegetation under the water and allowing the stumps to decay but the vegetation should be uprooted thoroughly.

Sewage disposal from the catchment area should be diverted to some other areas, to establish sewage treatment plant/stabilization of ponds.

The conservation and management is neither a simple job nor it can be acheived by the efforts of an individual or a single agency. It has to appreciated and a joint effort has to be initiated by many agencies to reach to goal. Department of Tourism/Science and Technology of J and K Government/Department of Environment of Government of India, should create a scientific unit to monitor the health of these lakes through continuous limnological investigations and population dynamics of aquatic biota.

ACKNOWLEDGEMENT

The author is indebted to Head, Department of Bio-Sciences, University of Jammu for Laboratory facilities and Prof. Y.R. Malhotra for valuable suggestions.

REFERENCES

APHA (1976). *Standard Methods for the Examination of Water and Waste water.* Am. Public Health Assoc. New York.

Kant, S (1985). Limnology of the Himalayan lakes problem of resourse use and management. In studies in Eco-Development Himalayas Mountains and Men (Eds. T.V. Singh and J. Kaur) Print house, Lucknow : 165-195.

Kaul, V and Zutshi, D.P. (1966). Some ecological considerations of floating Islands in Srinagar lakes. Proc. Natl. Acad. Sci , India *36* : 273-280.

Khan, M.A. and Zutshi D.P. (1980). Contribution to high altitude limnology of Himalayan system. I. Limnology and Primary Productivity of the plankton community of Nilnag lake, Kashmir. *Hydrobiologia, 75* : 103-112.

Kumar, S. (1991). Limnological studies on Mansar lake. I Physico-Chemical features. (Communicated).

Loffler, H. (1969). High altitude lakes in Mt. Everest region, *Verh. Internat. Verein. Limnol. 17* : 373-385.

Sehgal, H.S. (1980). *Limnology of lake Surinsar, Jammu with reference to Zooplankton and fishery prospects.* Ph.D. Thesis University of Jammu.

Welch, P.S. (1948). *Limnological Methods.* McGraw Hill Book Co Inc. New York.

Welch, P.S. (1952). *Limnology,* McGraw Hill Book. Co. Inc. New York.

Zutshi, D.P. (1985). The Himalayan lake ecosystems. Reprinted from Environmental regeneration in Himalaya—concepts and strategies (ed. J S Singh) : 325-338.

Zutshi, D.P. Kaul, V. and Vass, K.K. (1972). Limnology of high altitude Kashmir lakes. *Verh. Internat. Verein. Limnol. 18* : 599-604.

Zutshi, D.P. Subla, B.A., Khan, M.A. and Wanganeo, A. (1980). Comparative Limnology of nine Lakes of Jammu and Kashmir Himalayas. *Hydrobiologia 72* 101-112.

Zutshi D.P. and Vass, K.K. (1970). High altitude lakes of Kashmir. *Ichthyologica 10* : 12-15.

———and———(1973). Variation in the water quality of some Kashmir lakes. *Trop. Ecol. 14* (2) : 182-196.

5

STUDIES ON THE ECOLOGY AND FISH FAUNA OF AN OLIGOTROPHIC LAKE-HIMAYAT SAGAR, HYDERABAD, ANDHRA PRADESH

M. Babu Rao

Departmcnt of Fishery Sciencc
A.P. Agricultural University
Kakinada—533007, A.P.

and

Peetambar Pan

Zoological Survey of India, Freshwater Biological Station,
Ashoknagar, Hyderabad, Andhra Pradesh

Abstract—Himayat sagar lake is situated towards the south western suburbs of Hyderabad city. It has an area of about 19.7 sq. kms. (Permanent storage level : 19.684 sq. Kms) and serves as a reservoir for suppling drinking water to the city of Hyderabad. It was bunded on eastern side and northern and southern side have a rocky profile. On Western side there is a feeding channel which enters the lake through the fields. No known source of pollution enters the water body and hence water remains oligotrophic during the most part of the year. About 33 ecological parameters were studied from this lake, besides the routine planktonic studies. Altogether 32 species of 11 families of fishes were recorded from this lake.

INTRODUCTION

Himayatsagar lake is situated towards the south western suburbs of Hyderabad city in Andhra Pradesh at latitute 17°22′ N, longitude 78°27′ E at an elevation of 510 m above MSL. It is at about 26 Kms. from Hyderabad city (Fig. 1.1). It was built in 1926 by constructing a bund across a small tributory Esa of river Musi to control the floods and also serve

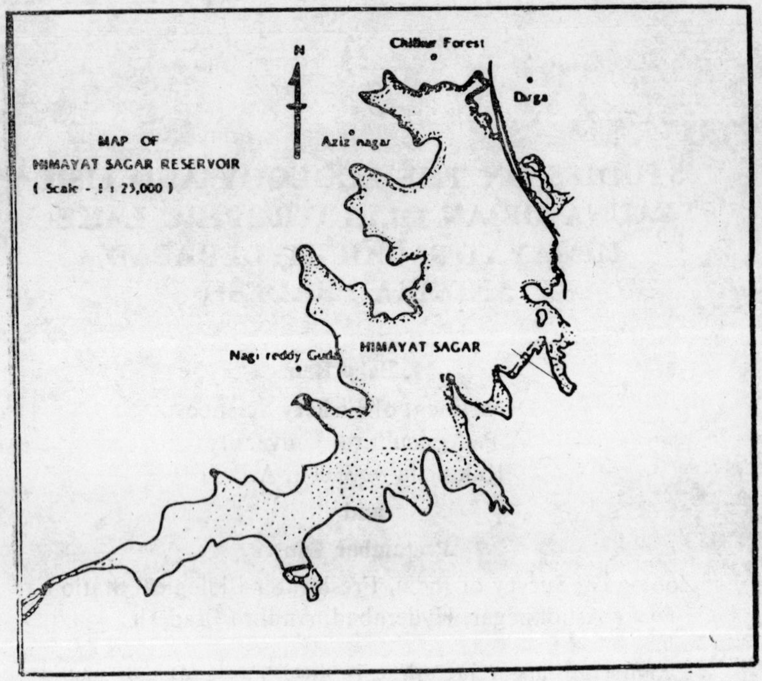

Fig. 1.1. Outline map of Himayat Sagar, Hyderabad as also sampling stations in the reservoir.

as an irrigation source. Presently in conjuction with Osman sagar reservoir it supplies drinking water to Hyderabad city. It was bunded on eastern side, northern and southern side have a rocky profile, where as the main feeding channel on western side, which enters the lake through the fields. Its catchment area is very large and measures 1308 Kms^2. Its permanent storage level is 19.684 Kms. No known source of pollution enters the water body and hence water remains oligotrophic during the most part of the year.

Morphometry of Himayat sagar reservoir

The following are the important features of the morphometry of Himayat Sagar reservoir.

Total catchment area	: 1308 Sq. Kms.
Total volume	: 102.1 Million m^3
Surface area	: 21.5 Sq. Kms.
Maximum depth	: 29.3 m
Mean depth	: 4.74 m
Maximum effective length	: 7.8 Kms
Maximum effective width	: 4.2 Kms
Lenght of shore line	: 35.44 Kms.
Permanent storage level	: 19.684 Sq. Kms.

Material and methods

Collection of water, plankton and fish collections were obtained from the reservoir monthly from Oct. 1985 to March

Himayat Sagar Reservoir

An over view of Himayat Sagar

Fishing Activity in Himayat Sagar

Shore Line Fauna of Himayat Sagar

An over view of Himayat Sagar Lake

Gastopoda Fauna (Shore Fauna) in Lake

1987. The physico-chemical analysis of water samples was done following A.P.H.A. (1985) etc., A total of 33 parameters of monthly average physico-chemical analysis data were presented from Oct. 1985 to March 1987 in Table I. Plankton samples were collected with the help of a nylon satin plankton net (200 mesh/sq. cm) and preserved in 5% formaldehyde soln. for quantitative analysis. The quantitative estimation of zooplankton was made with the help of 'Sedgwick Rafter Counting Cell'. Fish samples were obtained from fishermen catches monthly and were identified from Day (1889), Misra (1976), Rahimullah (1943), Jayaram (1981) and Babu Rao & Siva reddy (1984).

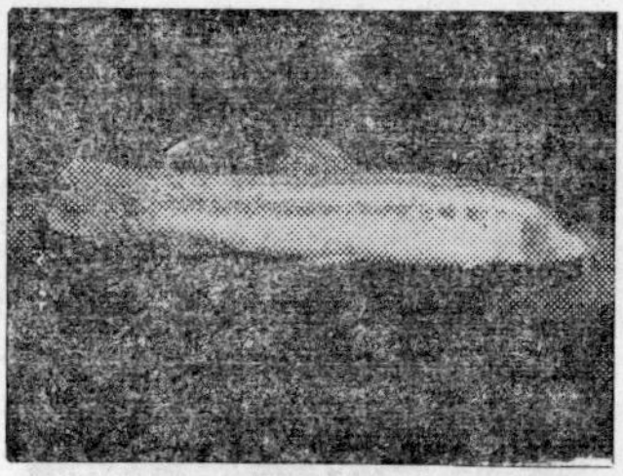

Fig. 1.

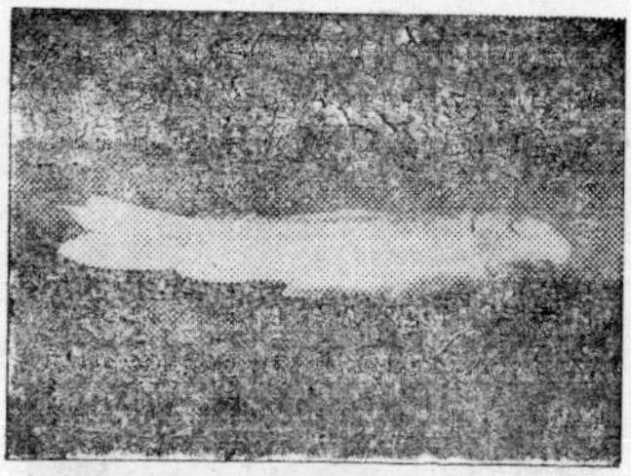

Fig. 2.

Fig. 3.

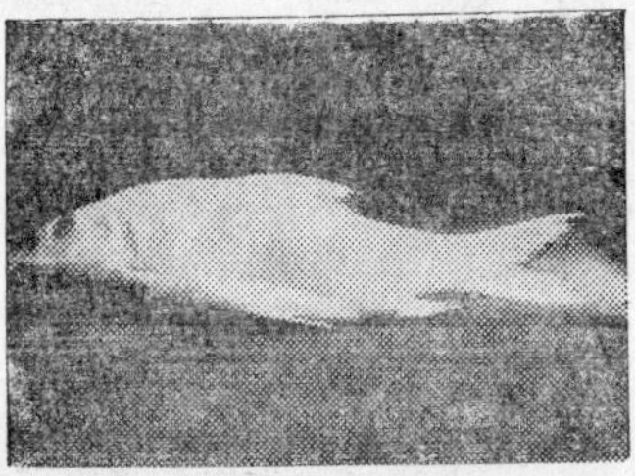

Fig. 4.

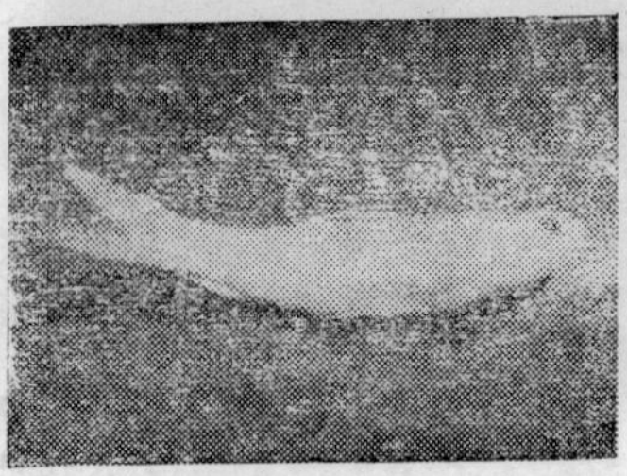

Fig. 5.

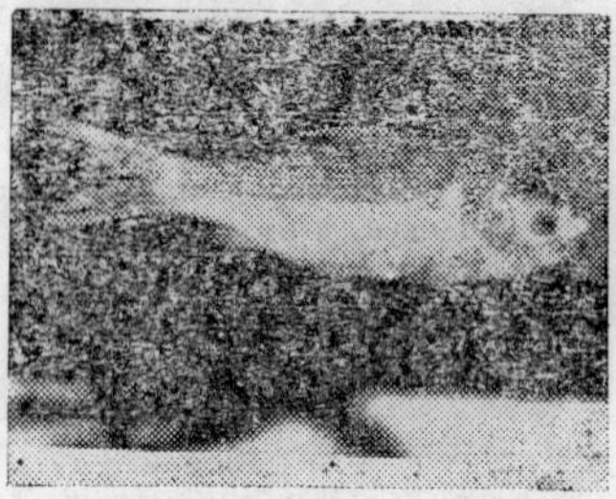

Fig 6.

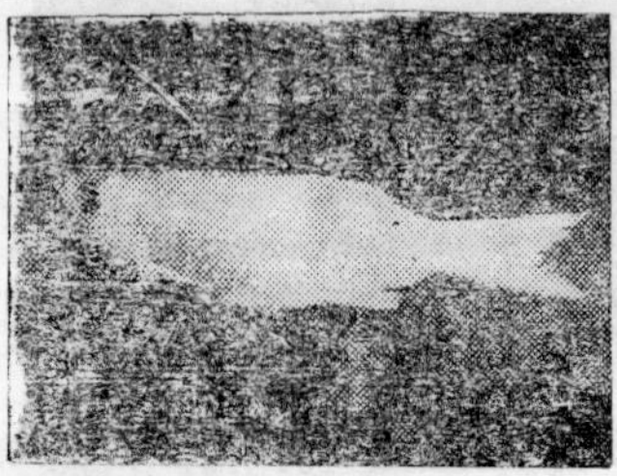

Fig. 7.

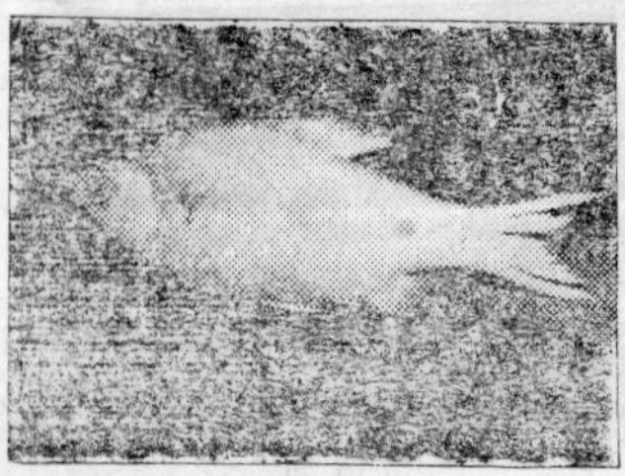

Fig. 8.

Fig. 9.

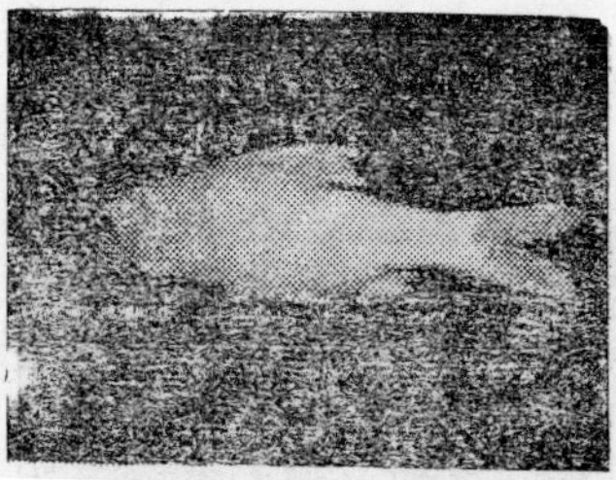

Fig. 10.

Fig. 11.

Fig. 12.

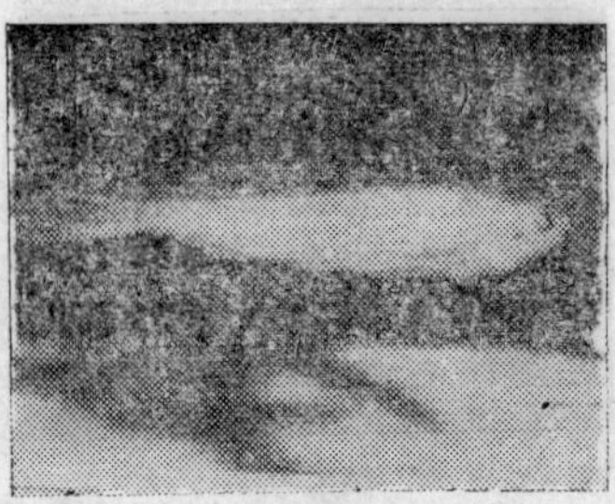

Fig. 13.

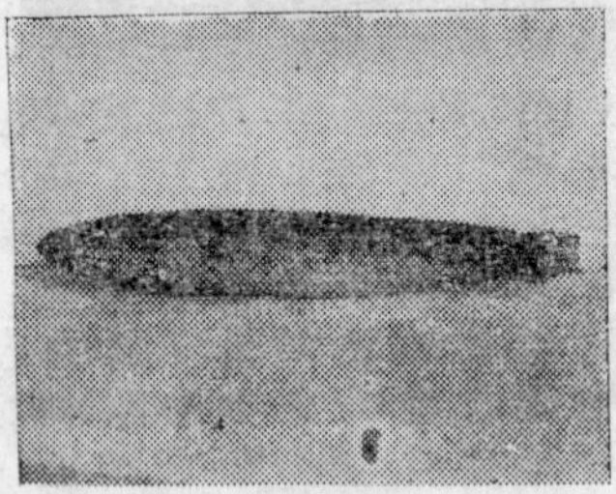

Fig. 14.

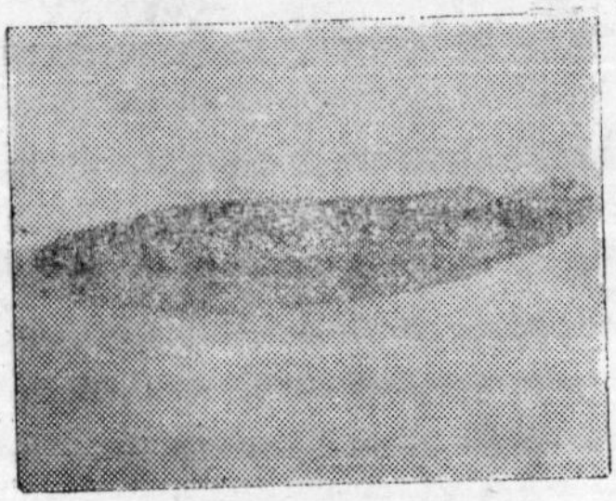

Fig. 15.

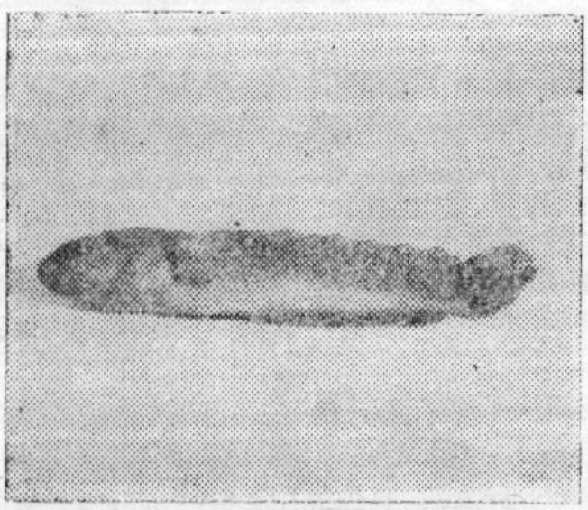

Fig. 16.

Fig. 17.

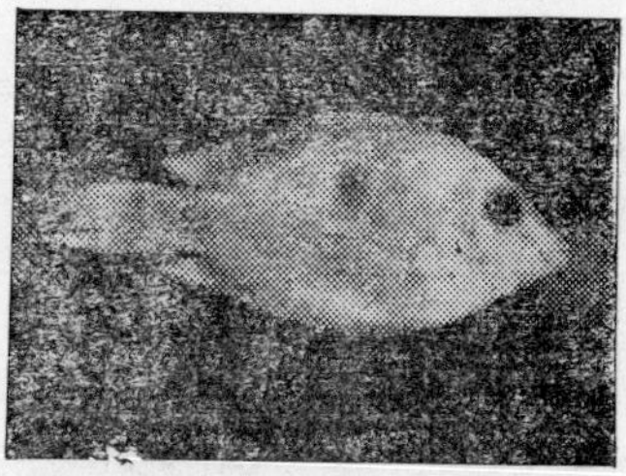

Fig. 18.

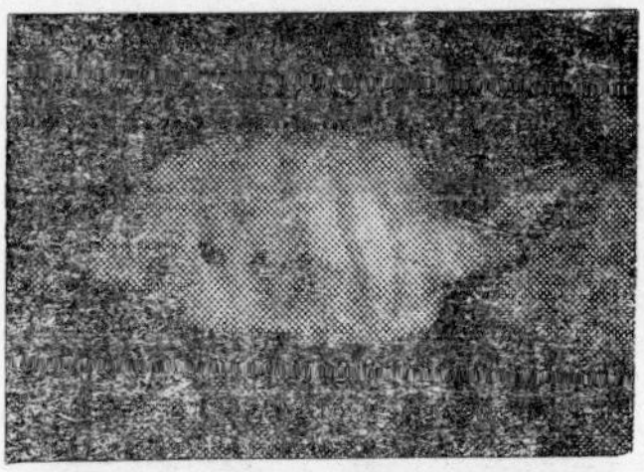

Fig. 19.

Fig. 20.

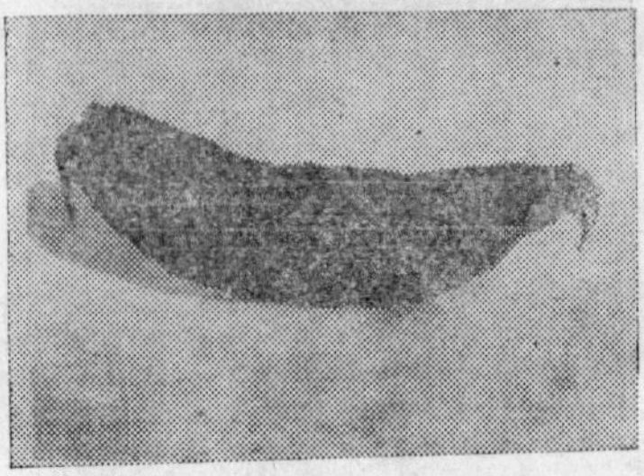

Fig. 21.

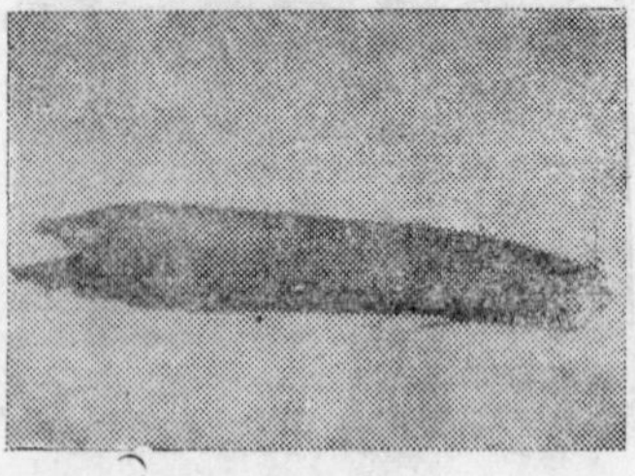

Fig. 22.

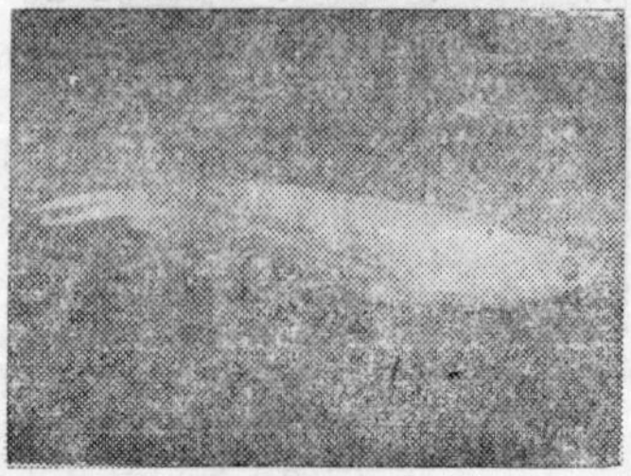

Fig. 23.

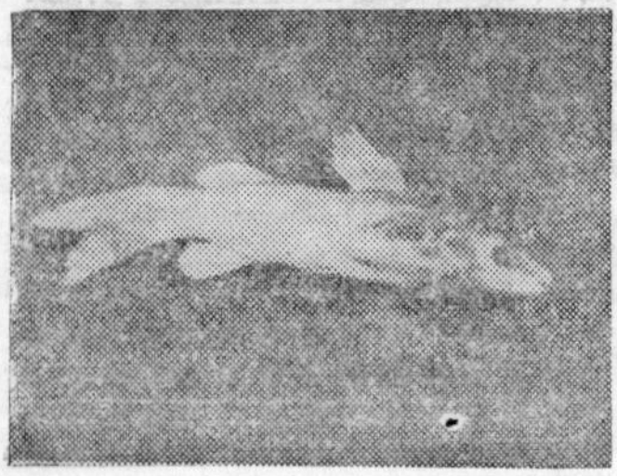

Fig. 24.

Fig. 25.

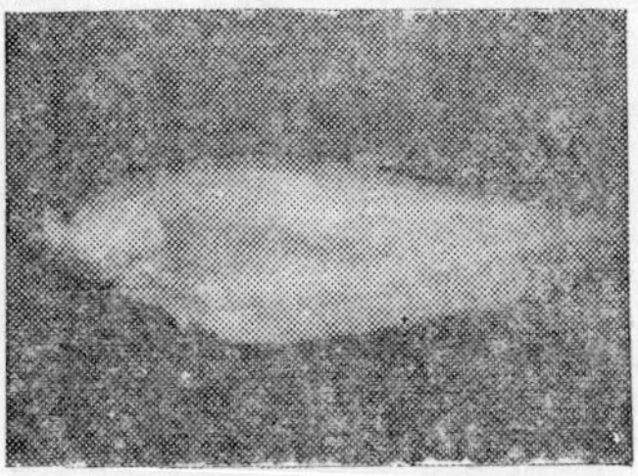

Fig. 26.

Fig. 27.

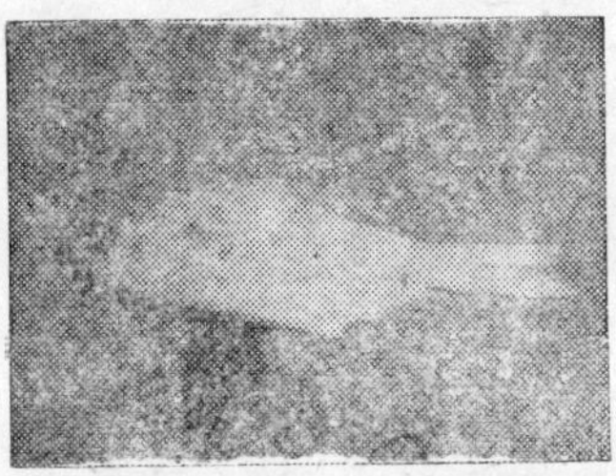

Fig. 28.

Table I. Showing the Physico-Chemical Characteristics of Himayat Sagar Reservoir during October 1985 to March 1987.

Parameters	*O/85* (*1*)	*N/85* (*2*)	*D/85* (*3*)	*J/86* (*4*)	*F/86* (*5*)	*M/86* (*6*)	*A/86* (*7*)	*M/86* (*8*)	*J/86* (*9*)
Air Temp. C	24.5	23.0	21.0	25.0	30.0	28.5	32.0	34.5	32.0
Water Temp. C	22 0	20.0	19.0	23.5	24.5	26.0	29.0	31.0	30.0
Rain fall (mm)	98.10	0.00	4.70	36.4	16.5	0.00	52.3	0.00	161.8
Relative Humidity (8.30 a.m./7.30 p.m. %)	69/53	50/37	68/41	75/43	74/39	53/30	56/29	47/22	73/54
pH	7.2	7.4	7.3	7.5	7.5	7.5	8.0	7.8	7.4
Conductivity Umos/cm	330	370	340	416	430	420	470	500	551
Dissolved Oxygen mg/l	7.4	7.9	8.0	7.4	6.0	6.3	6.5	6.2	7.4
Free carbon-dioxide mg/l	0.3	0.00	0.00	0.00	0.60	0.00	0.00	0.00	0.00
Carbonate mg/l	0.00	52.0	60.0	40.0	0.00	61.00	75.00	70.00	43.00

(Contd.)

Parameters	*J/86* (*10*)	*A/86* (*11*)	*S/86* (*12*)	*O/86* (*13*)	*N/86* (*14*)	*D/86* (*15*)	*J/87* (*16*)	*F/87* (*17*)	*M/87* (*18*)
Air Temp. C	31.0	28.0	26.0	23.5	24.0	20.0	26.0	31.0	30.0
Water Temp. C	28.5	25.0	25.0	20.5	22.0	18.0	22.0	28.0	28.5
Rain fall (mm)	109.3	218.5	65.2	12.7	28.2	15.5	7.8	0.0	61.3
Relative Humidity (8.30 a.m./7.30 p.m. %)	73/59	74/69	72/59	63/42	69/47	74/45	75/41	58/28	57/27
pH	7.5	7.9	7.6	7.7	7.6	7.3	7.7	7.6	7.6
Conductivity Umos/cm	510	430	370	410	390	390	360	380	440
Dissolved Oxygen mg/l	6.1	6.3	7.9	6.6	7.9	7.4	6.4	7.2	6.1
Free carbon-dioxide mg/l	0.00	0.00	0.00	0.00	0.00	0.00	0.40	0.00	0.00
Corbonate mg/l	30.00	39.00	48.00	51.00	44.00	63.00	0.00	46.00	63.00

(*Contd.*)

Table I. *(Contd.)*

Parameters	*O/85* (1)	*N/85* (2)	*D/85* (3)	*J/86* (4)	*F/86* (5)	*M/86* (6)	*A/86* (7)	*M/86* (8)	*J/86* (9)
Bicarbonates mg/l	162.0	181.0	187.0	162.0	115.0	173.0	220.0	208.0	120.0
Total alkalinity mg/l	162.0	223.0	247.0	202.0	151.0	234.0	295.0	278.0	163.0
Chloride mg/l	7.3	5.6	5.0	6.3	10.2	13.4	26.3	27.1	31.5
Calcium hardness mg/l	33.4	40.0	29.0	31.6	30.7	43.2	52.3	65.4	80.0
Magnesium hardness mg/l	3.7	4.2	2.5	5.2	3.9	6.4	7.2	10.4	11.6
Total hardness mg/l	37.1	44.2	31.5	36.8	34.6	49.6	59.5	75.8	91.6
Organic matter mg/l	1.6	0.9	0.9	1.2	0.9	1.0	2.1	3.7	3.2
S.R. Phosphate mg/l	0.00	0.016	0.013	0.00	0.00	0.02	0.02	0.02	0.018
Total phosphate mg/l	0.01	0.024	0.022	0.016	0.019	0.025	0.025	0.030	0.035

Parameters	*J/86* (*10*)	*A/86* (*11*)	*S/86* (*12*)	*O/86* (*13*)	*N/86* (*14*)	*D/86* (*15*)	*J/87* (*16*)	*F/87* (*17*)	*M/87* (*18*)
Bicarbonates mg/l	110.0	140.0	163.0	174.0	152.0	180.0	140.0	137.0	162.0
Total alkalinity mg/l	140.0	179.0	211.0	225.0	196.0	243.0	140.0	183.0	225.0
Chloride mg/l	30.2	16.9	18.0	13.4	9.3	8.4	8.7	11.5	16.4
Calcium hardness mg/l	73.5	68.0	68.0	72.0	59.5	37.4	44.6	50.4	53.2
Magnesium hardness mg/l	9.9	8.2	8.3	7.8	6.4	5.2	6.1	6.5	7.4
Total hardness mg/l	83.4	76.2	76.3	79.8	65.9	42.6	50.7	56.9	60.6
Organic matter mg/l	2.6	3.0	2.0	1.8	1.7	2.0	1.8	1.7	2.6
S.R. Phosphate mg/l	0.018	0.016	0.01	0.013	0.017	0.019	0.01	0.01	0.02
Total phosphate mg/l	0.038	0.025	0.024	0.024	0.024	0.028	0.01	0.02	0.026

Table I. (*Contd.*)

Parameters	*O/85* (*1*)	*N/85* (*2*)	*D/85* (*3*)	*J/86* (*4*)	*F/86* (*5*)	*M/86* (*6*)	*A/86* (*7*)	*M/86* (*8*)	*J/86* (*9*)
Nitrates mg/l	0.00	0.01	0.01	0.00	0.00	0.02	0.03	0.04	0.04
Silicates mg/l	0.32	0.41	0.40	1.20	2.10	2.06	1.98	2.63	2.84
Sulphates mg/l	28.10	26.30	24.00	25.16	27.90	26.30	32.60	35.60	34.30
Dissolved solides mg/l	106.0	98.0	102.0	122.0	130.0	166.0	190.0	198.0	210.0
Suspended solides mg/l	153.0	140.0	138.0	167.0	133.0	210.0	230.0	270.0	300.0
Total solides mg/l	259.0	238.0	240.0	289.0	313.0	376.0	420.0	468.0	510.0
Total Iron (ppm)	0.25	0.18	0.20	0.26	0.23	0.29	0.30	0.25	0.23
Cadmium (ppm)	0.000	0.000	0.003	0.000	0.001	0.001	0.003	0.004	0.002
Nickel (ppm)	0.00	0.00	0.00	0.00	0.00	0.00	0.00	0.00	0.00
Cobalt (ppm)	0.00	0.00	0.010	0.030	0.050	0.040	0.060	0.070	0.060
Lead (ppm)	0.002	0.00	0.00	0.001	0.004	0.001	0.002	0.00	0.000
Copper (ppm)	0.00	0.00	0.010	0.020	0.00	0.040	0.030	0.60	0.060
Zinc (ppm)	0.010	0.00	0.020	0.020	0.00	0.030	0.020	0.030	0.040
Manganese (ppm)	0.00	0.00	0.00	0.00	0.00	0.00	0.00	0.00	0.00

Parameters	J/86 (10)	A/86 (11)	S/86 (12)	O/86 (13)	N/86 (14)	D/86 (15)	J/87 (16)	F/87 (17)	M/87 (18)
Nitrates mg/l	0.02	0.01	0.01	0.03	0.01	0.02	0.02	0.03	0.03
Silicates mg/l	3.39	3.10	2.63	2.14	0.52	0.43	0.96	1.80	1.23
Sulphate mg/l	29.87	30.10	28.19	26.0	26.80	25.40	29.30	27.80	30.18
Dissolved solides mg/l	206.0	200.0	193.0	187.0	163.0	172.0	136.0	149.0	153.0
Suspended solides mg/l	320.0	310.0	292.0	284.0	216.0	239.0	183.0	192.0	206.0
Total solides mg/l	526.0	510.0	485.0	371.0	379.0	411.0	319.0	341.0	359.0
Total Iron (ppm)	0.23	0.28	0.19	0.10	0.12	0.09	0.11	0.13	0.19
Cadmium (ppm)	0.005	0.000	0.000	0.000	0.000	0.001	0.002	0.003	0.001
Nickel (ppm)	0.00	0.00	0.00	0.00	0.00	0.00	0.00	0.00	0.00
Cobalt (ppm)	0.080	0.070	0.050	0.040	0.040	0.00	0.020	0.00	0.030
Lead (ppm)	0.005	0.002	0.003	0.002	0.00	0.00	0.001	0.002	0.003
Copper (ppm)	0.040	0.040	0.40	0.030	0.010	0.00	0.020	0.002	0.003
Zinc (ppm)	0.040	0.040	0.020	0.010	0.00	0.00	0.00	0.020	0.030
Manganese (ppm)	0.00	0.00	0.00	0.00	0.00	0.00	0.00	0.00	0.00

OBSERVATIONS

A total of 33 physico-chemical parameters were obtained every month from October 1985 to March 1987 and presented in Table I. During the period of observation the range of water temperature was 1800°C to 31.00°C., dissolved oxygen 6.0 to 8.0 mg/l, Carbon dioxide 0.0 to 0.6 mg/l, pH 7.2 to 8.0, total alkalinity 140.0 to 295.0 mg/l, chlorides 5.0 to 31.5 mg/l, total hardness 31.5 to 91.4 mg/l, total phosphates 0.01 to 0.038 mg/l, nitrates 0.0 to 0.04 mg/l, dissolved solids 98.0 to 210.0 mg/l, suspended solids 133.0 to 310.0 mg/l, cadmium 0.C00 to 0.005 ppm, Nickel 0.00 ppm, cobalt 0.00 to 0.08 ppm, lead 0.00 to 0.005 ppm, copper 0.00 to 0.40 ppm, zinc 0.00 to 0.04 ppm, Manganese 0.00 ppm.

Number of plankton organisms per liter of water was counted for plankton biomass studies. Monthly plankton biomass values and gross and net productivity results of the lake were presented in Table II. It can be seen that there is a direct correlation between productivity and plankton biomass.

The phytoplankton species occuring in the lake were listed in table III. 18 species of chlorophyceae, 18 species of Bacillariophyceae, 11 species of Myxophyceae, 7 species of Euglenophyceae and 3 species of Dinophyceae were recorded from the lake during the period of study.

The zooplanktonic species recorded from the lake during the period of study were listed in Table IV, 5 species of protozoans, 15 species of ciliates, 3 species of sponges, 2 species of oligochetes, 10 species of rotifers, 8 species of cladocerans, 7 species of copepods and one species of ostracod were recorded.

The fishes occurring in the lake during the period of study were listed in the Table V. Altogether 32 species of fishes belonging to 11 families were recorded from the lake.

DISCUSSION AND CONCLUSION

In general the waters of the reservoir are clean and the secchi disc reading being above 2 meters. Nitrates and phosphates were in very low quantities during the course of

Table II. Biomass and Primary Productivity of Himayat Sagar Reservoir, During October 1985 to March 1987.

Month/ Year	*Biomass (Planktons/l)*	*Gross Production (G)**	*Net Production (N)**	*Resperation (R)**	*G/R*
Oct./85	190	0.0297	0.0148	0.0148	2.006
Nov./85	250	0 0372	0.0223	0.0148	2.513
Dec./85	370	0.0446	0.0297	0.0148	3.013
Jan./86	290	0.0148	0.0074	0.0074	2.000
Feb./86	410	0.0595	0.0372	0.0223	2.668
Mar./86	340	0.0223	0.0148	0.0074	3.013
Apl./86	470	0.0669	0.0372	0.0297	2.252
May./86	540	0.0744	0.0520	0.0223	3.336
Jun./86	380	0.0520	0.0297	0.0223	2.331
Jul./86	270	0.0446	0.0297	0.0148	3.013
Aug /86	110	0.0148	0.0074	0.0074	2.000
Sep./86	170	0.0297	0.0223	0.0148	2.006
Oct./86	160	0.0148	0.0074	0.0074	2.000
Nov./86	230	0.0223	0.0148	0.0074	3.013
Dec./87	390	0.0520	0.0372	0.0148	3.513
Jan./87	390	0.0520	0.0297	0.0223	2.331
Feb./87	430	0.0595	0.0297	0.0297	2.003
Mar./87	480	0.0669	0.0446	0.0223	3.000

*Primary productivity (G, N, R) mg/c/l/6 hrs.

study indicating that the low specific conductivity values support this lake is not very productive. Chlorides were also low indicating no organic pollution in the lake. The trace elements are below the ISI levels. Suspended solids and hardness also show low values.

Productivity was feund to be high during summer months. Plankton biomass has shown low values during monsoon months and high values during summer months.

Table III. List of Phytoplankton

Chlorophyceae

Chlamydomonas sp.
Phacotus lenticularis (Ehr.) Stein.
Volvox sp.
Spondylomorusm sp.
Miractinium sp.
Schroederia setigera (Schroder) Lemm.
Pediastrum boryanum (Turp.) Menegh.
Coelastrum sp.
Chlorella vulgaris Beyerinck
Ankistrodesmus sp.
Closteridium sp.
Selanastrum sp.
Tetraedron minimum (A Braun) Hansg.
Scenedesmus acuminatus
Scenedesmus dimorphus (Turp) Kuetz.
Closterium sp.
Staurastrum tetracerum Ralfs
Euastrum sp.

Bacillariophyceae

Melosira granulata (Ehr.) Ralfs.
Diatoma sp.
Synedra acus (Kuetz)
Cocooneis placentula Ehr.
Navicula gracilis Ehr.
Navicula mutica
Navicula pupula
Pinnularis sp.
Caloneis sp.
Anomoeoneis sp.
Diplonesis elliptica (Kuetz) Cleve.
Gyrosigma sp.
Gomphonema sp.
Cymbella affinis Kuetz.
Amphora Ovalis Kg.
Natzschia sp.
Nitzschia Palea
Nitzschia Sigmoidea Ehr.

Myxophyceae

Chroococcus minutus Kuetz Nag.
Mycrocystis sp.
Merismopedia elegans A. Br.
Synechocystis sp.
Spirulina sp.
Oscillatona amphigranulata Van Goor
Oscillatoria sp.
Phormidium sp.
Lyngbya sp.
Anabaenopsis reciborski Wolosz
Raphidiopsis mediterranea Skuja

Euglenophyceae

Euglena acus Ehr.
Euglena polymorpha Dang
Euglcna sp.
Phacus longicauda Ehr. Duj.
Trachelomonas volvocina Ehr.
Astasia sp.
Notosolenus sp.

Dinophyceae

Gymnodium sp.
Peridinium sp.
Ceratium hirundinella (O.F.M.) Schronk.

Table IV. List of Zooplankton

Protozoa

Cryptomonas sp.
Amoeba sp.
Arcella sp.
Difflugia sp.
Actinophrys sol. Ghr.

Ciliata

Holophrya sp.
Coleps hirtus (Muller)
Coleps octospinus (Noland)
Chilodonella dentata Fougue
Colpodas sp.
Tetrahymena sp.

Colpidium colpoda Ehr.
Paramecium sp.
Halteria grandinella Muller
Aspidisca lunceus Ehr.
Oxgtricha sp.
Uroleptus sp.
Euplotes sp.
Verticella sp.
Cerchesium polypinum (L.)

Porifera

Porifera
Eunapiacs carteri Bowerbank
Carvospongilla lapidosa Annandale
Radiospongilla hemiphydatia Annandale

Oligocheta

Nais sp.
Aulophorus sp.

Rotifera

Keratella tropica Apstein
Keratella cochlearic Gosse
Anuraeopsis fissa De Beuchamp
Brachionus angularis Gosse
Brachiorus quadridentatus Hermann
Lecane sp.
Monostyle sp.
Proales sp.
Cephalodella sp.
Filinia langiseta Ehrb.

Cladocera

Diaphanosoma sarsi Richard
Chydorus sp.
Indialona Ganapati Petkovaki
Alona sp.
Macrothrix spinosa Baird
Ceriodophnia cornuta Richard
Ceriodophnia reticulata (Jurine)
Moina micrura (Jurine)

Copepoda

Heliodiaptonus cinctus (Gurney)
Allodiaptomus sp.
Tropodiaptomus sp.
Cletocampu sp.
Neodiaptomus sp.
Mesocyclops leuckarti Claus
Mesocyclops hyalinus Rehberg.

Ostracoda

Stenocypris hislopi Ferguson

Table V. Fish Fauna of Himayat Sagar Reservoir and its Taxonomic Position

Class	: Pisces	
Sub class	: Teleostomi	
Super order	: Acanthopterygii	
Order	: Cypriniformes	
Family : 1	: Cobitidae	
Species : 1	: *Lepidocephalus (Lepidocephalichthys guntea)* (Hamilton)	Fig. 1
Family : 2	: Cyprinidae	
Sub family	: Garrinae	
Species : 2	: *Garra mullya (Skyes)*	Fig. 2
Sub family	: Cyprininae	
Species : 3	: *Catla catla* (Hamilton)	
Species : 4	: *Cyprinus carpio* (Linneus)	
Species : 5	: *Cirrhinus reba* (Hamilton)	Fig. 3
Species : 6	: *Labeo bata* (Hamilton)	Fig. 4
Species : 7	: *Labeo boggut* (Skyes)	Fig. 5
Species : 8	: *Labeo rohita* (Hamilton)	Fig. 6
Species : 32	: *Osteobrama Vigorsii* (Sykes)	Fig. 30
Species : 9	: *Puntius sophore* (Hamilton-Buchnan)	Fig. 7
Species : 10	: *Puntius ticto* (Hamilton-Buchnan)	Fig. 8
Species : 11	: *Puntius chola* (Hamilton-Buchnan)	Fig. 9
Species : 12	: *Puntius sarana* (Hamilton)	F ig. 10
Subfamily	: Rasborinae	
Species : 13	: *Rasbora daniconius* (Hamilton)	Fig. 11
Species : 14	: *Amblypharyngodon mola* (Hamilton)	Fig. 12
Sub family	: Cultrinae	

Species : 15	: *Chela bacaila* (Hamilton)	Fig. 13
Species : 16	: *Chela labuca* (Hamilton)	Fig. 14
Order	: Channiformes	
Family : 3	: *Channidae*	
Species : 17	: *Channa marulia* (Hamilton-Buchnan)	Fig. 15
Species : 18	: *Channa Punctata* (Bloch)	Fig. 16
Species : 19	: *Channa striata* (Bloch)	Fig. 17
Species : 20	: *Channa orientalis* (Hamilton)	Fig. 18
Species : 21	: *Channa gachua* (Hamilton)	Fig. 19
Order	: Perciformis	
Family : 4	: Cichlidae	
Species : 22	: *Etroplus maculatus* (Bloch)	Fig. 20
Species : 23	: *Etroplus suratensis* (Bloch)	Fig. 21
Family : 5	: Gobiidae	
Species : 24	: *Glossogobius giuris* (Hamilton)	Fig. 22
Family : 6	: Claridae	
Species : 25	: *Clarias batrachus* (Linnacus)	Fig. 23
Order	: Siluriformes	
Family : 7	: Heteropneustidae	
Species : 26	: *Heteropneustes fossills* (Bloch)	Fig. 24
Family : 8	: Siluridae	
Species : 27	: *Wallago attu* (Schneider)	Fig. 25
Family : 9	: Bagridae	
Species : 28	: *Mystus vittatus* (Bloch)	Fig. 26
Species : 29	: *Mystus bleekeri* (Day)	Fig. 27
Order	: Osteoglossiformes	
Family : 10	: Notopteridae	
Species : 30	: *Notopterus notopterus* (Pallas)	Fig. 28
Order	: Mastacembeliformes	
Family : 11	: Mastacembelidae	
Species : 31	: *Mastacembelus armatus* (Lacepede)	Fig. 29

Both phytoplankton and zooplankton have shown a variety of species. Zooplankton dominated over phytoplankton. Among phytoplankton diatoms were better represented. Fishes show all varieties including sensitive genera like *Catla*, *Labeo*, *Glosogobius*, *Puntius* etc., and hardy genera like *channa*, *clarias*, *Mystus* etc.

All the above facts go to show that the lake is oligotrophic and not subjected to any sort of pollution.

SUMMARY

Hydrobiological studies on an oligotrophic lake, Himayat sagar, Hyderabad (A.P.) were made for the period October 1985 to March 1987. 33 physico-chemical parameters were obtained month wise during study period. Plankton biomass and productivity studies were also made month wise. Phyto-plankton and zooplankton organisms occurring in the lake were listed. 32 species of fishes belonging 11 families were recorded from the lake. Ecological parameters and faunistic composition indicates that this lake is an oligotrophic lake.

ACKNOWLEDGEMENTS

Authors are thankful to the Director, Zoological Survey of India and Officer-In-Charge, Freshwater Biological Station, Zoological Survey of India, Hyderabad for their interest and giving facilities to carry out this work.

REFERENCES

APHA. 1985. *Standard methods for the examination of water and waste water.* 15th Ed. Washington, D.C. 1268 P. American Public Health Association.

Babu Rao, M. and Y. Siva Reddy 1984. Studies on fishes of Hussain sagar, *Jantu, 2 : 1-21.*

Day, F. 1989. *Fauna of British India, including Ceylon and Burma, Fishes. 1.* 548 pp. London Tayler and Francis.

George, M.G. 1966. Comparative plankton ecology of five fish tanks in Delhi, India. *Hydrobiologia. 27(1-2)* : 81-108.

Jayaram, K.C. 1981. *The Freshwater Fishes of India.* A Handbook, Zoological Survey of India, Calcutta 475 p.

Jhingran, V.G. 1975. *Fish aud Fisheries of India.* Hindustan Publishing Corpn., India.

Misra, K.S. 1976. *The Fauna of India and the adjacent countries, Pisces.* 2nd Ed. 3. Teleostomi : Cylopriniformes, Siluri, 367 pp.

Michael, R.G. 1968. Studies on the Zooplankton of tropical fish pond. *Hydrobiologia,* 31-31 : 47-48.

Munawar, M. 1974. Limnological studies on freshwater pond of Hyderabad. *Hydrobiologia,* 44(1) 13-27.

Patil, S.G. & Panda, P. 1986. Preliminary Limnological survey of a freshwater tank Patan, Andhra Pradesh. *Geobios New Reports* 5 : 171-172.

Qureshi, T.A. & Qureshi, N.A. 1983. *Indian* fishes. *Brij Brothers, Bhopal.*

Rahimullah, M. 1943. Fish survey of Hyderabad state. Part II—Fishes of Hyderabad city and its suburbs. *Journ. Bombay Nat. Hist. Soc.* 44(1) : 88-91.

Reddy, S. 1984. Limnological studies of the Himayat Sagar and Hussain Sagar Lakes. ***Ph. D. Thesis Osmania University.***

Zafar, A.R. 1966. Limnology of the Hussain Sagar lake, Hyderabad, India—*Phykos* 5(1 & 2) : 115-126.

6

STUDIES ON THE AQUATIC PLANTS FOUND IN AND AROUND HYDERABAD

V. Bhavani Devi

and

Kaza V. Rama Rao

Freshwater Biological Station,
Zoological Survey of India, 1-1-300/B,
Ashoknagar, Hyderabad-500 020.

Abstract—A total of 47 tanks have been surveyed for their biological communities along with physico-chemical parameters of water. A careful examination of the aquatic plants collected during these surveys revealed a total of 15 species from 19 localities. Out of the 15 species, 2 are emergent, 6 are floating and 7 are submerged. The distribution of the aquatic plants found in and around Hyderabad in relation to physico-chemical parameters of water is discussed in the present communication.

INTRODUCTION

A total of 47 tanks located in and around Hyderabad upto a distance of 150 km have been surveyed mainly for listing of animal communities with physico-chemical parameters and while doing so the aquatic plants also have been collected from these tanks with a view to assess the trophic stage of the tanks. This study resulted in the identification of 15 plants which are tabled (1-5) under aquatic plants recorded tankwise, ecological classification of aquatic plants, the range of physico-chemical parameters in relation to occurrence of plants along with their abundance and physico-chemical parameters pertaining to the stations where the plants were collected/not collected in the present communication.

MATERIAL AND METHODS

The plants and animals collected from tanks were sorted out, labeled and preserved in 10% formalin. The water samples were analysed in the field as well as in the laboratory after following the methodology given in APHA (1985). The aquatic plants have been identified using the keys given by Fassett (1940), Subramanyam (1974), Gupta (1979), Riemer (1984) and Santhanam, *et. al.*, (1989).

OBSERVATIONS

15 species of aquatic plants were encountered during the period of the study from 11-4-88 to 26-4-90 and they are as follows :

Family : Characeae
1. *Chara* sp.

Family : Pontederiaceae
2. *Eichhornia crassipes* (Mart.) Solms

Family : Hydrocharitaceae
3. *Hydrilla verticillata* (L.f.) Royle
4. *Vallisneria spiralis* L.
5. *Ottelia alismoides* (L.) Pers.

Family : Lemnaceae
6. *Spirodela polyrhiza* (L.) Schleid

Family : Araceae
7. *Pistia stratiotes* L.
8. *Cryptocoryne retrospiralis* (Roxb.) Fischer

Family : Najadaceae
9. *Najas minor* (Pers.) All.
10. *Najas* sp.

Family : Cyperaceae
11. *Fimbristylis* sp.

Family : Nymphaeaceae
12. *Nymphaea stellata* Willd
13. *Nymphaea* sp.

Family : Gentianaceae
14. *Nymphoides cristatum* (Roxb.) O. Kuntze

Family : Convolvulaceae
15. *Ipomea aquatica* Forsk.

Out of 47 tanks surveyed only 19 tanks were found mainly with aquatic vegetation (Table 1) as mentioned above. The aquatic plants *Eichhornia crassipes, Ipomea aquatica* are collected from six localities ; *Hydrilla verticillata, Vallisneria spiralis, Najas minor* are from three localities ; *Najas* sp., *Nymphaea* sp., *Nymphoides cristatum* from two localities and remaining *Chara* sp., *Ottelia alismoides, Spirodela polyrhiza, Pistia stratiotes, Cryptocoryne retrospiralis, Fimbristylis* sp., and *Nymphaea stellata* each from single station. Among the aquatic plants 2 are found to be emergent, 7 are submerged and 6 are floating (Table 2). The range of Physico-chemical parameters where plants were noticed is given in the Table 3. The physico-chemical parameters of tanks where plants were collected and where plants were not collected are given separately in the table 4 and 5 respectively.

The given data (the range of parameters from tanks where plants were not collected is given in parenthesis) indicate that the aquatic plants thrive under these physico-chemical parameters which range from 22°C—27°C temperature (24°C—33°C) ; 7.0—9.0 pH (6.5—9.1) ; 220—3843 specific conductance (200—5185 mg) ; 0.6—15.5 dissolved oxygen (6.0—12.6 mg/l) ; 60—720 mg/l total hardness (60—450 mg/l) 58—234 mg/l calcium (10—160 mg/l) ; 63—486 mg/l magnesium (8.5—290 mg/l) ; Nil—31 mg/l carbonates (10—48 mg/l) ; 73—612 mg/l bicarbonates (16—335 mg/l) ; 25—436 mg/l chlorides (12—130 mg/l) and Nil—12 mg/l Carbondioxide (Nil).

The occurrence of *E. crassipes* in Nalla cheruvu, Pitlam cheruvu, Ramchandrapuram tank, Nacharam tank, Pedda cheruvu and Hussain sagar indicate the eutrophic status of above tanks.

ACKNOWLEDGEMENTS

The authors are thankful to Prof. Mohammad Shamim Jairajpuri, Director, Zoological Survey of India, Calcutta and Dr. A.N.T. Joseph, Joint Director, Zoological Survey of India, Madras, for encouragement and providing necessary facilities. We are also thankful to the officers and staff of the Freshwater

Table 1. Aquatic plants recorded tankwise. + = Present, — = Absent

Sl. No.	Name of the aquatic plant	Sadashivpet pond 11-4-88 (1)	Patan Cheruvu 11-4-88 (2)	Vikarabad reservoir 20-4-88 (3)	Jeedimetla tank 27-4-88 (4)	Shameerpet stream 4-5 88 (5)	Shameerpet tank 4-5-88 (6)	Water bodies around NFC, Nacharam, ECIL & Uppal 10-5-88 (7)	Indra Sagar, 10-5-88 (8)	Pedda Cheruvu 4-1-90 (9)	Kaverammapeta tank 4-1 90 (10)	Ketireddypalli tank 4-1-90 (11)	Sangam Cheruvu 11-1-90 (12)	Nagaya Kunta, 11-1-90 (13)	Vikarabad tank, 11-1-90 (14)	Ibrahimpatnam tank, 18-1-90 (15)	Masab tank, 18-1-90 (16)
1.	*Chara* sp	—	—	—	—	—	—	—	—	—	—	+	—	—	—	—	—
2.	*Eichhernia crassipes* (Mart) Solms	—	—	—	—	—	—	—	—	—	—	—	—	—	—	—	—
3.	*Hydrilla verticillata* (L.f.) Royle	—	—	—	—	—	—	—	—	—	—	—	—	—	—	—	—
4.	*Vallisneria spiralis* (L.)	—	—	—	—	+	—	—	—	—	—	—	—	—	—	—	—
5.	*Ottelia alismoides* (L.) Pers.	—	—	—	—	—	—	—	—	—	—	—	—	—	—	—	—
6.	*Spirodela polyrhiza* (L) Schleid	—	—	—	—	—	—	—	—	—	—	—	—	—	—	—	—
7.	*Pistia stratiotes* L,	—	—	—	—	—	—	—	—	—	—	—	—	—	—	—	—
8.	*Cryptocoryne retrospiralis* (Roxb.) Fisher	—	—	—	—	—	—	—	—	—	—	—	—	—	—	—	—
9.	*Najas minor* (Pers.) All.	—	—	—	—	—	—	—	—	—	—	—	—	—	—	—	—
10	*Najas* sp.	—	+	—	—	—	—	—	—	—	—	—	—	—	—	—	—
11.	*Fimbristyles* sp.	—	—	—	—	—	—	—	—	—	—	—	—	—	—	—	—
12.	*Nymphaea stellata* Willd.	—	—	—	—	—	—	—	—	—	—	—	—	—	—	—	—
13.	*Nymphaea* sp.	—	+	—	—	—	—	—	—	—	—	—	—	—	—	—	—
14.	*Nymphoides cristatum* (Roxb.) O. Kuntze.	—	—	—	—	—	—	—	—	—	—	—	—	—	—	—	—
15.	*Ipomea aquatica* Forsk	+	+	—	—	—	—	—	—	—	—	—	—	—	—	—	—

Table 1. (*Contd.*)

Sl. No.	Name of the Aquatic Plant	*Pangal tank, 25-1-90* (17)	*Singaram Cheruvu, 25-1-90* (18)	*Kambala Kunta, 30-1-90* (19)	*Raigiri tank, 31-1-90* (20)	*Bhongiri tank, 30-1-90* (21)	*Bibinagar tank, 30-1-90* (22)	*Nalla Cheruvu, 30-1-90* (23)	*Narmala Cheruvu, 20-2-90* (24)	*Kasipalli tank, 20-2-90* (25)	*Chintal Cheruvu, 20-2-90* (26)	*Lakshmi Chernvu, 26-2-90* (27)	*Shameerpet tank, 26-2-90* (6)	*Fish breeding pond, 27-2-90* (28)	*Fish Stocking pond, 27-2-90* (29)	*Fish growth tank, 27 2-90* (30)	*Pitlam Cheruvu, 27-2-90* (31)	*Amangal tank, 15-3-90* (32)
1.	*Chara* sp.	—	—	—	—	—	—	—	—	—	—	—	—	—	—	—	—	—
2.	*Eichornia crassipes* (Mart.) Solms	—	—	—	—	—	—	+	—	—	—	—	—	—	—	—	+	—
3.	*Hydrilla verticillata* (L f.) Royle	—	—	—	—	—	+	—	—	—	—	—	—	—	—	—	—	—
4.	*Vallisneria spiralis* L.	—	—	—	—	—	—	—	—	—	—	—	—	—	—	—	—	—
5.	*Ottelia alismoides* (L.) Pers.	—	—	—	—	—	—	—	—	—	—	—	—	—	—	—	—	—
6.	*Spirodela polyrhiza* (L.) Schleid	—	—	—	—	—	—	—	—	—	—	—	—	—	—	—	—	—
7.	*Pistia stratiotes* L.	—	—	—	—	—	—	—	—	—	—	—	—	—	—	—	—	—
8.	*Cryptocoryne retrospiralis* (Roxb.) Fisher	—	—	—	—	—	—	—	—	—	—	—	—	—	—	—	—	—
9.	*Najas minor* (Pers.) All.	—	—	+	—	—	—	—	—	—	—	—	—	—	—	—	+	—
10.	*Najas* sp.	—	—	—	+	—	—	—	—	—	—	—	—	—	—	—	—	—
11.	*Fimbristylis* sp.	—	—	—	—	—	—	—	—	—	—	—	—	—	—	—	—	—
12.	*Nymphaea stellata* Willd	—	—	—	—	—	—	—	—	—	—	—	—	—	—	—	—	—
13.	*Nymphaea* sp.	—	—	—	—	—	—	—	—	—	—	—	—	—	—	—	—	—
14.	*Nymphoides cristatum* (Roxb.) O. Kuntze	—	—	—	—	—	—	—	—	—	—	—	—	—	—	—	—	—
15.	*Ipomea aquatica* Forsk	—	—	—	—	—	—	—	—	—	—	—	—	—	—	—	—	—

Table 1. (*Contd.*)

Sl. No.	Name of the Aquatic Plant	*Govindpalli tank, 15-3-90* (33)	*Keesarguta, 20-3-90* (34)	*Shameerpet Stream, 20-3-90* (35)	*Sadashivpet tank, 22-3-90* (36)	*Ramchandra puram tank (B.H.E.L.) 22-3-90* (37)	*Venuka Cheruvu, 22-3-90* (38)	*Narsapur town, 28-3-90* (39)	*Malkal Cheruvu, 28-3-90* (40)	*Dairy effluent plant, 5-4-90* (41)	*Nacharam tank, 5-4-90* (42)	*Pedda Cheruvu, (Nacharam) 5-4-90* (43)	*Nalla Cheruvu, (Uppal tank) 5-4-90* (44)	*Suuraram tank, 12-4-90* (45)	*Loyola tank, 12-4-90* (46)	*Kukatpalli Stream, 19-4-90* (47)	*Hussain Sagar, opp. (Kukatpalli nala) 24-5-90* (48)	*Kotula Kunta, 26-4-90* (49)	*Bhadrakali tank, 26-4-90* (50)
1.	*Chara* sp.	—	—	—	—	—	—	—	—	—	—	—	—	—	—	—	—	—	—
2.	*Eichhornia crassipes* (Mart) Solms	—	—	—	+	+	—	—	—	—	+	+	+	—	—	—	+	—	—
3.	*Hydrilla verticillata* (L.f.) Royle	—	—	—	—	—	+	—	—	—	—	—	—	—	—	—	—	—	+
4.	*Vallisneria spiralis* L.	—	—	+	—	—	—	—	—	—	—	—	—	—	—	+	—	—	+
5.	*Ottelia alismoides* (L.) Pers.	—	—	—	—	—	—	—	—	—	—	—	—	—	—	+	—	—	—
6.	*Spirodala polyrhiza* (L.) Schloid	—	—	—	—	—	—	—	—	—	—	—	+	—	—	—	—	—	—
7.	*Pistia stratiotes* L.	—	—	—	—	—	—	—	—	—	—	—	—	—	—	—	+	—	—
8.	*Cryptocoryne retrospiralis* (Roxb.) Fisher	—	—	—	—	—	—	—	—	—	—	—	—	+	—	—	—	—	—
9.	*Najas minor* (Pers.) All.	—	—	—	—	—	+	—	—	—	—	—	—	—	—	—	—	—	—
10.	*Najas* sp.	—	—	—	—	—	—	—	—	—	—	—	—	—	—	—	—	—	—
11.	*Fimbristylis* sp.	—	—	+	—	—	—	—	—	—	—	—	—	—	—	—	—	—	—
12.	*Nymphaea stellata* Willd	—	—	—	—	—	—	—	—	—	—	—	—	+	—	—	—	—	—
13.	*Nymphea* sp.	—	—	—	—	+	—	—	—	—	—	—	—	—	—	—	—	—	—
14.	*Nymphoides cristatum* (Roxb.) O. Kuntze	—	—	—	—	+	—	—	—	—	—	—	—	—	—	—	—	—	+
15.	*Ipomea aquatica* Forsk	—	—	+	+	+	+	+	+	—	—	—	—	—	—	—	—	—	—

Table 2. Ecological classification of Aquatic Plants.

Sl. No.	*Aquatic Plant Names*		*Ecological Distribution*		
	Scientific Names	*English Common Names*	*Emergent*	*Floating*	*Submerged*
1.	*Chara* sp.	Chara	—	—	×
2.	*Eichhornia crassipes* (Mart.) Solms	Water Hyacinth	—	×	—
3.	*Hydrilla verticillata* (L.f.) Royle	Hydrilla	—	—	×
4.	*Vallisneria spiralis* L.	Eelgrass	—	—	×
5.	*Ottelia alismoides* (L.) Pers.	—	—	—	×
6.	*Spirodela polyrhiza* (L.) Schleid	Gaint duckweed	—	×	—
7.	*Pistia stratiotes* L.	Water lettuce	—	×	—
8.	*Cryptocoryne retrospiralis* (Roxb.) Fisher	Crypt	—	—	×
9.	*Najas minor* (Pers.) All.	Brittle naiad	—	—	×
10.	*Najas* sp.	Naiad	—	—	×
11.	*Fimbristylis* sp.	Joria	×	—	—
12.	*Nymphaea stellata* Wild	Water lily	—	×	—
13.	*Nymphaea* sp.	Water lily	—	×	—
14.	*Nymphoides cristatum* (Roxb.) O. Kuntze.	—	—	×	—
15.	*Ipomea aquatica* Forsk.	Water spinach	×	—	—

Table 3. The range of physico-chemical parameters in relation to occurrence of plants and their abundance.

Sl. No.	Name of the Aquatic Plant	Plants Coll. Stn. Nos.	Physico-chemical parameters range										
			Temp°C	pH	Specific Cond. in μ s at 25°C	mg/l							
						DO	T. Hardness	Ca^{++}	Mg^{++}	CO_3^{--}	HCO_3^-	Cl^-	CO_2
1.	*Chara* sp.	11	25	8.3	549	7.1	120	—	—	14	230	67	Nil
2.	*E. crassipes*	23, 31, 35, 40, 41, 45	22-33	7.2-8.5	610-3843	N-12.0	120-720	98-234	210-486	N-18	153-612	50-436	N-12
3.	*H. verticillata*	22, 36,47	26-37	7.8-9.0	570-1037	7.4-11.8	205-332	80-113	219-230	16-31	103 256	60-112	Nil
4.	*V. spiralis*	5, 44, 47	26-37	7.5-9.0	400-2200	8.0-15.5	120-565	113-122	63-219	N-16	114-451	42-316	Nil
5.	*O. alismoides*	44	28	7.5	2200	15.5	565	122	63	Nil	451	316	—
6.	*S. pelyrhiza*	23	33	8.0	3538	0.6	561	213	348	Nil	572	318	8
7.	*P. stratiotes*	45	—	—	—	—	—	—	—	—	—	—	—
8.	*C. retrospiralis*	42	29	8	671	4.1	140	68	72	18	73	60	Nil
9.	*N. minor*	19, 26, 36	25-26	7.5-9 0	570-830	8.4-11.8	120-205	64	140	N-29	84-256	37-60	Nil
10.	*Najas* sp.	2, 20	26	9.0	960	7.2	196	58	138	31	74	50	Nil
11.	*Fimbristylis* sp.	5	26	7.7	400	10.8	120	—	—	Nil	170	42	—
12.	*N. stellata*	42	29	8	671	4.1	140	68	72	18	73	60	Nil
13.	*Nymphaea* sp.	2, 35	26	7.4	860	10.0	255	—	—	Nil	384	50	—
14.	*N. cristatum*	35, 47	26-37	7.4-9.0	860-1037	8.9-10	255-332	113	219	N-16	114-384	50-78	Nil
15.	*I. aquatica*	1, 2, 35, 36, 37, 38	26-30	7-9.0	380-860	7.8-11.8	125-255	—	—	N-18	128-384	50-70	—

Table 4. Physico-chemical parameters pertaining station where plants were collected.

Sl. No.	*Name of the location (Where plants were collected)*	Physico-chemical parameters										
					mg/l							
		Temp. °C	*pH*	*Specific Cond. in μs at 25°C*	*DO*	*T. Hardness*	Ca^{++}	Mg^{++}	CO_3^{--}	HCO_3^-	Cl^-	CO_2
1.	Sadashivpet pond, 11-4-88	—	—	—	—	—	—	—	—	—	—	—
2.	Patan Cheruvu, 11-4-88	—	—	—	—	—	—	—	—	—	—	—
3.	Shameerpet stream, 4-5-88	—	—	—	—	—	—	—	—	—	—	—
4.	Ketireddypalli tank, 4-1-90	25	8.3	549	7.1	120	—	—	14	230	67	Nil
5.	Kambala kunta, 30-1-90	26	8.5	830	8.4	204	64	140	29	84	37	Nil
6.	Raigiri tank, 30-1-90	26	9.0	960	7.2	196	58	138	31	74	50	Nil
7.	Bibinagar tank, 30-1-90	27	7.8	671	7.4	310	80	230	31	102	112	Nil
8.	Nalla Cheruvu, 30-1-90	22	8.5	2840	Nil	720	234	486	Nil	612	436	1.2
9.	Pitlam Cheruvu, 27-2-90	25	7.5	610	12.0	120	—	—	Nil	213	142	—

(Contd.)

10. Shameerpet stream, 20-3-90	26	7.7	400	10.8	120	—	—	Nil	170	42	—
11. Sadashivpet tank, 22-3-90	26	8.2	800	8.3	250	—	—	Nil	275	113	—
12. Ramchandrapuram tank, 22-3-90	26	7.4	860	10.0	255	—	—	Nil	384	50	—
13. Venuka Cheruvu, 22-3-90	26	9.0	570	11.8	205	—	—	18	256	60	—
11. Narsapur town, 28-3-90	28	7.0	380	8.0	125	—	—	Nil	128	60	—
15. Malkal Cheruvu, 28-3-90	30	8.0	610	7.8	225	—	—	Nil	256	70	—
16. Nacharam tank, 5-4-90	23	7.5	3843	Nil	396	178	218	Nil	234	216	12
17. Pedda Cheruvu, 5-4-90	29	7.2	1281	1.3	308	98	210	18	153	103	Nil
18. Nalla Cheruvu, 5-4-90	33	8.0	3538	0.6	561	213	348	Nil	572	318	8
19. Suuraram tank, 12-4-90	29	8.0	671	4.1	140	68	72	18	73	60	—
20. Kukatpalli stream, 19-4-90	28	7.5	2200	15.5	565	122	63	Nil	451	316	—
21. Hussain Sagar, 25-4-90	—	—	—	—	—	—	—	—	—	—	—
22. Bhadrakali tank, 26-4-90	37	9.0	1037	8.9	332	113	219	16	114	78	Nil

Table 5. Physico-chemical parameters pertaining station where plants were not collected.

Sl. No.	Name of the location (Where Plants were not Collected)	Physico-chemical parameters										
		Temp. °C	pH	Specific Cond. in µs at 25°C	mg/l							
					DO	T. Hardness	Ca^{++}	Mg^{++}	CO_3^{--}	HCO_3^-	Cl^-	CO_2
1.	Vikarabad reservoir, 20-4-88	—	—	—	—	—	—	—	—	—	—	—
2.	Jeedimetla tank, 27-4-88	—	—	—	—	—	—	—	—	—	—	—
3.	Shameerpet tank, 4-5-88	—	—	—	—	—	—	—	—	—	—	—
4.	Water bodies around NFC, ECIL, Nacharam and Uppal, 10-5-88	—	—	—	—	—	—	—	—	—	—	—
5.	Indra Sagar, 10-5-88	—	—	—	—	—	—	—	—	—	—	—
6.	Pedda Cheruvu, 4-1-90	24	8.6	854	7.3	156	—	—	12	180	40	Nil
7.	Kaverammapeta tank, 4-1-90	24	8.4	488	6.9	184	—	—	10	240	53	Nil

(Contd.)

Table 5. (*Contd.*)

8. Sangam Cheruvu, 11-1-90	29.5	8.1	732	8.1	250	70	180	20	31	58	Nil
9. Nagayakunta, 11-1-90	30	8.0	610	6.9	178	40	138	17	21	57	Nil
10. Vikarabad tank, 11-1-90	26	7.6	366	6.0	160	30	130	11	16	50	Nil
11. Ibhrahimpatnam, 18-1-90	31	8.0	793	9.1	290	80	210	30	70	60	Nil
12. Masab tank, 18-1-90	30	8.5	1098	8.7	450	160	290	35	96	97	Nil
13. Pangal tank, 25-1-90	29	8.5	538	8.6	182	38	144	18	24	42	Nil
14. Singaram Cheruvu, 25-1-90	26	8.5	730	8.9	235	82	153	27	63	34	Nil
15. Bhongiri Cheruvu, 30-1-90	26	7.5	810	6.8	193	53	140	48	92	130	Nil
16. Narmala Cheruvu, 20-2-90	25	8.9	215	8.6	60	10	8.5	12	79	21	—
17. Kasipalli tank, 20-2-90	25	9.1	240	10.4	70	12	9.7	18	104	39	—
18. Chintal Cheruvu, 20-2-90	24	8.5	220	7.2	60	12	7.3	12	91	25	—
19. Lakshmi Cheruvu, 26-2-90	24	8.0	665	7.2	65	—	—	Nil	244	120	—

(*Contd.*)

Table 5. (*Contd.*)

20. Shameerpet tank, 26-2-90	24	8.5	380	8.8	75	—	—	12	146	67	—
21. Fish breeding pond, 27-2-90	25	7.8	460	8.0	70	—	—	Nil	213	74	—
22. Fish Stocking pond, 27-2-90	25	8.8	510	8.6	65	—	—	18	104	81	—
23. Fish growth tank, 27-2-90	25	7.8	740	8.0	165	—	—	Nil	335	100	—
24. Amangal tank, 15-3-90	27	8.0	610	7.3	178	—	—	26	7.9	122	Nil
25. Govindpalli tank, 15-3-90	26	8.0	620	8.5	164	—	—	32	87	116	Nil
26. Keesargutta, 20-3-90	25	7.9	200	12.6	60	—	—	Nil	91	21	—
27. Dairy effluent plant, 5-4-90	33	6.5	5185	Nil	—	—	—	—	—	—	—
28. Loyola tank, 12-4-90	30	8.2	488	8.3	168	30	138	10	18	12	Nil
29. Ketula kunta, 26-4-90	29	8.0	854	6.4	161	68	93	10	68	59	Nil

Biological Station, Zoological Survey of India, Hyderabad for assistance.

REFERENCES

APHA, AWWA and WPCP. (1985). *Standard methods for examination of water and waste water*. American Public Health Association, Washington, D.C., (17th edition).

Fassett, N.C, (1940). *A Manual of Aquatic Plants*. McGraw-Hill Book Company, Inc,, New York and London, vii, 382. pp.

Gupta, O.P. (1979). *Aquatic Weeds : Their Menace and Control*. Today and Tomorrow's Printers and Publishers, New Delhi, 272 pp. 14 figs.

Riemer, D.N. (1984). *Introduction to Freshwater Vegetation*. Avi Publishing Company, Inc. Westport, Connecticut, vii 207 pp., 14 figs , 14 tabs.

Santhanam, R., P. Velayutham and G. Jegatheesan (1989). *A Manual of Freshwater Ecology (An aspect of Fishery Environment)*. Daya Publishing House, Delhi, ix 134 pp., 101 figs.

Subramanyam, K. 1962 (1974). *Aquatic Angiosperms. A systematic Account of Common Indian Aquatic Angiosperms*. Council of Scientific and Industrial Research, New Delhi, viii 190 Pp , 63 figs.

7

FOOD AND FEEDING ANALYSIS OF CATFISH FROM KSHIPRA RIVER

Sharad Shrivastava
School of Studies in Zoology,
Vikram University, Ujjain (M.P.)

Abstract—Food and feeding habits of freshwater catfish *viz. Heteropneustes fossilis, Mystus seenghala Mystus bleekeri* and *Ompok bimaculatus* of Kshipra river have been studied. The fish food organisms have been classified into 16 food groups with their frequency of occurrence in the stomach of catfish. Their food consisted mainly of Diptera, Ephemeroptera, Coleoptera, Odonata, Mollusca and fish. The utilization of fish food groups in different seasons have been correlated with their occurrence. Preferential food spectrum with reference to fish size is also pointed out. The present result describe that catfishes feeding on similar diet with overlapping food organism groups indicate interspecific food competion among them in Kshipra river.

INTRODUCTION

Since the beginning of fishery biological studies, the nutrition of fishes has been an important theme. Nutrition is one of many factors which regulate or atleast influence the occurrence, growth and migration of fishes and therefore any information concerning the mechanisms involved would aid in the development of an optimal stock management programme. Inadequate attention was devoted to the subject in India till 1955, when Das and Moitra in a series of papers (1955, 1956 a, b, c and 1958), elucidated and quantitative correlation of the food for freshwater fishes and the type and the length of alimentary canal. This was further substantiated by a series of original observations on plankton peaks, bottom fauna peaks and hydrology of Indian freshwater (Das and Moitra 1963 ; Kapoor *et. al.* 1975 ; Pandian and Vivekanandan, 1985). Several aspects concerning the nutrition of catfish in Kshipra river, have been examined in the present study.

MATERIAL AND METHODS

Gut contents analysis was performed on freshwater catfish *H. fossilis, M. Seenghala, M. Bleekeri* and *O. bimaculatus,* which are commonly found in Kshipra river. Catfishes were collected from five sampling stations by cast netting between September 1987 and August 1988 at monthly intervals. The fish were killed immediately after capture, the length and weight recorded and alimentary tract removed, labelled and preserved in 10% formalin. The individual food organisms were categorized into 16 major food groups (Teleost, Diptera, Epheroptera, Coleoptera, Odonata, Placoptera, Hemiptera, Trichoptera, Algae, egg of Invertebrate, Copepods and Daphnids, Mollusca, Ostracoda, Orthoptera, Decapoda and Miscellaneous) with their frequency of occurrence in the gut of catfish.

RESULTS AND DISCUSSION

In the present investigation from September 1987 to August 1988, stomachs of *M. seenghala* were examined and nine major food groups were observed (Table 1). The percentage occurrence of fish as a food item in *M. seenghala* was found to be dominant (18.39 to 78.1%) throughout the year except in May and June were miscellaneous group (51.38 to 57.53%) observed as a most dominant food group. The other food groups *viz.*, Diptera 6.3 to 18.62%, Coleoptera 1.6 to 15.17%, Odonata 1.5 to 9.78%, Placoptera 1.5 to 4.78%, Hemiptera 1.2 to 2.27%, Decapoda 3.2 to 10.61%, Ephemeroptera 1.8 to 9.28% and miscellaneous 4.4 to 57.33% were observed from the gut of *M. seenghala. M. seenghala* feeds on fishes as reported by earlier workers (Alikunhi and Rao, 1948 ; Saigal and Motwani, 1961 ; Sehgal, 1967 ; Bhatt, 1970 ; Vinci, 1986). Bhatt (1970), in their investigation reported that *M. seenghala* is a carnivorous fish with cannibalistic habit. In the present study cannibalism was also occurred at only few instances. *O. bimaculatus* feeds mainly on ten major food groups (Table 1). The percentage occurrence of forage fishes as a food very high 18.36 to 73.2% in all the

Table 1. Average percentage contribution of different prey items in four catfish in Kshipra river.

S. No.	*Food groups*	*Mystus seenghala*	*Ompok bimaculatus*	*Mystus bleekeri*	*Heteropneustes fossilis*
1.	Teleosts	50.70	42.77	—	—
2.	Diptera	13.00	16.56	26.15	28.78
3.	Ephemeroptera	2.31	3.70	1.76	1.96
4.	Coleoptera	6.26	7.59	3.41	4.75
5.	Odonata	3.71	5.59	3.42	4.29
6.	Placoptera	1.61	2.53	—	—
7.	Hemiptera	0.580	3.26	0.106	0.938
8.	Trichoptera	—	—	0.22	0.865
9.	Algae	—	—	7.94	7.100
10.	Eggs of invertebrates	—	—	10.37	10.440
11.	Copepods and Daphnids	—	—	5.64	7.11
12.	Mollusca	—	—	13.79	9.36
13.	Ostracods	—	—	3.13	5.250
14.	Orthoptera	—	0.90	—	—
15.	Decapoda	6.27	6.57	1.76	1.87
16.	Miscellaneous	15.52	10.49	22.20	16.87

months except in August were dominated food group was miscellaneous 35.39%. The insect order Diptera occurred as a second most dominated food group ranged between 5.9 to 28.30% and occurred throughout the year. The other important food groups which were next in the order of importance *i.e.* Miscellaneous 2.10 to 35.39%, Coleoptera 3.47 to 12.10%, Odonata 2.1 to 12.5%, Decapoda 2.10 to 14.10%, Hemiptera 2.30 to 6.60%, Placoptera 1.7 to 8.73%, Ephemeroptera 2.00 to 6.8% and Orthoptera 4.4 to 6.4%. *O. bimaculatus* feeds mainly on forage fishes and insects as reported by Qayyum and Qasim (1964), and Shivakami (1987). Similar observations were made in the present study.

Gut contents examination *M. bleekeri* reveals that it feeds mainly on 13 food groups (Table 1). Diptera 13.62 to 34.10%

was the most dominant food group were observed throughout the year. The Mollusca 2.18 to 36.81% was observed as a third dominant food group. The Cladocera and Copepoda 1.62 to 16.46%, were recorded throughout the year except in two months January and August, Algae 2.15 to 24.20% shows steady occurrence throughout the year except in October and November, Ostracoda were 1.24 to 12.24% recorded for nine months. The eggs of invertebrates 2.24 to 38.76% were recorded throughout the year, Decapoda 2.18 to 4.7%. Observed only for six months, Ephemeroptera 2.14 to 6.20% observed only for six months, Odonata 1.4 to 7.81% were recorded throughout the year except in three months and miscellaneous group 6.61 to 41.52% were also recorded throughout the year from the gut of *M. bleekeri.* Das and Moitra (1963) and Pandey (1983) reported that *M. bleekeri* feeds mainly on insects larvae, insects, crustacean, Molluscs and other planktonic organisms. Findings of the present study was consistent with the observation made by Das and Moitra (1963) and Pandey (1983). Gut content examination of *H. fossilis* reveals that it feeds mainly on 13 food groups (Table 1). In the present study *Heteropneustes fossilis,* feeds mainly on insects, insect larvae, mollusca Cladocera and Copepoda, Algae, invertebrate eggs, Ostracoda and Decapoda. The highest percentage of Diptera 12.50 to 36.58% were observed throughout the year except in January were the dominanted food group was eggs of invertebrates 34.00%. The Mollusca 1.21 to 36.8% were observed as a dominant food group with maximum percent 36.8% in August, Cladocera and Copepoda 1.58 to 16.46% recorded only for ten months, eggs of invertebrates 2.26 to 34.00% were occurred throughout the year. Ostracoda 1.38 to 15.16% were recorded for 9 months, Algae 3.18 to 12.68% were occurred throughout the year except in June and July. Coleoptera 2.10 to 9.14% were consumed throughout the year except in July, Ephemeropteran 2.11 to 6.8% were observed only for seven months. Trichoptera 2.00 to 4.68% were observed only during October, November and December. Hemiptera 1.2 to 6.0% consumed only for three months, Dacapoda 1.20 to 10% were consumed only in five months, miscellaneous 8.0 to 44.58% were observed throughout the study period respectively. *H. fossilis* is not markedly piscivorous as reported by Alikunhi

(1957). Hora and Pillay (1962) wrote that *H. fossilis* is a bottom feeder feeds on protozoa, Crustacean and insects which contributes 70% of diet, the rest being the algae and higher plants. Chaudhary (1971) observed that Juvenile and adult *H. fossilis* feeds on insects, insect larvae, crustacean, Ostracods and also on Algae, higher aquatic plants and organic debris. They are predominantly insectivorous and though predatory in habit, are not markedly piscivorous. Bhatt (1968) reported that *H. fossilis*, does not have any regular food preference he noted that gastropods during August and September form the chief food of the fish. Thakur and Das (1985 b) observed Mollusca 15.5% *Cypris* 15.6%, *Chironomid* larvae 13.4%, fish 16.5%, debris 41% and Algae 26.6% as a food items of *H. fossilis*. These findings of earlier workers confirm the observations of present study.

In the present study it was observed that fish fingerlings are generally abundant in the gut of *M. seenghala* and *O. bimaculatus* during post breeding season. This variation to a large extent, seems correlated with the breeding season of other fish. Fish fingerlings are generally abundant in rivers during the postmonsoon months and therefore their availability in the gut of *M. seenghala* and *O. bimaculatus* correspond with their availability in the environment. A gradual decline of these food items from the guts during subsequent months suggests a decreasing effect of predation on them as probably they become too large to remain susceptible to be seized by *M. seenghala* and *O. bimaculatus*. Similar observations were made by Qayyum and Qasim (1964) and Bhatt (1970). In *M. seenghala* and *O. bimaculatus* forage fishes were recorded from July to November and February to March. From April to June the occurrence of fish in the gut progressively falls. This may be connected with the maturation of gonads because during these months fish become less active and fails to catch other fishes. Qayyum and Qasim (1964) in their study also represented similar reasons. In the present study maximum percentage of insects and insect larvae were occurred from December to April corresponds with their abundant in the environment. The minimum percentage of insect and insect larvae were observed from May to August may be due to the reduction in feeding intensity and reduction of insect population in the environment due to flood events in July and August respectively. Govind (1969), describe December to April as productive period of fish food, in general mayfly nymphs, chironomids and other insects were present in high densities in this period. Occurrence of sand and gravel in males (parents) signifies that while guarding the youngones the males (parents) feeds mainly on debris and do not go about hunting for fish and other organisms, leaving the larvae unguarded.

In the *M. bleekeri* and *H. fossilis* the insect and insect larvae occurred throughout the year and the only time when their percentage showed considerable reduction was during July and August. These being the spawning months. There was a marked seasonal variation in the presence of gastropods in the guts. Their occurrence was maximum in May and August in *Mystus bleekeri* while August and September in *H. fossilis*. Field observations revealed that the gastropods particularly breed during the monsoon months which leads to their maximum density in rivers during the rainy season and form food of these fishes (Bhatt, 1968).

It is expected that the composition of the food consumed will changed as the fish length increases. Since catfish Juvenile and adult were represented in the study, their food groups were correlated with the fish length. In *M. seenghala* it was observed that 15-21 cm long fishes feed on four food groups, 21-25 cm long fishes feed on eight food groups and 25-40 cm long fishes feed on two food groups. In *O. bimaculatus* fish size 10-15 cm, 15-20 cm, and 20-26 cm long feed on four, eight and three food groups respectively (Fig. 1 and 2). In *M. bleekeri* it was observed that 8-11 cm long fishes feed on three food groups, 11-15 cm long fishes feed on four food groups, 15-18 cm long fishes feed on nine food groups and 18-21 cm long fishes feed on four food groups (Fig. 1 and 2). In *H. fossilis* it was observed that youngest fishes 9-15 cm long feed on four food groups while the 15-20 cm long feed on twelve and 20-.5 cm long fish feed on four food groups (Fig. 1). In the present study it was established that broadest nutritional spectrum was

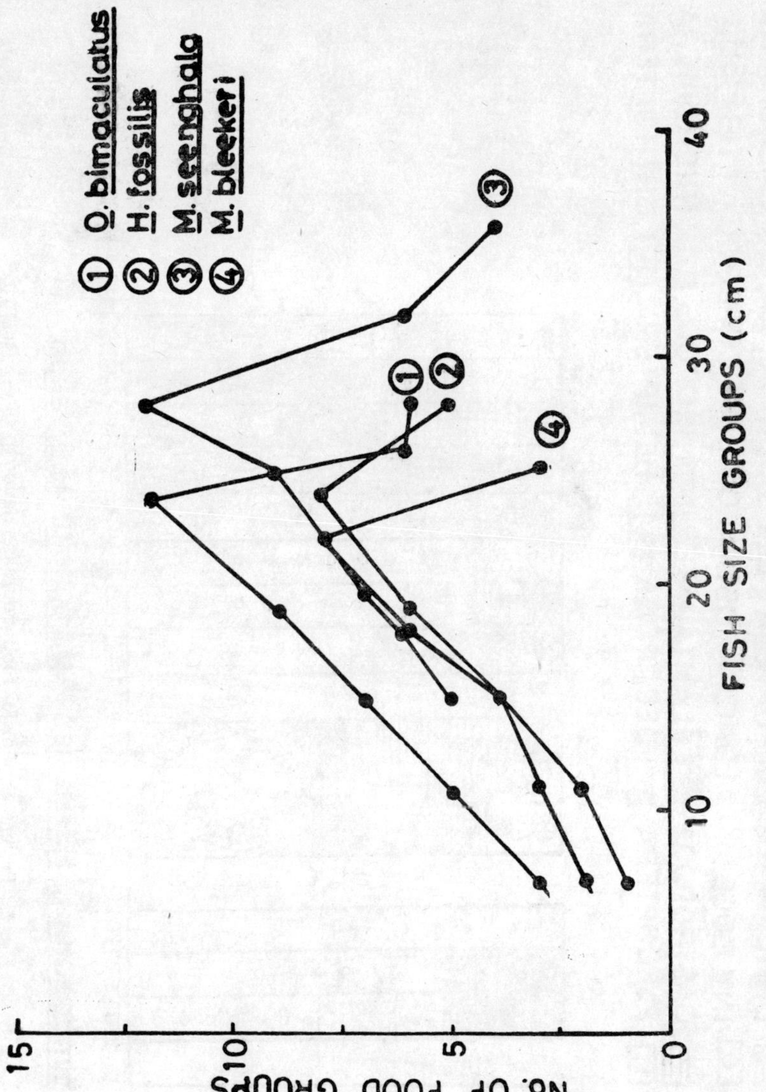

Fig. 1. Number of food groups in relation to catfish length.

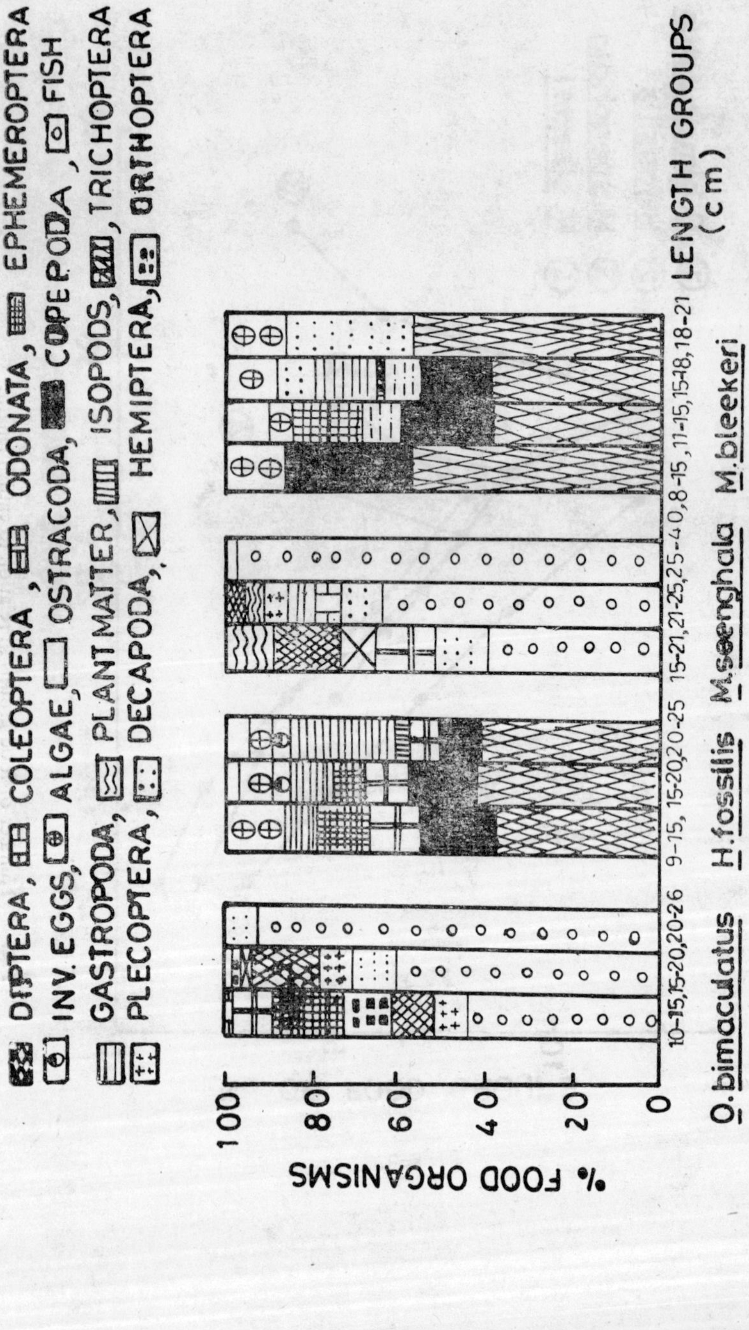

Fig. 2. Percentage of major food groups in relation to catfish length.

evident in medium size fishes consuming 8-12 food group in *M. seenghala*, *O. bimaculatus*, *M. bleekeri* and *H. fossilis* while the smallest fish and largest fishes had only 3-4 food groups in their stomach (Fig. 1 and 2). It was observed that catfish feeding on similar diet with overlapping food organisms type, indicate food competition among them in Kshipra river (Fig. 1). The results of the feeding and size relationship were similar to Worthmann and Oliveira (1987) in *P. squamosissimus* and *P. monti*.

REFERENCES

Alikunhi, K.H. and Rao Nagaraja. S. (1948). An investigation into the food and feeding habits of some of the common freshwater fishes of Madras. *Pro. 13th Indian Sci.* Cong. pp. 201.

Bhatt, V.S. (1968). Studies on the biology of some freshwater fishes. Part VII. *Heteropneustes fossilis* (Bloch). *Ind. J. Fish.* 15. 99-115.

Bhatt, V.S. (1970) Studies on the biology of some freshwater fishes Part IV *Mystus seenghala* (Sykes). *J. Bombay. Nat. His. Soc. 67* (*2*), 194-211.

Chaudhary, H. (1971). Culture of airbreathing fishes. *In first workshop report of all India coordinated Research Project on culture of air-breathing fishes held at Cuttack*, 6 p.

Das, S M and Moitra, S K (1955). Feeding habits of freshwater fishes of Uttar Pradesh India Pt. I. The surface feeders, the mid feeders and the bottom feeders. *Proc. Nat. Acad. Sci. 25* (B) (I and II): 1-6.

Das, S.M. and Moitra, S K. (1956 a), The Surface, the mid and the bottom feeding fishes of Uttar Pradesh. *Proc. Ind. Sci. Cong. Pt. III*: 307.

Das, S.M. and Moitra, S.K. (1956 b), Studies on the food of some common fishes of Uttar Pradesh Pt. II : On the types of fish food and the variations in the relative length of the alimentary canal *Proc. Nat. Acad. Sci. 26 (B) IV* : 213-223.

Das, S.M. and Moitra, S.K. (1956 c) Studies on the food of some common fishes of Uttar Pradesh India III : Comparative anatomy of the alimentary tract in some freshwater fishes and their modification in relation to feeding habit. *Proc. Nat. Acad. Sci. 26 (B)* IV—224-233.

Govind, M.S. (1969), Bottom fauna and macro-vegetation in Tunghabhadara reservoir and their role in the food chain of food communities. *Pro. of Seminar on the Ecology and freshwater reservoir, Barrackpore* : 91-123 pp.

Hora, S.L. and Pillay TVR (1962). Hand book on fish culture in the Indo-Pacific region *FAO Fish Biol. Tech. Pap* : *14* : 204.

Kapoor, B.G , Smit, H. and Verighina IA (1975). The alimentary canal and digestion in teleosts *Adv. Mar. Biol. 13* : 109.

Karekar, P.S. and Bal, D.V. (1958). Food and feeding habits of *Polynemus indicus. Ind. J. Fish. 5* : 77-94.

Pandey, K.C. (1983). *Concepts of Indian fisheries* (K. Nath and Company, Meerut India).

Pandian, T.J. and Vivekanandan, E. (1985). Energetics of feeding and digestion In Tytler Peter and Peter Calow (eds) *Fish Energetics New Perspective.* 98-124 pp. Croomhelm London.

Qayyum, A. and Qasim, S.Z. (1964). Studies on the biology of some freshwater fishes Part I. *J. Bombay Nat. Hist. Soc. 61 (3)* : 627-650.

Saigal, B M and Motwani, M.P. (1961). Studies on the Fisheries and biology of the commercial catfishes of the Ganga river system. Early life history bionomics and breeding of *M. seenghala. Indian J. Fish. 8 (1)* : 61-74.

Sehgal, Prem (1967). Food feeding habits of *Mystus seenghala. Research Bulletin of Punjab University 18* (I-II) : 149-155.

Shivakami (1987). Length weight relationship and relative condition in *Ompok bimaculatus* from Bhavanisagar Reservoir *Indian J. Fish 36* : 202-207.

Thakur, N.K. and Das, P. (1985 a) Synopsis of Biological data on *Magur*. Pulbished by *Central Inland capture Fisheries Research Institute Barrackpore Bulletin 41* : 1-80.

Thakur, N.K and Das, P. (1985 b). Synopsis of the biological data on Singhi Published by *Central Inland Capture Fisheries Research Institute Barrackpore. Bulletin 39* : 1-32.

Vinci, G.K. (1986). of Biology *Mystus seenghala* (Sykes) from Nagarjuna Sagar Reservoirs, Andhra Pradesh. *Indian J. of Anim. Sci. 56, (7)* : 814-821.

Worthmann, Heino and Oliveriva De L. Jacinta (1987). Comparative nutritional analysis of two Sciaenidian Species, the Pescades *Plagioscion squamosissimus* and *Plagioscion monti* from different water system of the Central Amazon. *Anim. Res.* and *Dev. 25* : 7-34.

8

STUDIES ON FISHERY POTENTIAL OF THE WESTERN NARMADA BASIN

S.N. Chatterji, K.S. Rao and Anil K. Singh
School of Studies in Zoology
Vikram University, Ujjain

INTRODUCTION

Fisheries potentials of various rivers are on record. The fisheries of Cooum river (Abraham, 1962), Bhavani, Tungbhadra, and Cauvery rivers (Chacko and Dinamani, 1949 : Chaoko and Jiriyan 1948 ; Chaoko *et. al.* 1954 and Ranganathan and Natarajan, 1969). Data with regard to Hooghly river (David, 1954), river Godavari and Krishna (David, 1963 a), river Gandak (David 1963 b), Tungbhadra river (David *et. al.* 1959), Chambal river (Dubey and Mehra 1959), Rihand river (Hora, 1949), Mahanadi river (Job *et. al.* 1955), Damodar river (Job *et. al.* 1952 ; Pantulu *et. al.* 1966) Tapti river (Karamchandani and Pisolkar, 1967), Narmada river (Karamchandani *et. al.*, 1967), Brahmaputra river (Motwani *et. al.*, 1962) are reported. Besides these, comprehensive works on riverine fisheries by Saha (1970), Sen Gupta and Jacob (1968), CIFRI report (1966), M.P. Fisheries Directorate report (1971), National commission of Agriculture report (1978), IIMA report (1983) through light on the gross fishery potentials and its fluctuations in the riverine fisheries of our country.

River Narmada is the fifth largest west flowing river of India. This river originates from Amarkantak traverses a length of 1312 km, out of which it flows for 1077 km in the state of M.P. About 87% of Narmada basin is present in M.P.

***Director of Fisheries, Government of M.P. Bhopal.**

wherein Tawa and Bergi reservoirs on its tributaries have recently been constructed, and in the western basin, the Indira Sagar Reservoir, Omkareshwar reservoir, Maheshwar reservoir and Sardar Sarovar reservoir construction have already been initiated. As many as 29 major, 450 medium and more than 3500 minor reservoirs have been planned to be constructed. Of the 6 major reservoirs mentioned above, Barna reservoir (7705 hactares) and Tawa (20050 hactares), Bergi reservoir (27290 hactares), fish catch data is available in a gross manner. The Narmada sagar (91349 hactares) and Sardar Sarovar (39134 hactares) when completed shall become the biggest reservoirs in the country. The Omkareshwar (9393 hactares) and Maheshwar (4856 ha) reservoirs are comparatively smaller. Fisheries data in the above 4 reservoirs of Western Narmada basin is fragmentary and not available. It is the aim of the present study to furnish a preimpoundment survey and fish out put report giving the present status of these reservoirs and envisage their future fishery potential which shall help in drawing the fisheries management method.

MATERIALS AND METHODS

As a part of the preimpoundment limnological studies in some reservoirs of the Narmada basin, the fisheries survey, fish marketing survey, and catch statistics records for the year 1989-90 done under the N.V.D.A. scheme are presented. The physico-chemical and biological data are not included in the present report. Monthly data recordings as mentioned above for the year 1989-90 have been compiled, evaluated and presented. Wherever felt necessary, experimental fish catching operations were conducted, with the help of local fisherman by using cast nets and gill nets. Fish catch efficiency was calculated by computing the fishing effort hours and the fish catch. Available information about the fish catch statistical surveys conducted earlier by Indian Institute of Management, Ahmedabad, CIFRI, Barrackpore and M.P. Fisheries Department, Bhopal was utilized for comparative purposes.

RESULTS AND DISCUSSION

Table 1 denotes a qualitative list of all the fishes collected in the western zone of Narmada river in the present study.

Table 1. List of Fishes Recorded from Narmada River.

1. *Hilsa ilisha* (Ham.)
2. *Gonialosa manmina* (Ham.)
3. *Notopterus notopterus* (Pallas)
4. *Oxygaster laubuca* (Ham.)
5. *O. clupeoides* (Bl.)
6. *Barilius barila* (Ham.)
7. *B. bendelisis* var, *chedra* (Ham.)
8. *B. evezardi* (Day)
9. *B. radiolatus* (Gunther)
10. *Danio* (Danio) *aequipinnatus* (Mc Clelland)
11. *D.* (Danio) *devario* (Ham.)
12. *D.* (Brachydanio) *revio* (Ham.)
13. *Esomus danricus* (Ham.)
14. *Rasobora daniconius* (Ham.)
15. *Amblypharyngodon mola* (Ham.)
16. *Tor khudree* (Sykes)
17. *T. putitora* (Ham.)
18. *T. Tor* (Ham.)
19. *Puntius conchonius* (Ham.)
20. *P. guganio* (Ham.)
21. *P. sarana* (Ham.)
22. *P. sophore* (Ham.)
23. *P. ticto* (Ham.)
24. *P. ambassis* (Day)
25. *P. amphibins* (Cuv. and Vol)
26. *P. Chrysopoma* (cuvier)
27. *P. dorsalis* (Jordon)
28. *P. pinnanratus* (Day)
29. *P. titius* (Ham.)
30. *Oreichthys cosuatus* (Ham.)
31. *Catla catla* (Ham.)
32. *Cirrhinus mrigala* (Ham.)
33. *C. reba* (Ham.)
34. *Garra gotyla* (Gray)
35. *G. lamta* (Ham.)
36. *G. mullya* (Sykes)
37. *Labeo bata* (Ham.)
38. *L. boggut* (Skyes)
39. *L. calbasu* (Ham.)
40. *L. dycheillus* (Mc Celland
41. *L. fimbriatus* (Bloch)
42. *L. gonius* (Ham.)
43. *Labeo rohita* (Ham.)
44. *Osteobrama cotio* (Ham.)
45. *Crossocheilus latius* (Ham.)
46. *Parapsilorhynchns fentaculatus* (Ann)
47. *Nemacheilus botia* (Ham.)
48. *N. deyi* (Hora)
49. *N. evezardi* (Day)
50. *Lepidoccphalichthys guntea*
51. *Ompak bimaculatus* (Bloch)
52. *Wallago attu* (Block and Schneider)

53. *Mystus bleekeri* (Day)
54. *M. cavasius* (Ham.)
55. *M. vittatus* (Bloch)
56. *M. aor* (Ham.)
57. *M. seenghala* (Sykes)
58. *Rita pavimentata* (Gunther)
59. *Amblyceps mangois* (Ham.)
60. *Gagata itchkeea* (Sykes)
61. *Glyptothorax lonah* (Sykes)
62. *G. vibeiroi* (Sykes)
63. *Clupisoma garua* (Ham.)
64. *Heteropneustes fossilis*
65. *Clarias batrachus* (Linnaus)
66. *Xenentodon cancila* (Ham.)
67. *Channa gachna* (Ham.)
68. *C. marullus* (Ham.)
69. *C. punctatus* (Bloch)
70. *Chanda nama* (Ham.)
71. *C. ranga* (Ham.)
72. *Badis badis* (Ham.)
73. *Nandus nandus* (Ham.)
74. *Glossogobius giuris* (Ham)
75. *Mastacembelus armatus*
76. *M. pancalus*
77. *Anabas testudineus* (Block)
78. *Rita rita* (Ham.)
79. *Labeistis reticulatus*
80. *Notopterus chitala* (Ham.)
81. *Oxygaster bacaila* (Ham.)
82. *Eutropiichthya yacha* (Ham.)
83. *Rhinomugil corsula* (Ham.)
84. *Colisa fasciatus* (Bl. and Schn)

About 84 species (table 1) were recorded belonging to 45 genera, 20 families and 6 orders in the western zone of Narmada river from the 5 fish landing centres of Punasa, Omkareshwar, Mandleshwar, Maheshwar and Barwani. The Barwani fish landing centre data includes the fish recorded from Hiranfall area also. The location of these fish landing centres is shown in Fig. 1. Hora and Nayar (1941) collected 40 species from the hilly areas of Narmada of Satpura ranges in Hoshangabad district. The Central Inland Fisheries Research institute in their survey conducted from 1960 to 1968 recorded 77 species belonging to 19 families. The department of Fisheries, M.P. during their survey from 1960 to 1971 indicated the presence of 46 species. The variations of the fish species recorded by various survey parties cannot be explained. M.P. Department of fisheries in their research report of 1978-79

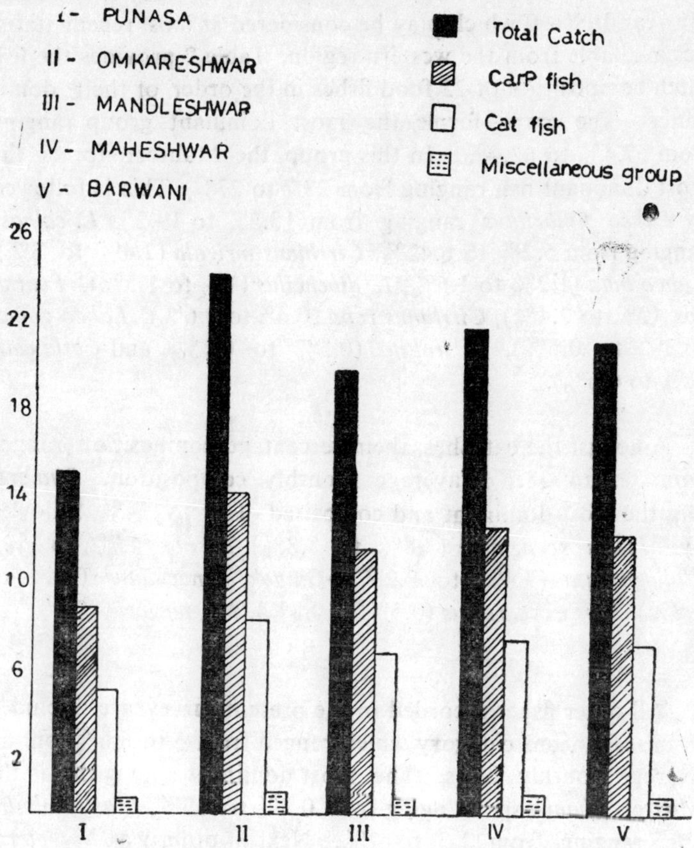

Fig. 1. Average catches (t) of 9 months of various fish groups from Narmada river.

describes 37 species, recorded from Barna. The present report systematically surveyed and recorded, the fish recorded from western zone of Narmada river. It can be broadly assumed that after considering all these reports, about 90 fish species seem to occur in river Narmada. But it is suggested that a detail fish survey of the entire river stretch should be necessarily considered. **Fish Catch Statistics.**

As mentioned earlier, the present fish catch statistics recorded from 5 fish catching centres *i.e.*, Punasa,

Omkareshwar, Mandleshwar, Maheshwar and Barwani during the year 1989-90 which may be considered as most recent statistics available from the western region. Table 2 provides the fish catch composition of 22 food fishes in the order of their dominance. The carps forms the most dominant group ranging from 57.4% to 62.4%. In this group, the Mahseer forms the most dominant fish ranging from 23.7 to 27%. This is followed by *Labeo fimbriatus* ranging from 13.2% to 19.2% *L. calbasu* ranging from 5.2% to 6.42%, *Cirrhinus mrigala* (2.4% to 3%), *Labeo bata* (1.2% to 1.6%, *L. diochcilus* (1% to 1.5%), *Puntius sps.* (2% to 2.4%), *Cirrhinus reba* (0.4%to 0.6%), *Labeo gonius* (0.3% to 0.6%), *L. rohita* (0.3% to 0.5%) and *Catla catla* (1% to 0.6%).

Amongst the catfishes, their percentage composition ranged from 34.6 to 38.1% average monthly composition. *Rita* sps. was the most dominant and comprised of 12 to 14%, followed by *Mystus seenghala* (7.8% to 8.8%), *M. aor* (4.7% to 5%), *Wallago attu* (7.4% to 8.2%), *Ompok bimaculatus* (1.4% to 1.8%), *Mystus carosius* (0.5% to 0.8%), *M. tengara* (0.2% to 0.5%).

All other fishes recorded in the present survey are included in miscellaneous category which ranged from 3 to 6.6% on an average monthly basis. The most dominant among this is the Murrel, *Channa* sps. ranging from 0.8% to 2.9%, *Mastacmbelus* sps. ranging from 1.3 to 2%. Next in order was *Notopterus* ranging from 0.4 to 1.5%. The other small fishes ranged from 0.22 to 0.4% on an average monthly basis. The CIFRI during their sample survey collected catch statistics from a stretch of 48 km from Narmada covering 2 important stations, Houshangabad and Shahganj. The sampling was done for about 21 days at Houshangabad and about 9 days at Shahganj in each month. During the period 1958-66, the average monthly fish weight ranged from 32.8 to 52.7 tons. Considering percentage composition, they reported *Tor tor* as the most dominant in the carp group, followed by *Labeo fimbriatus*, *L. calbasu*, *Cirrhinus mrigala*, *Labeo bata*, *L. diocheilus*, *Puntius sarana*, *Catla catla*, *Cirrhinus reba* and *Labeo gonius*. The present data from the western zone of Narmada follows the same pattern except that after *Puntius sarana*, the next dominant was *Catla catla cirrhinus reba* and *Labeo gonius* in the CIFRI report while in the present report after *Puntius*, the next dominant species were *Cirrhinus reba*, *Labeo gonius*, *L. rohita* and *Catla catla*. The average monthly catches of the species was slightly more in western zone when compared to the data of central Narmada region. Amongst the catfish, the same order of dominance existed in the eastern region showing a considerable increase in the monthly average catches of fish in western zone of Narmada. Sufficient populations of *Mystus tengara* ranging from 0.22 to 0.6% were also encountered in this region which was not reported from central Narmada region in the CIFRI survey. Further *Clupisoma forua* recorded in the central zone was encountered in very small numbers in the western zone.

The fish catch statistics of the western Narmada zone resemble those of the central zone in the order of dominance of the carps, Maheer and other food fishes. A similar statistics of food fishes were reported by the fish catches data presented by the Deptt. of Fisheries, M.P. 1967-1970. Their catch statistics from Maheshwar and Mandleshwar show similar results. Their statistics show the dominance of *Tor tor* at Houshangabad, while the Shahganj area was dominated by *L. fimbriatus*. It is interesting to observe that *Labeo rohita* was not reported in the two surveys of CIFRI (1958-66) and the Department of Fisheries (1967-78), through a reference of its presence has been made in the report of CIFRI. However, Barna reservoir has registered about 10.5% of *Labeo rohita* (CIFRI research report 1978-79). The present study has reported on average monthly catch of *Labeo rohita* 0.2 to 0.5%.

Fish Catch per Unit Effort

Fish catch statistics in Indira Sagar, Omkareshwar, Maheshwar and Sardar Sarovar area was done to a limited extent wherein experimental fish catch was done by using cast nets and gill nets at the study sites. Fish catch per day ranged upto 58.4 kg at Punasa (Indira Sagar), 91.8 kg. at Omkareshwar,

Table 2. Average % Composition of Monthly Catches of Various Species from Narmada River of Selected Study Sites During September 1989 to May 1990.

1	Sep. 2	Oct. 3	Nov. 4	Dec. 5	Jan. 6	Feb. 7	March 8	April 9	May 10	$\bar{X}$ 11
Carp group fishes	*57.6*	*62.4*	*62.2*	*60.2*	*60.1*	*59.5*	*58.7*	*58.1*	*57.4*	*59.53*
1. *Tor* sps.	25.4	27.0	26.8	26.4	26.3	26.2	26.0	25.1	23.7	25.87
2. *Labeo fimbriatus*	18.1	19.2	19.1	18.5	18.5	18.4	18.2	18.4	18.5	18.54
3. *Labeo calbasu*	5.2	5.6	5.6	5.5	5.5	5.4	5.3	5.3	5.4	5.42
4. *Cirrhinus mrigala*	2.4	2.8	2.8	3.0	3.0	2.8	2.6	2.5	2.5	2.71
5. *Labeo bata*	1.3	1.6	1.6	1.3	1.3	1.2	1.2	1.5	1.6	1.4
6. *Labeo dyocheilus*	1.4	1.5	1.5	1.3	1.3	1.1	1.0	1.2	1.5	1.31
7. *Puntius* sps.	2.1	2.3	2.4	2.2	2.2	2.1	2.0	2.0	2.1	2.15
8. *Cirrhinus reba*	0.4	0.6	0.6	0.5	0.5	0.4	0.5	0.5	0.4	0.48
9. *Labeo gonius*	0.4	0.6	0.6	0.5	0.5	0.6	0.6	0.3	0.3	0.48
10. *Labeo rohita*	0.2	0.4	0.4	0.4	0.3	0.5	0.5	0.3	0.3	0.36
11. *Catla catla*	0.7	0.8	0.8	0.6	0.7	0.8	0.8	1.0	1.1	0.81
Cat fish group	*35.80*	*34.6*	*34.7*	*35.2*	*35.4*	*36.2*	*36.5*	*37.2*	*38.1*	*35.92*
1. *Rita* sps.	12.5	12.0	12.1	12.0	12.2	12.6	13.0	13.7	14.0	12.60
2. *Mystus seenghala*	8.4	8.0	8.0	7.8	7.8	7.9	8.2	8.5	8.8	8.18

(*Contd.*)

1	*2*	*3*	*4*	*5*	*6*	*7*	*8*	*9*	*10*	*11*
3. *Mystus aor*	4.7	4.7	4.7	4.8	4.8	5.0	5.0	4.9	5.0	4.84
4. *Wallago attu*	7.6	7.4	7.4	7.6	7.6	7.6	7.9	8.0	8.2	7.7
5. *Ompok bimaculatus*	1.5	1.5	1.5	1.7	1.7	1.8	1.5	1.5	1.4	1.56
6. *Mystus cavasius*	0.8	0.7	0.6	0.8	0.8	0.8	0.6	0.6	0.5	0.68
7. *Mystus tengara*	0.3	0.3	0.4	0.5	0.5	0.5	0.3	0.3	0.2	0.36
Miscellaneous fishes	*5.6*	*3.0*	*3.1*	*4.6*	*4.5*	*4.3*	*4.8*	*4.7*	*4.5*	*4.44*
1. *Channa* sps.	2.9	0.8	0.9	1.4	1.9	1.8	2.6	2.4	2.3	1.88
2. *Mastacembelus* sps.	2.0	1.5	1.5	2.0	1.9	1.7	1.5	1.4	1.3	1.64
3. *Notopterus* sps.	1.5	0.7	0.7	1.0	0.5	0.5	0.4	0.5	0.5	0.7
4. Other small fishes	0.2	—	—	0.2	0.2	0.3	0.3	0.4	0.4	0.22

75.5 kg at Mandleshwar, 82.3 kg. at Maheshwar and 30.1 kg. at Barwani. This points out the comparatively lower potential at Indira Sagar when compared to Omkareshwar, Maheshwar and Barwani (Sardar Sarovar). The fish catch/hour relatively ranged up to 2.4 kg/h, 3.8 kg/h, 3.1 kg/h, 3.4 kg/h and 3.3 kg/h respectively at the different fish landing centres. Table No. 3 provides the preimpoundment fish catch statistics of all the 4 reservoirs. After the impoundments are build, the fish catches are supposed to increase several folds and the comparatively small reservoirs like Omkareshwar and Maheshwar seem to be highly promising for future development of fishery while the large reservoir like Indira Sagar and Sardar Sarovar offer high promise to provide areas for developing future fish conservation centres or possible prawn cultures centres.

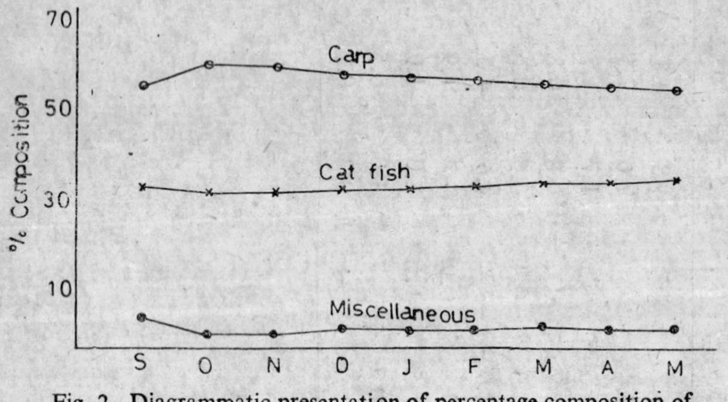

Fig. 2. Diagrammatic presentation of percentage composition of monthly catches of various groups of fishes of Narmada river during September 1989 to May 1990.

Present fish catch, statistics described above broadly agreed with the catch per unit effort studies made by CIFRI (1958-66) and M.P. Fisheries Department. While maximum catches of Mahseer were recorded by cast net and gill net operations. Highest densities of *Rita* sps. were also recorded. Data recorded from the fish market centres also support the same. Prawn catches eventhough negligible hold good promise in future for development. Migration of Mahseer during late monsoon (July to September) where it migrates upstream in the hilly

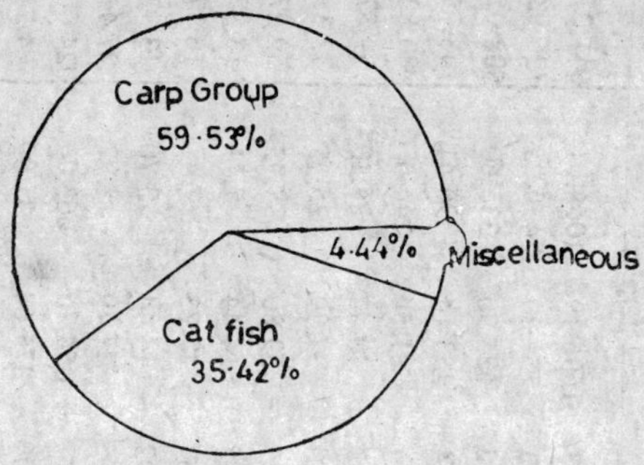

Fig. 3. Graph showing average percentage composition of various groups of fishes from Narmada river during September 1989 to May 1990.

areas of river Narmada for breeding Studies must immediately be taken to locate the Mahaseer breeding ground along the entire stretch of river and the future conservation areas must be identified before filling of these reservoirs start. Dubey (1969) pointed out the abundance of fry and fingerlings of Mahaseer along the river. Since successful artificial breeding of Mahaseer was done in Maharashtra (Dubey, 1990) by Kulkarni, 1971 and *Tor putitora* in Veem Tal reported by Tripathy 1978.

The present study could not record *Hilsa ilisha* from the present study sites eventhrough this fish is supposed to be occurring commonly in western part of Narmada. In this regard the present authors agree with Dubey, 1990 that by conservation method, this fish could be established and propagated in M.P. reservoirs also. This idea seems to be more promising since the CIFRI (in their news letter Vol. IV, September, October, 1981, No. 4) have reported the successful artificial breeding of *Hilsa ilisha* at Ukai centre through stripping like gravid fishes. They reared 5000 young *Hilsa* in Ukai reservoir and recorded gravid females of 600 g after 17 months confirming the natural breeding of this species in the reservoir.

ACKNOWLEDGMENTS

The authors are highly thankful to the Narmada Valley Development Authority, (NVDA) Narmada Bhawan, Tulsi-

Table 3. Average Monthly Catches Ranges (kg/month) of Various Species from Narmada River at Selected Sites during September 1989 to May 1990

	Punasa		*Omkareshwar*		*Mandleshwar*		*Maheshwar*		*Barwani*	
1	*2*	*3*	*4*	*5*	*6*	*7*	*8*	*9*	*10*	*11*
Carp group fishes	831.68	1174.93	1172.10	1840.81	1042.32	1637.48	1131.43	1696.19	1068.01	1641.21
1. *Tor* sps.	367.46	520.39	516 86	810 57	459.62	725.25	498.93	751.29	470.95	726.93
2. *Labeo fimbriatus*	261.85	364.27	368.31	569 25	327.52	507.67	355.53	525.90	335.60	508.05
3. *Labeo calbasu*	75.27	106.08	105.81	167.41	94 09	147.84	102.14	153.14	96.41	148.75
4. *Cirrhinus mgigala*	34.72	58.87	48 83	91.31	43.42	72.76	47.14	81.54	44.5	80.83
5. *Labeo bata*	18.80	27.04	26.45	44.70	23.52	38.58	25.53	40.42	24.10	39.87
6. *Labeo dyocheilus*	20 01	25 39	28.48	41.91	25.33	36.12	27.50	37.89	25.95	37.38
7. *Puntius sps.*	30.38	41.70	42 73	67.05	38.00	56 89	41.25	59.80	38.93	59.27
8. *Cirrhinus reba*	5.78	10.15	8.13	16.76	7.23	13.94	7.85	14 44	7.41	13.97
9. *Labeo gonius*	4.86	11.67	7.70	18.56	7.22	26.73	7.57	17 33	7.41	16.77
10. *Labeo rohita*	2.89	10 00	4.06	15.46	3.61	13.94	3.92	14.44	3.70	13.97
11. *Catla catla*	10,12	18.03	14.24	28.26	12.66	16 49	13.75	27.79	12.97	27.41
Cat fishes group	515.06	730 58	728 49	1119.95	647.83	1018.16	697.77	1054.70	663.00	1024.00
Rita sps.	179.81	260.17	254.36	389.81	226 98	362.61	242 00	375.64	231.09	364.71
2. *Mystus seenghala*	120.13	164 11	170.93	244 41	151.53	228.72	161 33	236.94	154.06	230 04
3. *Mystus aor*	67 61	100.06	95.64	154.69	85.04	139.46	92.32	144 48	87.14	140.27
4. *Wallago attu*	109 32	158,10	154.65	235,12	137.52	220.00	149 23	228 27	140.91	221.63

(Contd.)

1	*2*	*3*	*4*	*5*	*6*	*7*	*8*	*9*	*10*	*11*
5. *Ompok bimaculatus*	21.57	35.03	30.52	55.68	27.14	46.73	29.46	50.08	27.81	49.58
6. *Mystus cavasius*	8.11	15.56	12.84	24.75	11.98	20.79	12.63	22.25	12.46	22.03
7. *Mystus tengara*	3.24	9.73	5.13	15.46	4.81	12.99	5.05	13.91	4.98	13.77
Miscellaneous fishes	41.37	96.07	81.58	141.44	56.83	133.89	60.5	138.70	57.77	134.66
1. *Channa* sps.	12.0	52.0	21.75	76.61	15.14	72.47	16.13	75.12	15.40	72.94
2. *Mastacembelus*	21.08	37.39	33.40	59.88	28.4	49.10	30.25	51.64	28.80	51.17
3. *Notopterus* sps.	8.0	21.68	12.34	30.52	11.15	27.13	12.63	29.46	11.21	25.00
4. Other small fishes	2.89	7.18	4.06	30.37	3.61	9.90	3.92	10.46	3.70	10.03
	1443.89	2001.58	2034.90	3093.80	1809.6	2789.50	1964.3	2889.3	1854.2	2805.50

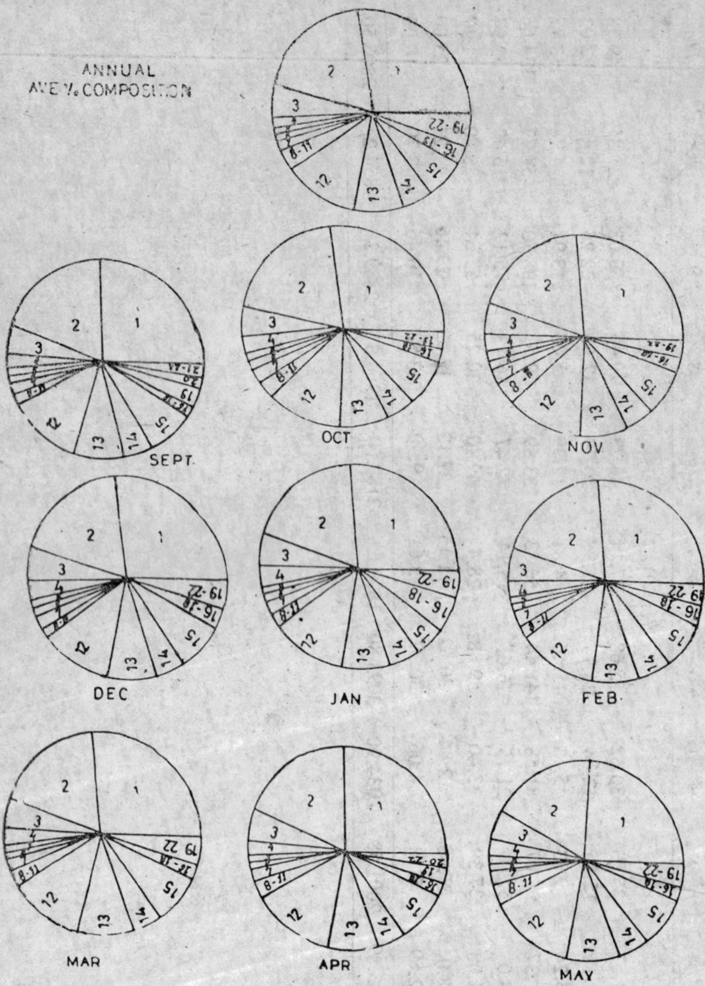

Fig. 4. Diagrammatic representation of Con parative commercial fishes of Narmada river during September 1989 to October, 1990 (Annual average percentage composition).

nagar, Bhopal for financial assistancc provided for this study. Thanks are also due to Dr. A.G.K. Menon, Emeritus Scientist Z.S.I. Madras for his valuable comments and discussions in identification of some of the fish species in this study. Laboratory facilities provided by the Head, School of Studies in Zoology, Vikram University, Ujjain is acknowledged. The assistance of Dr. Sharad Srivastava, Usha Chaubey, and B.S. Rao for the help they provided during the course of this study is gratefully acknowledged.

REFERENCES

Abraham, J.C. (1962). A survey of hydrobiological and fisheries of the Cooum river, Madras. *J. Fish* 1 : 50-69.

Chacko, P.I. and Dinamani, P. (1949). Preliminary report on the fishery resources of the Bhavani river in relation to the Lower Bhavani Project. *Proc. Indian Sci Congr.*, 36 (3) : 166.

Chacko, P.I. and Kuriyan, G.K. (1948). Survey of the fisheries of the Tungbhadra river *Proc. Indian Acad. Sci.* 28 (5) : 165-76.

Chacko, P.I., Kuriyan, G.K. and Thyagarajan, S. (1954). A survey of the fisheries of the Cauvery river. *Contr. Freshw. Fish Biol. Stn. Madras*, (12) : 19.

CIFRI Unit. Survey of Narmada river at Houshangabad, 1958-66.

David, A, (1954). A preliminary survey of the fish and fisheries of a five-mile stretch of the Hooghly river near Barrackpore. *Indian J. Fish.* 1 (1 and 2) : 231-35.

David, A. (1963 a). Studies on fish and fisheries of the Godavari and Krishana river system—Part I, *Proc, Nat. Acad. Sci. India.* 33 B (2) : 263-86.

David, A. (1963 b). Report on fisheries survey of River Gandak (North Bihar), *Sur. Rep. Cent. Int. Fish. Res. Inst. Barrackpore* [1 : 24 p. (Mimeo)].

David, A., Ray., P. Govind, B.V., Rajagopal, K.V. and Bauerjee, R.K. (1959). Linmology and fisheries of the Tungbhadra reservoir, *Bull. Cent. Int. Fish. Res. Inst, Barrackpore* (13) : 188 p. (Mimeo).

Dubey, G.P. and Mehra, R.K. (1959). Fish and fisheries of Chambal river, *Proc. All. India Congr. Zool.* 1 (2) : 647-65.

Fisheries Department, M.P. Fisheries Survey in Narmada River. 1967-71.

Bora, S.L. (1949). The Fish fauna of Rihand river and its zoogeographical significance. *J. Zool. Soc. India* 1 (1) : 1-7.

IIMA Report, 1983, Vol. I, page 29, exhibit 3,21 and page 36.

Job, T.J. David, A. and Das, K.N. (1955). Fish and Fisheries of the Mahanadi in relation to the Hirakund Dam, *Indian J. Fish.* 2 (1) : 1-36.

Job, T.J., David, A. and Motwani, M.P. (1952). Observations on the fish and fisheries of the Damodar basin with reference to multipurpose projects of the Valley, *J. Asist. Soc.* 18 (2) : 147-72.

Karamchandani, S.J. and Pisolkar, M.D. (1967). Survey of fish and fisheries of the Tapti river, *Surv. Rep. Cent. Inl. Fish. Res. Inst., Barrackpore* (4) : 29 p. (Mimeo).

Karamchandani, S.J., Desai, V.R., Pisolkar, M.D. and Bhatnagar. G.K. (1967). Biological investigations on the fish and fisheries of Narmada river (1958-66), *Bull Centr. Inl. Fish. Res. Inst. Barrackpore* (10) : 40 p. (Mimeo).

Motwani, M.P. Jayaram, K.C. and Sehgal, K.L. (1962). Fish and Fisheries of Brahmaputra river system, Assam. 1. Fish Fauna with observations on their zoo-geographical significances, *Trop, Ecol.* 3 (1-2) : 17-43.

Pantulu, V.R. Alagaraja, K, and Bhimachar, B.S. (1966). Fisheries of the Damodar Valley in relation to construction of dams, *Proc. Nat. Inst. Sci. Ibdla* 32 B (5 & 6) : 191-207.

Ranganathan, V. and Natarajan, V. (1969). Fisheries of Mettur reservoir, an artificial impoundment on the river Cauvery, Tamilnadu. In : Seminar on the Ecology and Fisheries of Freshwater Reservoirs. Sponsored by ICAR at CIFRI, Barrackpore, Nov. 27-29.

Report on National Commission of—Agriculture. 1978. Part VIII.

Saha, K.C. (1970). Fisheries of West Bengal, Calcutta : West Bengal Government Press, 138 p.

Sen, Gupta, J.M. and Jacob, P. (1968). Fishery statistics of India with special reference to estimation of inland fish catch. *J. Indian Soc. Agric. Statist.* 20 (2) : 79-89.

Survey of Central Inland Fisheries Research Institute Bulletin No. 10. 1958-66.

Survey of Narmada River, Directorate of Fisheries, M.P. 1967-71.

9

STATUS (PRIOR TO THE YEAR 1980) OF HYDROBIOLOGY, FISHERIES AND SOCIO ECONOMIC CONDITIONS OF FISHERMEN IN NARMADA BASIN

Dr. G.P. Dubey

Retd. Director of Fisheries, M.P.,
Ramchandra Nagar,
11/2, Bijasan Road,
INDORE

Abstract—Like the conservation and development of water resources in Chambal Valley long back, Madhya Pradesh, since its formation in 1956, has been planning to develop the huge water potential resources of Narmada Basin for power irrigation, fisheries etc. by creating 29 major 450 medium and innumerable minor reservoirs both on the main Narmada river and its 41 tributaries spreadover in 24 districts out of 45 districts in M.P. In a course of last 25 years in Narmada basin a few of the major reservoirs like Tawa and Barna, one on each of the important tributaries. Bergi reservoir on the main river, a good number of medium and large number of minor reservoirs have already been created providing good potential for irrigation, fisheries etc. The work on other remaining ones and the work of biggest reservoir of the country on main Narmada river—the Narmada Sagar has now been started.

As these reservoirs provide immense potentialities for fisheries development, thereby providing animal protein food and job potentialities to large people in the valley, certain important preliminary studies for scientific development of fisheries, were also started in 1956, simultaneously alongwith the planning and development of water resources in the Valley. The present paper, therefore, deals with the detailed status on hydrobiology fish fauna fish production trend, environmental effects on fishery in Narmada basin and the fluctuation of fishermen population and their socio-economic condition for the period from 1956-80 particularly at the time when the author also worked for most of the time as Director and Head of the Department of Fisheries, M.P. The study also includes the above status stated above of the two important reservoirs *i.e.* Tawa (20055 ha) and Barna (7705 ha) constructed on each one of the two important tributaries of Narmada river prior to the year 1980.

INTRODUCTION

The river Narmada is the fifth largest, west flowing river of India. It originates from Amarkantak in M.P. and meets the Gulf of Cambay in Arabian Sea, out of its total length of 1312 Kms. the river flows for 1070 Kms. in M.P. and the rest in Gujarat and partly forming the boundary of M.P. and Maharashtra. During its course, the river has 6 important falls and possess number of big and deep pools. The main river receives 41 tributaries, 20 from its left bank and 21 from its right bank.

As 87% of its basin lies in Madhya Pradesh, the state, since its formation in 1956, has been planning to develop most of the basins for construction of reservoirs. Among others, the Narmada valley is being developed in a systematic manner for integrated development. As many as 29 major, 450 medium and more than 3500 minor reservoirs have been planned to be constructed. These reservoirs, besides generating power, will yield benefits of irrigation and provide immenee scope for development of fisheries in them.

Fisheries development work in some of the reservoirs like Barna (7705 ha.) and Tawa (20050 ha) which were completed a few years ago, has been initiated. Similarly, Bergi reservoir (27296 ha.) which has also been completed recently in 1987, is now taken up for fiisheries development. And now the construction of the biggest man-made lake of the country both in area and capacity—Narmada Sagar (91340 ha.) alongwith others, are now being taken up. Sardar Sarovar, the last in the series on the Narmada River, is also being constructed in Gujrat State but its major portion (20000 ha.) would be living in Madhya Pradesh.

On completion of these reservoirs, Madhya Pradesh will take big strides in reservoir fishery resources in the country. It will yield an estimated production of about 10,000 m. tonnes of fish from these reservoirs alone. Further, large command area of each reservoir will provide scope for many ponds, tanks etc. there yielding at least two times more production of fish from them. Above all, such a vast water resource, besided providing sufficient protein food, will generate immense possibilities in job orientation to fishermen, tribals, rural people and others in the valley.

HYDROBIOLOGY & FISHERIES STATUS OF RIVER NARMADA

Background

Emphasis has always been laid on pre-impoundment survey of the valley prior to the construction of any dam. Pre-impoundment survey of fish and fisheries of Chambal valley was done as a pre-requisite for proper development of Gandhisagar reservoir (66,000 ha.)—a multipurpose project. The detailed survey was done for hydrobiology, fishery status and environmental aspects of the river (Dubey & Mehra 1959). After about 3 decade, the value of such preimpoundment survey has now been found to be of great importance.

Keeping in view the future prospects of fisheries development in the Narmada Valley. The hydro-biological studies of Narmada river were taken up immediately after the formation of the new State of Madhya Pradesh (1956) with the initiative taken up the author who was the head of the then Fisheries Section. Accordingly, a Research and Investigation Unit of Central Inland Fisheries Research Institute, Barrackpore was established at Hoshangabad in 1956 with the main objective to undertake hydrobiological studies, fishery status, fishermen population, etc. of the River Narmada. The unit carried out a detailed study of the river Narmada from 1958 to 1966.

After the closure of the unit in 1966, the State Fisheries Department, again with the interest taken by the author, continued certain investigations on the river Narmada to review the fish fauna, production trend ; environmental effects on fishery and fluctuations in fishermen population (1967-71). In late 70's when two reservoirs Tawa and Barna were completed on two different tributaries of the river Narmada, the State Fisheries Department also took up investigations on their hydrobiology and fish behaviour. This provided preliminary information about the trends in fisheries development of the large number of the reservoirs proposed to be constructed and will provide facilities for experimenting upon the methodology for developing most of the reservoirs with regard to behaviour of fish and development of fisheries on the main river.

Physico-Chemical Characteristics

Physico-Chemical characteristics of water was observed at 4 sampling stations at the river Narmada, *i.e.* Jamtara, Hoshangabad, Mandleshwar and Barwaha (sometimes during 1970—80) is given in Annexure—I. Some of the important indicators identified are as follows :

(1) The average temperature of the river ranges from 20°C. to 32°C the dam site.

(2) The pH of the river varies from 7.3 to 8.5 at various sampling stations, the minimum being in November and December, while maximum has been recorded in April and May.

(3) Dissolved oxygen content vary between 5.8 and 10.9 mg/l, with minimum values during winter (November to January) and maximum during the monsoon (July to September.

(4) The total alkalinity varies from (70 mg/l. July to 280 mg/l February).

(5) The total solid content of the river shows a fluctuating trend. The values were recorded in the range of 149 mg/l to 1780 mg/l, the lowest being at Jamtara. At Hoshangabad, it ranged from 208 mg/l to 401 mg/l BOD values varied from 1.2 mg/l to 16.8 mg/l.

The above data on physico-chemical characteristics of the water are conductive to fish growth. These characteristics will however, alter after the creation of the reservoirs and affect the fish.

Plankton

During the survey (1967—71) plankton samples from important water pools (Deh.) were collected and analysed. The results indicated that the river is rich in plankton, both of zooplankton and phytoplankton (Annex. 2).

Macrophytes

As the river is subjected to floods, macrophytes are largely eliminated. However, they are limited to the river banks. Common weeds found in different parts of the river are *Ceratophyllum*, *Najas*, *Hydrilla*, *Chara*, and *Vallisneria*.

Fish Fauna

During the investigations conducted by Central Inland Fisheries Research Institute, (1958-66), 77 species (Annex. 3) belonging to 19 families have been reported in the river Narmada. (9 marked with asterisk are from earlier record Hora and Nair, 1941). Hora and Nair (1941) collected 40 species from hill strains arising from Satpura ranges in Hoshangabad district and joining river Narmada. The Department of Fisheries of M.P. during their survey (1967-71) indicated the presence of 46 species (Annex 19). No definite reasons for the fluctuation in fish fauna of the river in both the surveys can be given. About 37 species of fishes were recorded from Barna by Department of Fisheries (Research Report 1978-79). However, to ascertain the facts, a comprehensive survey of fish fauna of the river Narmada and of existing reservoirs is considered necessary.

Source. Directorate of Fisheries, Madhya Pradesh, Research Report, Project Reports, 1978-80.

HYDRO-BIOLOGY OF THE PRESENT RESERVOIRS IN THE BASIN

Primary Productivity of Reservoirs

Primary productivity investigations of the two reservoirs in the basin *i.e.* Tawa and Barna were carried out. Cross production generally varied from 20 mg/cm³/hr to 120 mg/cm³/hr for Tawa and 20 mg/cm³/hr to 100 mg/cm³/hr for Barna and the net production ranges from 20 mg/cm³/hr. to 100 mg/cm³/hr. and 20 mg/cm³/hr. to 80 mg/cm³/hr. respectively and the community respiration ranges from zero to 50 mg/cm³/hr. In both the reservoirs. However, in Barna the primary productivity is estimated to be 2534.465 kg/yr/ha and has a capacity to maintain an yield of 45 kg/ha. (M.P. Research Report 1978-79).

Physico-Chemical Characteristics

The inflow from the catchment in the form of foliage, dung, decay matter and basin soil run off generally effect the water quality of a reservoir. The productivity of reservoirs, therefore, is greatly influenced by its hydrobiological features.

In Tawa and Barna, pH values are within desirable range and dissolved Oxygen content is also relatively high. The details are as under :

Table 1

	TAWA		*BARNA*	
	Max.	*Min*	*Max.*	*Min*
1. Temeprature (°C)	14	25.5	34	20
2. Saachi disc (Transparency) CM	80	32	151	70
3. Dissolved Oxygen (ppm)	8.8	6.0	9.6	6.4
4. Carbondioxide (mg/lit)	4	—	8	—
5. Total Alkalinity (ppm)	128	54	94	54
6. pH	8.4	7.4	8.4	7.5

Plankton

In Barna, about of species of plankton were recorded (Annex 15). The density of plankton ranged from 0.02 to 0.2

mg, 50 L of water. Zooplankton mostly comprised of Protozoans, Crustaceans and Rotifors, while Phytoplankton were represented by Myxophyceae, Chlorophyceae and Bacillariophyceae.

Macrovegetation

Ipomea, Ceratophyllum, Hydrilla, Vallisneria, Offelia, Lemna and Potamageton are found in both the reservoirs.

CATCH STATISTICS AND PERCENTAGE COMPOSITION

The Central Inland Fisheries Research Institute during their sample survey collected catch statistics from a stretch of 48 km, of Narmada covering two important centres *i.e.* Hoshangabad and Shahganj. The sampling was done for about 21 days at Hoshangabad and about 9 days at Shahganj in each month.

During the period of eight years from 1958 to 1966, the average annual landings and average monthly catch for these years are as under : (Table 2)

Table 2

YEAR (Sept. to August)	*(Fish landings in tonnes)* Annual weight	Average monthly weight
1958-59	34.7	2.9
1959-60	39.4	3.8
1960-61	35.0	2.9
1961-62	43.4	3.8
1962-63	57.2	4.8
1963-64	56.8	4.7
1964-65	32.3	2.7
1965-66	32.8	2.7

The fishery shows fluctuations in the river Narmada as in the other rivers. Catch was lowest in monsoon months (July to September) which is attributed to suspension of cast nets and gill nets operations during high floods. During the observations of CIFRI, the fish landings have been found to be maximum in post monsoon and summer months, except in the months of December and May when the catches were comparatively low.

The commercial catches were represented by about 23 species of fishes ; 10 belonging to carp group ; 8 to cat fish group ; 2 to murrel group ; 2 to spinyeel and 1 to feather-back. The average annual catches of various species and their percentage composition in the annual landing of eight years (1958-66) are given below : (Table 3)

Source. Central Inland Fisheries Institute, Barrackpore Bulletin No. 10, 1958-66.

Table 3

Species	Annual Landings 1958-59 to 1965-66	
	Average weight in tonnes	Percentage composition
Tor tor	7.60	28.0
Labco fimbriatus	5.35	19.7
Labeo calbasu	1.11	4.4
Cirrhinus mrigala	0.69	2.5
Labeo bata	0.45	1.7
Labeo dyocheilus	0.42	1.6
Puntius sarana	0.38	1.4
Catla catla	0.15	0.6
Cirrhinus reba	0.14	0.5
Labeo gonius	0.07	0.3
CARP GROUP		60.4
Rita pavimentata	2.78	10.2
Mystus seenghala	2.45	9.0
Mystus aor	1.29	4.7
Wallago attu	2.08	7.7
Clupisoma garua	0.48	1.8
Ompok bimaculatus	0.12	0.4
Mystus cavasius	0.07	0.3
CATFISH GROUP		34.1

Channa spp.	1.00	3.7
Mastacembelus ssp.	0.35	1.3
Notopterus notopterus	0.14	0.5
MISCELLANEOUS FISHES		5.5

Source. Survey of Narmada CIFRI, Barrackpore Bulletin No. 10, 1958-66.

Carps (60%) including major carps, *Tor tor* (28%) formed the main bulk of fishery, followed by *Labeo fimbriatus* about 20%) and *Labeo calbasu* (4%), Cat fishery (34%) was mainly constituted of *Rita pavimentata* (10.2%), *Mystus spp* (4% to 9%) and *wallago attu* (7%). Amongst miscellaneous species, *Channa sps.* was maximum (3.7%).

A similar statistics of fish catches were collected by the Department of Fisheries (1967-68). They also observed that from 46 species, the commercial catches represented the same 23 species and the *Tor tor* dominated in catches. Their catch statistics from Maheshwar and Mandleshwar also revealed more of less similar percentage of carps (62%), callishes (32%) and others (6%) (Annex 18%).

The Department also carried-out comparative study of catches at Hoshangabad and Shahganj. *Tor tor* has higher percentage in the catches at Hoshangabad than at Shahganj where as *L. fimbriatus* figured better in Shahganj catches. *Rita pavementata* has higher percentage in the catch composition recorded at Shahganj.

It has been interesting to note that *Labeo rohita* was not reported in any of catch composition of the two surveys of CIFRI (1958-66) and the Department of Fisheries (1967-78) as indicated in Table 3 above and 5 below. Though a reference of its presence has been made in the report of CIFRI while giving suggestions on major carp stocking. It is very likely that during floods it might have entered into the river from some ponds in the Narmada river basin where fish culture of major carps have been in vogue since 3 to 4 decades. However, Barna which is a reservoir of the Narmada basin and is stocked with major carps since 6-7 years, later registered as much as 10.8% of *Labeo rohita* (Research Report 1978-79). Like *L. rohita*, *Catla Catla*, which is the fastest growing major carp is hardly 0.56% in the catches. It may, however, be indicated that Catla fishery, more or less had the same percentage in Chambal valley (1959) before the construction of Gandhisagar. But, later on, after sustained development of reservoir by stocking major carps, it registered as much as 75-80% in catches during 1978-79 and 1982-83.

Catch Per Unit Effort

In order to determine the fluctuations in the abundance of fish in 48 km. stretch of river Narmada near Hoshangabad, observations on catch per unit effort in respect of long line, cast net and gill net operations were made at Shahganj (1958-64) and at Bhilpura and Dongarwara (Hoshangabad) from 1961 to 1966 by CIFRI Unit. The data of various years is presented below : (Table 4)

Table 4

	Catch per unit of effort (in Kg.)			
Year	*Long line*	*Cast net*	*Gill net*	*Combined*
1958-59	0.078	0.017	—	0.048
1959-60	0.208	0.353	—	0.281
1960-61	0.173	0.355	—	0.264
1961-62	0.194	0.370	0.249	0.271
1962-63	0.342	0.451	0.358	0.384
1963-64	0.231	0.581	0.638	0.483
1964-65	0.241	0.540	0.480	0.420
1965-66	0.170	0.509	0.519	0.399

Source. CIFRI, Barrackpore—Bulletin No. 10 (1958-66).

It is noteworthy that the fluctuations in catch-per-unit-effort from 1958-59 to 1965-66 followed a trend similar to that of annual landings recorded during the corresponding years.

The percentage composition of catches from long line, cast net and gill net operations are given in Table 5.

Table 5

Species	Percentage composition (by weight)		
	Long line	Cast net	Gill net
Tor tor	10.0	30.2	34.2
Labeo fimbriatus	3.4	31.3	32.6
Labeo calbasu	1.0	7.6	6.8
Cirrhinus mrigala	0.6	2.6	3.5
Catla catla	0.2	0.5	2.3
Labeo gonius	—	0.4	—
Labeo dyachlilus	—	0.9	1.2
Labeo bata	0.1	3.6	1.2
Cirrhinus reba	0.1	1.7	0.1
Puntius garana	1.3	2.7	1.1
CARP GROUP	16.7	81.5	83.0
Mystus seenghala	2.4	4.2	9.0
Mystus aor	6.1	3.2	1.4
Wallago attu	6.1	7.2	5.4
Rita pavimentata	52.9	0.4	0.3
Clupisoma garua	7.0	0.1	—
Ompok bimacuatus	1.4	—	—
Mystus cavasius	1.7	—	—
CATFISH GROUP	77.6	15.1	16.1
Channa spp.	1.6	2.4	0.8
Notopterus notopterus	0.6	0.3	0.1
Mustacembelus armatus	4.1	0.1	—
Small fish	—	0.6	—
MISCELLANEOUS	5.7	3.4	0.9

Generally, catches are good at Hoshangabad and Maheshwar. The cast nets are found to be common white gill nets are plied in deep pools during winter season. It is also observed that percentage composition of catches of long line, cast net and gill nets show maximum catch of Mahseer, through cast net and gill net whereas highest catch of *Rita pavimentata* (53%) was recorded in long line.

The giant prawn fishery (*M. rosanburgii*) which forms an important fishery in the estuary of river Narmada in Gujrat State was said to be present in traces in the Narmada river in M.P. territory.

MIGRATORY FISHES

In true sense of the word, there exist no migratory fish in the river Narmada, in State territory.

Source. Survey of river Narmada Bulletin No. 10 Cultural Inland Fisheries Research Institute 1958-66.

The Mahseer *i.e. Tor tor* known as the Game fish of India forms a predominant fishery in commercial catches of Narmada. It constituted about 46.0% of the carp group and about 28% of the total catch.

Mahseer is the only local migratory fish. Generally during late monsoon (July to September) it migrates upstream in hilly areas of river Narmada or its tributaries as they provide excellent breeding ground, since Mahseer prefers deep moving waters for their breeding and depositing their eggs.

It may be indicated here that river Narmada is the only river in the country where lakhs of fry and lingerlings of mahseer could be collected in a year from certain collection centres (Dubey 1969). Moreover, *Tor Khudree* has been successfully bred artificially at Lonawala reservoir in Maharashtra (Kulkarni 1971) and *Tor Pulitora* in *Bhimtal* in Himalayan rivers (Tripathi 1978). In Madhya Pradesh generally *Tor tor* has established in most of the reservoirs as all the age-groups of Mahseer are found in them. Thus, all the important species of Mahseer, a local migratory fish is now completely established in the different waters and it can no more be considered as a migratory fish in true sense of the word.

RORE SPECIES

The *Hilsa ilisha,* earlier considered a typical long distance inadromous estuarine fish is now reported to be present only in the lower estuarine part of Narmada beyond M.P. territory. Probably the fish seems to have found it difficult to ascend in Narmada river on account of certain rapids in the long gorge of 113 Kms. in lower reaches. Though it does not form fishery in M.P. but like any other fish *Hilsa ilisha* could possibly be established and propagated in reservoirs. Interestingly CIFRI in their news letter (Vol. 4, September—October, 1981, No. 4) have reported that scientists at Ukai centre in Gujrat have achieved success in artificial breeding of *Hilsa ilisha* through stripping live gravid fishes from river. They, reared them in nylon happa for 17 days and then stocked the 5000 young Hilsa in Ukai reservoir. Further, after 17 months they got a gravid female of 600 grams with 0.65 lakhs eggs. After 2 months a Hilsa fingerling was caught in the lentic sector of the reservoir confirming the natural breeding of this species in reservoir.

Though more study is required in this respect and it is verified and confirmed, the remarkable phenomenon of this only migratory fish will also come to stay in reservoirs.

EXOTIC FISH

Exotic fishes like common carp, grass carp and silver carp have been introduced in India. But it does not seem to have been reported in the river basin.

ESTIMATES OF FISH CATCH

Fishing is free in the rivers of M.P. Therefore to evaluate the total fish production from river Narmada in Madhya Pradesh (CIFRI 1958-66 and DFMP 1967-71) information was collected from various sources *i.e.*, markets, fishermen and contractors.

For, this purpose, the Department of Fisheries divided the river stretch in to three fishery zones :

Eastern zone	Mandal to Gadarwara
Central zone	Gadarwara to Handia
Western zone	Handia to Barwaha

Details of monthly fish production is given below :

Table 6. Estimated monthly fish production

	Eastern Zone (tonnes)	*Central Zone (tonnes)*	*Western Zone (tonnes)*	*Total estimated fish production in tonnes*	*Total length of river stretch surveyed in M.P.*
November	7.0	8.2	6.0		
December	9.4	10.0	8.7		
January	11.5	12.0	11.2		
February	14.0	15.7	14.3	361.40	732 Kms.
March	15.3	16.4	18.4		
April	22.2	23.3	20.5		
May	20.4	21.0	22.0		
June	16.2	18.3	19.4		
TOTAL	116.0	124.9	120.5	361.40	

An average fish production of river Narmada per km. length of river can be estimated to be 0.2 to 0.5 M. tonne per year. The National Commission on Agriculture (1976) indicated the average annual catch per km. length of the river Narmada near Hoshangabad as 0.364 tonne per km. This production is much lower than other river basins in India (NCA 1976).

The Central Inland Fisheries Research Institute also during their survey estimated the total fish production of Narmada and covered 720 kms. of the river. They had also collected statistics of catch from weekly market days and other days. They also divided the river into three zones each covering 240 kms. each (*Annex 20*). But the CIFRI collected fish statistics only for 5 months *i.e.* from November, 1960 to March, 1961. This does not give any idea to fish production per km. However, the species composition of fish catches given by CIFRI confirm that each zone indicated the highest catch of *Tor tor* (25%–30%) followed by *Labeo fimbriatus* (21.5% to 24.4%) as shown in table below :

Source. Survey of Narmada river, Directorate of Fisheries, M.P. (1967-71).

FISH LADDERS

In view of the foregoing, construction of fish-ladders for migration of fish in reservoirs of Narmada basin is not required. Also fish ladders constructed earlier elsewhere in India, have not only proved to be of any utility but even of adverse effect.

SOCIO-ECONOMIC STATUS OF FISHERMEN

Fish Market and Marketing

The fish caught from Narmada is generally brought to local markets. Except for big cities the village markets are weekly and this effects the fishing trend. The main fish markets are listed below : (Table 7)

Table 7

District	*Location*	*Fishing area*
1. Mandla	Mandla	Rampuraghat, Nanighat Maharajpura.
2. Jabalpur	Gulandibazar Jabalpur	Gurighat, Lamehtaghat etc.

Source. Survey of Central Inland Fisheries Research Institute Bulletin No. 10–1958-66.

3. Narsinghpur	Barmanghat	Barman area
4. Hoshangabad	Piparia	Sandiya area
—	Hoshangabad	Tawa confluence to Hasalpur
5. Sehora	Nasurullaganj	Miladia area
6. Dewas	Nimawar	Nimawar area
7. Khandwa	Omkareshwar	Area beyond Omkareshwar
8. Khargone	Badwaha	Billora to All area
9. Khargone	Mardana	Pitamali to peetnagar

Source. Survey of Narmada river, Directorate of Fisheries, M.P. (1967-71).

1. Report on National Commission of Agriculture 1978 Part VIII.

Fishermen

The CIFRI Unit, Hoshangabad during their survey (1958-66) covered 945 kms. of the river ; 800 kms. in M.P. and 145 kms. in Gujrat. Nearly 712 villages were covered in their survey in both the States. Out of 69, 152 fishermen residing in 349 villages (285 in M.P. and 64 in Gujrat fishing villages) only 3082 fishermen (2600 in M.P. and 482 in Gujrat) were reported to be actively engaged in fishing throughout the year (Annex 21). The rest of the fishermen population is either wholly engaged in cultivation and manual labour or engaged only occasionally. Later on, the Fisheries Department, M.P. during survey of 732 Km. stretch of Narmada in M.P. (1967-71) in 322 villages, observed 191 fishing villages having a population of 11,502. Out of this, they located 4504 fishermen actively engaged in fishing throughout the year. District-wise list of fishermen and their equipment is given in *Annex 22*.

According to CIFRI, the population of active fishermen is 3.2 fishermen/km. in 1958-66 whereas according to the Fisheries Department report it is 6.1/km. in 1967-71 period. Thus the estimation of active fishermen needs fresh survey.

Economic Status of Fishermen

The fishermen, in general, belong to the weaker section of the society. They are socially, economically and educationally backward. In view of the fishing intensity and population per kilometre, wherein they hardly get 2 to 3 kg. fish per day per head, their income is fairly low and they live much below the poverty line. The community as a whole is still under the clutches of money-lenders and fishing contractors. The Indian Institute of Management, Ahmedabad during their survey (1983, Vol. I) has observed that the average income of fishermen from fish business was found to be Rs. 1008/- and over all income is Rs. 1302/- per year. Both of them were found to be lowest in the country[1].

1. IIMA Report, Vol. I, 1983, page 29 exhibit 3.21 and page 36.

Annex 1. Physico-Chemical Characteristics of River Narmada

S. No.	Characteristics	Units	Sampling Stations Jamtara	Hoshangabad	Mandleshwar	Barwaha
1.	Temperature	0°C	17.5 to 29.5	—	21 to 30	20 to 32
2.	Turbidity	mgl	11 to 2000	10 to 320	4.0 to 11.50	6.12 to 1520
3.	Colour	pt Co. Sc	Colourless	Muddy clear	Greenish Muddy	Muddy brown
4.	Odour	T. No.	Odourless	Odourless	—	To clear
5.	PH	PH units	7.3 to 8.4	7.0 to 8.5	7.5 to 8.5	7.5 to 8.5
6.	Specific conductivity	Mohs.	18 to 410	152 to 350	—	—
7.	Total Solids	Mgl	149 to 1780	208 to 401	250 to 1610	220 to 1600
8.	Dissolved Solids	Mgl	126 to 500	92 to 108	240 to 340	190 to 342
9.	Suspended Solids	,,	23 to 1280	62 to 283	10 to 1330	20 to 1300
10.	Ammonical Nitrogen	,,	0.1 to 1280	62 to 283	10 to 1330	20 to 1300
11.	Nitrite	,,	Traces to 0.32	Nil to 0.003	Nil to 0.003	Nil to 0.4
12.	Nitrate	,,	Traces to 0.85	Traces to 0.02	Nil to 0.92	Nil to 0.34
13.	Phosphate	,,	Traces to 3.3	Traces to 0.05	Nil to 2.0	Nil to 0.36
14.	Chloride	,,	8 to 20	10 to 27	12 to 28	12 to 48

(Contd.)

Annex. 1. *(Contd.)*

15.	Total Alkalinity	Mgl	70 to 100	86 to 220	82 to 260	84 to 280
16.	Total Hardness	,,	72 to 160	80 to 210	42 to 130	30 to 160
17.	Calcium	,,	40 to 68	60 to 120	22 to 80	20 to 110
18.	Magnesium	,,	6.2 to 8.7	12 to 90	22 to 48	10 to 100
19.	Dissolved Oxygen	,,	6.2 to 10.5	5 to 10.5	6.9 to 10.8	1.2 to 10.9
20.	BOD	,,	.1 to 1.7	2.2 to 3.0	1.2 to 1.8	0.5 to 2.0
21.	COD	,,	8.4 to 16.4	12.5 to 15.3	8 to 42	12 to 35
22.	Sodium	,,	—	9.0 to 30	Nil to 19	Nil to 34
23.	Potassium	,,	—	Nil to 60	—	120 to 2400

Source. M P. Pollution Prevention Board, Bhopal.

Annex 2. A list of planktons both zoo and phyto from the Narmada River

Blue green algae (Myxophyceae)

1. Oscillatoria
2. Merismopedia
3. Anabaena
4. Microcystia
5. Ulothrix
6. Spirulina
7. Polycystis
8. Tetrapedia

Green algae (Chlorophyceae)

1. Spirogyra
2. Eremospheera
3. Pediastrum
4. Zygnema
5. Ulothrix
6. Sphaeroplea
7. Cladophora
8. Microspora
9. Tetrastora
10. Protococcus
11. Botryococcus
12. Clasterium
13. Pilliginora
14. Cocyalayte
15. Playrosivma

Desmids

1. Desmidium
2. Cosmarium
3. Sparotaenia
4. Docidium
5. Cenicularia

Diatoms

1. Nitzschia
2. Synedra
3. Navicula
4. Diatoma
5. Fragilaria
6. Amphora
7. Tabellaria
8. Cyclotella
9. Epithemia
10. Rhizosolenia
11. Comphonema
12. Cymbella
13. Brelissonia

Rotifers

1. Keratella
2. Notholca
3. Distimus
4. Mytilina
5. Distomus
6. Cynchaeta
7. Colurella
8. Diurella
9. Limnias
10. Trichocerea
11. Barachionus sps.
12. Polyarthra sps.
13. Dinocharis

Protozoa

1. Chlamydomonas
2. Paramaecium

Arthropodes

1. Nauplius
2. Bosmina

3. Arcella
4. Amoeba
5. Euglena
6. Furcuraria
7. Actinosphaerium
8. Edudorina

3. Simocephalus
4. Cyclops
5. Moina
6. Diaptomis
7. Daphnia
8. Cadis fly larvae
9. Dragon fly nymph
10. Diptera larvae (Mostly Chironomous)
11. Water bugs

Source. Survey of River Narmada
Directorate of Fisheries M.P.

Annex 3. Classification of the fishes from Narmada river

Class TELEOSTOMI
Subclass ACTINOPTERYCII
Order CLUPEIFORMES
Suborder CLUPEOIDEI
Family CLUPEIDAE
Subfamily CLUPEINAE

1. *Hilsa ilisha* (Hamilton)
2. *Conialosa manminna* (Hamilton)

Suborder NOTOPEEROIDEI
Family NOTTPTERIDAE

3. *Notopterus notopterus* (Pallas)

Order CYPRINIFORMES
Division Cyprini
Suborder CYPRINOIDEI
Family CYPRINIDAE
Subfamily ABRAMIDINAE

4. *Chela laubuca* (Hamilton)
5. *Oxygaster clupeoides* (Bloch)

Subfamily RASBORINAE

6. *Barilius barila* (Hamilton)
7. *Barilius bendelis* var. *Chedra* (Hamilton)
8. Barilius evezardi Day
9. Barilius radiolatus Gunther
10. Danio (Danio) aequipinnatus (McClelland)
11. Danio (Danio) devario (Hamilton)

12. Danio (Brachydanio) rerio (Hamilton)
13. *Esomus danrica* (Hamilton)
14. Rasbora daniconius (Hamilton)

Subfamily CYPRININAE

15. Amblypharyngodon mola (Hamilton)
16. Tor Khudree Sykes**
17. Tor putitora (Hamilton)
18. Tor tor (Hamilton)
19. Puntius ambassis (Day)
20. Puntius amphibius (Cuvier and Valenciennes)
21. Puntius chrysopoma Cuvier**
22. Puntius conchonius (Hamilton)
23. Puntius dorsalis (Jerdon)
24. Puntius gugania (Hamilton)**
25. Puntius pinnauratus (Day)**
26. Puntius sarana (Hamilton)
27. Puntius sophora (Hamilton)
28. Puntius ticto ticto (Hamilton)
29. Pintius titius (Hamilton).
30. Oreichthys cosuatis (Hamilton)
31. Catla catla (Hamilton)
32. Cirrhinus mrigala (Hamilton)
33. Cirrhinus rebu (Hamilton)
34. Garra gotyla (Gray)**
35. Garra lamta (Hamilton)
36. Garra mullya (Sykes)**
37. Labeo bata (Hamilton)
38. Labeo boggut (Sykes)
39. Labeo calbasu (Hamilton)
40. Labeo dyocheilus McClelland)
41. Labeo fimbriatus (Bloch)
42. Labeo gonius (Hamilton)
43. Labeo rohita (Hamilton)
44. Osteobrama cotio cotio (Hamilton)

Subfamily CORRINAE

*The classification adopted here is mainly after LS Berg, "Classification of fishes, both, recent and fossil." Irav. *Inst. Zool. Acad. Sci.*, USSR, 5(2), 1940.

**Recorded by Hora & Nair (1941).

45. Crossoc eilus latius latius (Hamilton)
Family PSILARHYNCHIDAE
46. Parapsilorhynchus tentaculatus (Ann.)**
Family COBITIDAE
47. Noemacheilus botia (Hamilton)
48. Noemacheilus dayi Hora
49. Noemacheilus evezardi Day**
Subfamily COBITINAE
50. Lepidocephalichythes guntea (Hamilton)
Division Siluri
Suborder SILUROIDEI
Family SILURIDAE
51. Ompok bimaculatus (Block)
52. Wallago attu (Block and Schneider)
Family BAGRIDAE
53. Mystus (Mystus) bleekeri (Day)
54. Mystus (Mystus) cavasius (Hamilton)
55. Mystus (Mystus) vittatus (Bloch)
56. Mystus (Osteobagrus) aor (Hamilton)
57. Mystus (Osteobargus) seenghala (Sykes)
58. Rita pavimentata Gunther
Family AMBLYCIPITIDAE
59. Amblyceps mangois (Hamilton)**
Family SISORIDAE
60. Gagata itchkeea (Sukes)
61. Glyptothorax lonah (Sykes)
62. Glyptothorax ribeiroi (Sykes)
Family SCHILBEIDAE
63. Clupisoma garua (Hamilton)
Family HETEROPNEUSTIDAE
64. Heteropneustes fossilis (Bloch)
Family CLARIDAE
65. Clarias batrachus (Linnaeus)
Order ANGUILLIFORMES
Suborder ANGUILLOIDEI
Family ANGUILLIDEA

**Recorded by Hora & Nair (1941).

67. Xenentodon cancila (Hamilton)
Order OPHIOCEPHALIFORMES
Family CHANNIDAE
68. Channa gachua (Hamilton)
69. Channa marulius (Hamilton)
70. Channa punctatus (Bloch)
Order PERCIFORMES
Suborder PERCOIDEI
Family CENTROPOMIDA
71. Chanda nama (Hamilton)
72. Chanda ranga (Hamilton)
Family NANDIDAE
73. Badis badis (Hamilton)
74. Nandus nandus (Hamilton)
Suborder GOBIODEI
Family GOBIIDAE
75. Glossogobius giuris giuris (Hamilton)
Order MASTOCEMBELIFORMES
Family MASTOCEMBELIDAE
76. Mastocembelus armatus armatus (Lacepede)
Mastocembelus pancalus (Hamilton)

Source. CIFRI Barrackpore Bulletin No. 10. 1958-66.

10

AQUATIC MICROBIOLOGY—A NEW PERSPECTIVE IN HYDROBIOLOGICAL STUDIES

S.D. Tripathi

Central Institute of Freshwater Aquaculture
Kausalyaganga, Bhubaneswar-751002, Orissa

INTRODUCTION

Production, consumption and decomposition are the three important eco-regulatory and balancing processes in an aquatic habitat, the interplay between which determine and regulate the output levels. Numerous microbial communities are involved in each of these ecological processes and their activities constitute the basis of trophic interactions. Exhaustive reviews of Indian limnological research as presented by Gulati and Wurtz-Schulz (1980) and Michael (1980) indicate that studies on freshwater impoundments present only short term pictures of hydrobiological conditions during known intervals of time or record occurrences of biotic components. While there is much published literature on aspects of hydrography, plankton and primary production (Ganapati, 1957; Das and Srivastava, 1959; Sreenivasan, 1963, 1976; George, 1966; Sumitra, 1969; Mathew, 1975; Saran and Adoni, 1985), there is paucity of information regarding the aquatic microbiological processes. The need for hydrobacterial research was stressed as early as in 1927 by Thieremann, who inferred that the greatest need of limnological studies is bacteriological information, which also has immense applications in fishesies. In aquatic habitats, without adequate data on bacterial populations and their activities, chemical investigations, however detailed they may be, do not solve the mystery of biotransformation of matter (Kuznetsov, 1970; Rodina, 1972).

Most of the Indian aquatic microbiological studies that are only a few, pertain to the marine environment, including those on microflora of sea's, estuaries and beaches (Velankar, 1950, 1957; Santhakumary, 1966; Gore, 1971; Nair, 1979; Parimala and Sridhar, 1982). Only in the recent years, attempts are being made at studying the microbiology of natural freshwater lakes and fish ponds (Adoni, 1975; Jana *et al.*, 1980, 1982; Bagde and Varma, 1982; Jana and Roy, 1983, 1986), signifying the importance of bacterial communities in trophic interactions. In this direction, systematic aquatic microbiological investigations with reference to production potentials have been initiated at the Central Institute of Freshwater Aquaculture (ICAR), Kausalyaganga, Bhubaneswar and the present account highlights the different aspects being studied and their potential applications in improving aquaculture productivity.

Microbial coenoses of natural and managed waters

Fifteen bacterial groups involved in nutrient cycling in the water and sediment media of a peninsular tank were analysed with reference to the prevailing hydrobiological conditions during 1981-1984 (Ayyappan, 1987). The aerobic heterotrophic bacterial counts were low in water and high in the sediment, commensurate with their organic contents (Table 1). A similar pattern was observed with regard to populations of phospholytic, proteolytic, amylolytic, aerobic and anaerobic cellulolytic methanogenic and iron bacteria. The bacterial groups in the nitrogen cycle (ammonifiers, urea decomposers, aerobic and anaerobic nitrogen fixers, nitrifiers-I and II) and coliforms were in the moderate ranges. The heterotrophic food chain was observed to be playing an important role in the trophic dynamics of the tank, which was classified as mesotrophic in comparison to other habitats in the area (Parimala and Ayyappan, 1989).

The aerobic heterotrophic bacterial population in the water and sediment media and primary production levels with reference to the physico-chemical conditions of carp polyculture ponds under different management practices, *viz.*, fortnightly application of cowdung in heaps, spray-application

Table 1. Bacterial coenoses in the weter and sediment media of a peninsular tank (1981-84)

Group	*Water (no/ml)*	*Sediment (no/g)*
Aerobic heterotrophs	80-5580	$22\text{-}800 \times 10^3$
Ammonifying bacteria	100-1400	$12\text{-}76 \times 10^3$
Ureolytic bacteria	2-250	$0.10\text{-}22.00 \times 10^3$
Aerobic nitrogen fixers	5-80	$0.24\text{-}9.00 \times 10^3$
Anaerobic nitrogen fixers	2-45	75-1400
Nitrifying bacteria-I	1-40	8-2200
Nitrifying bacteria-II	1-280	11-2800
Inorganic phosphorus solubilisers	5-170	$0.29\text{-}14.20 \times 10^3$
Proteolytic bacteria	5-277	$0.48\text{-}91.10 \times 10^3$
Amylolytic bacteria	7-186	200-3040
Aerobic cellulose decomposers	1-15	10-450
Anaerobic cellulose decomposers	1-110	15-1400
Methanogenic bacteria	1-20	10-200
Iron bacteria	1-45	20-1100
Total coliforms	10-720	100-450

of cowdung on alternate days, gradual leaching of inorganic fertilizers, manual raking of pond bottom and bioturbation by common carp, were studied along with observations on the growth characteristics of bacterial populations under simulated conditions. The respective ranges of bacterial counts in the water medium (no $\times 10^3$/ml) were 1.02—64.67, 1.55—157.07, 0.71—26.32, 0.51—39.89 and 0.90—3.65. The corresponding ranges in the sediment (no. $\times 10^6$/g) were 0.39—15.03, 1.89—5.10, 0.40—5.00, 0.43—12.27 and 0.36—1.78. The generation time and generation number of the populations in the water and sediment media were in the ranges of 0.9—24.8 h, 0.7—21.2 h and 0.2—5.3, 0.2—6.4 respectively, indicating intense bacterial activities. Primary production levels in the range of 2.01—6.01 g C/m³/d comprised high proportions of respiration (24.10—139.82%) of the gross production. The results brought out the significance of the heterotrophic food chain in fish pond ecosystems and demonstrated the positive effects of simple pond environmental manipulation measure like bottom ranking on sediment-water interactions (Ayyappan *et. al.*, MS). Microbiological studies are valuable in determining the types, doses and periods of fertilizer application. Though in preliminary stages of investigation, selective enrichment of microflora is a potential tool in optimising ecological processes.

Trophic efficiencies

Quantification of production efficiencies considering the pathways of energy transfer is important from the point of view of optimising production processes. In the peninsular tank studied, the annual average photosynthetic efficiency values ranged from 0.81 to 1.42% and the fish production efficiency worked out to 0.43%. In the farm ponds under different management practices, the photosynthetic efficiencies were in the range of 0.91—2.69% of the incident light energy. The fish production efficiencies based on photosynthetic energy fixation and total carbon input were 0.97—2.94% and 0.90—2.33% respectively (Ayyappan *et. al.*, 1990 a).

High organic accumulations were observed in non-drainable rural ponds of Orissa (Table 2). Community respiration formed a large portion of the primary production (1.76—4.79 g C/m^2/d). The sediment oxygen consumption comprised upto 30% of the total respiration, with dominance of chemical uptake followed by bacterial and macroinvertebrate respiration. The significance of differences between actual and potential sediment respiration rates and limiting effect of oxygen availability on the uptake rates were demonstrated. The studies showed that the deep sediment layers rich in organic matter played a limited role in the community metabolism of non-drainable fish ponds. Certain measures such as manual raking of pond bottom and introduction of bottom feeding fishes were suggested for improved sediment-water interactions and balanced transport mechanisms (Olah *et al.*, 1986, 1987 a).

Table 2. Fish food resources, production and decomposition levels in rural non-drainable fish ponds of Orissa (1985-87)

Bacterioplankton, no. $\times 10^6$/ml	0.20—12.90
Planktonic detritus, >10 μ, no. $\times 10^6$/ml	1.90—19.30
Seston, >60 μ, mg/l	0.40—232.00
Seston, >150 μ, mg/l	19.00—115.00
Gross primary production, g C/m^2/d	1.76—4.79
Community respiration, g C/m^2/d	1.66—4.69
Sediment oxygen consumption, g O_2/m^2/d	0.39—3.39
Decomposition loss of water hyacinth, %/day	0.01—10.99
Oxygen consumption by decomposing macrophytes, mg O_2/g/h	0.08—2.65

Culture of fish food organisms (*Spirulina*, *Brachionus*, *Moina*) and provision of periphytic substrates are the aspects of trophic enrichment. Large scale production of *Spirulina* using different media and its incorporation in carp diet are being standardised. Increase in productivity levels on periphytic enrichment, even in terms of fish, was demonstrated in an experiment on monoculture of rohu, *Labeo rohita*. The production rate was 13% higher in ponds provided with periphytic substrates (bamboo mats, 10% of pond area) over those without substrates. Diatoms and Conchostracans formed the major part of periphyton.

Organic decomposition and microbial processing

The degradation of organic matter regulates a number of important functions like nutrient recycling, food production for detritus feeders, production of regulatory ectocrine substances and modification of inert materials. The International Conference on detrital systems for aquaculture held at Italy in 1985 emphasised the role of detritus food chain in sustaining aquatic productivity and stressed the need for detailed research on these aspects to maximise fish yields (Moriarty and Pullin, ed., 1987). In the Institute's programmes, decomposition of organic substrates was studied both as a means to assess the decomposition potentials of the different types of ponds as well as to recycle the organic matter, both plant and animal, into carp culture. Seasonal and vertical variations in the decomposition rates of water hyacinth and its ingredient-simulating model substrates of filter paper cellulose and silk protein were monitored in a peninsular tank and a rural pond in Orissa. An *in situ* procedure to measure the weight losses was developed to provide a simple means of obtaining quick estimates. Of the three substrates, water hyacinth showed maximum decay rates of 1.1—6.3% and 4—11% per day in the two waters. The corresponding values for cellulose and protein were 0.1—3.5, 0—3.6 and 0—1.2, 0—3% per day. Weight losses were higher in bottom waters and at the sediment-water interface (Ayyappan *et al.*, 1986).

Detritus-associated oxygen consumption of different macrophytes was quantified during different stages of its formation. While it was in the low ranges of 0.26 – 0.44 mg O_2/g/h for *Salvinia*, high values of 0.12—2.31 mg O_2/g/h were measured in *Najas* (Olah *et al.*, 1987 b). Further, the decomposition pattern of leaves of water hyacinth was measured in response to nutrient enrichment with nitrogen and phosphorus. The maximum total decay loss was observed in the natural water medium. A compensatory nutrient flux was observed during the decomposition and at high nutrient level, macrophytes served more as a base for bacterial activity than as a substrate for decomposition (Ayyappan *et al.*, 1988).

Experiments on the processing and utilisation of the problem weed, water hyacinth, with a view to recycle it into carp culture ponds as manure and/or feed were conducted with inocula of 10% and 20% cowdung and 20% poultry manure along with 2% urea in each case A period of 2-3 weeks was observed to be adequate for treatment, considering the changes in its nutrient compartments, bacterial population and the effects on water quality (Table 3). High organic processing potentials (5 g C/m^2/d) and primary production sustenance capabilities (5 g C/m^2/d) were demonstrated. With fish pro-

Table 3. Bacterial populations in processed water hyacinth with three treatments at four weeks, No. $\times 10^4$/g

Bacterial populations	*Water hyacinth +10% cowdung*	*Water hyacinth +20% cowdung*	*Water hyacinth +20% poultry manure*
Aerobic heterotrophic bacteria	13.51	15.29	9.21
Anaerobic bacteria	8.60	6.36	8.96
Ureolytic bacteria	0.88	0.07	5.01
Euterobacteria	0.56	4.20	0.10
Lactic acid bacteria	4.30	3.15	7.34
Aerobic cellulolytic bacteria	3.71	1.51	0.15
Anaerobic cellulolytic bacteria	6.35	7.61	1.82

duction levels of 2.43 and 1.13 g/m^2/d at high stocking densities of 5 and 19/m^2 respectively, the production efficiencies considering the total carbon input were 2.68 and 1.45. The conversion ratios were 2.02—3.72 : 1 on dry weight basis. Biomanipulation of the environment of shortening food chains achieved through both the application of processed hyacinth and maintenance of high fish densities increased the efficiency of the vegetation-fed and detritus-based carp culture system (Ólah *et al.*, 1990).

Integrated fish farming combining fish culture with livestock farming and horticulture provides for all channels of recycling wastes into fish production. Though raw cowdung is traditionally used as a manure in fish ponds, intermediary measures are being introduced for processing the organic matter and reducing the in-pond oxygen demand. An improvement in the nutrient quality of the cowdung through microbial processing of the added water hyacinth in the biogas plant was demonstrated, with the daily input of water hyacinth and cowdung (4 : 1 ; 25 kg) into a 2-m^3 biogas plant (Table 4). The mean net fish production levels in three sets of ponds with six species of carps stocked at 6000/ha and treated with cowdung (10 t/ha/a) and biogas slurry at 15 and 30 t/ha/a were 3711.60, 2226.60 and 3269.85 kg/ha/10 months. With corresponding mean primary production levels of 3.29, 3.19 and 3.21 g C/m^2/d, the respective production efficiencies were 3.76, 2.33 and 3.40%. Based

Table 4. Chemical composition of cowdung and biogas slurry

Parameter	*Cowdung*	*Biogas slurry*
Moisture, %	81.30	92.24
Organic carbon, %	6.62	2.96
Total nitrogen, %	0.38	1.18
Total phosphorus, %	0.20	0.82
C : N ratio	17.42	2.50

on the total carbon input, the efficiencies worked out to 3.11, 1.94 and 1.75% (Tripathi *et al.*, 1989 a).

The dominant decomposing bacterial population in two types of integrated fish farming systems, *viz.*, with grass carp as the main component fed on aquatic vegetation and biogas slurry as manure were studied. The first system showed an initial spurt in the heterotrophic bacterial activity (540—2-5.000/ml water and 14.32 $\times$ 10^4—23.95 $\times$ 10^4/g sediment), gradually tending to stabilisation. The other bacterial groups in the water and sediment media were aerobic cellulolytic bacteria 167—3363/ml, 2400—13933/g ; pectinolytic bacteria 1—110/ml, 35—1400/g, being significantly higher than those in natural waters. The guts of grass carp, full with decomposing vegetation, recorded high counts of the above. The second system showed much higher heterotrophic bacterial loads of 1.1—9.7 $\times$ 10^3/ml water and 1.54—112 $\times$ 10^4/g sediment. The results indicate a sustained bacterial activity in such organic-rich culture systems (Ayyappan and Tripathi, 1989).

In attempts at demonstrating the trophic significance of microbial communities in carp rearing, the nutritive qualities of microorganisms were assessed using both pure cultures and substrates. The crude protein levels (percentages on dry weight basis) in the feeds, *viz.*, bacteria (*Bacillus subtilis*), fungus (*Aspergillus flavus*), plankton, wheatbran-groundnut oilcake mixture, fermented bran-cake and processed water-hyacinth were 43.63, 47.27, 42.91, 29.09, 36.37 and 14.54 respectively. The corresponding energy values (k cal/g dry weight) were 2.98, 4.03, 3.78, 3.02, 3.68 and 3.01. The specific growth rates of the carp fry fed on the above feeds for four weeks ranged between 0.02 and 3.40. Providing comparative evaluation of the natural and processed foods, the study showed the possibilities of better utilisation of macrobial production for carp rearing through medium enrichment or diet incorporation (Ayyappan *et al.*, 1990 b).

Nitrogen fixation and biofertilization

Biological nitrogen fixation has been shown to be a significant process in the functioning of aquatic ecosystems (Horne and Viner, 1971 ; El Samra and Olah, 1979). While a clear evidence of nitrogen fixation in lake water was first reported by Dugdale *et al.* (1959) using ^{15}N tracer technique, the discovery by Dilworth (1966) and Schollhorn and Burris (1966) that nitrogenase catalyses the reduction of acetylene to ethylene provided the basis for a more sensitive and practical assay of nitrogen fixation activity. Considering the lacuna of information on the *in situ* aquatic nitrogen fixation levels in the country, comprehensive investigations were initiated at the Institute since 1988. In three sets of farm ponds with different sets manured with cowdung (@ 10 t/ha/a), treated with inorganic fertilizers (@ 100 kg N and 500 kg P/ha/a) and the third untreated (providing information on prevailing conditions), the monthly variations of acetylene reduction activity (n mol C_2H_4/l/h) were 0.12—2.85, 0.08—1.96 and 0.06—1.57 respectively. The corresponding ranges in the sediment (n mol C_2H_4/kg/h) were 2.24—27.45, 2.22—17.47 and 0.13—11.44. The associated bacterial population levels are presented in Table 5. In terms of nitrogen, the mean fixation levels in the water and sediment media (kg N/ha-m/a) in the three sets of ponds were 0.7280, 0.6743, 0.3973 and 1.6841, 1.3835, 0.8606 respectively. These account for the first estimates of nitrogen fixation rates of Indian freshwater fish ponds (Ayyappan and Pandey, 1989).

Table 5. Bacterial populations in the water and sediment media of the experimental fish ponds (1988-89)

Group	*Water (no/ml)*	*Sediment (no/g)*
Aerobic heterotrophs	$0.12-29.19\times10^3$	$0.79-43.88\times10^4$
Ammonifiers	141—2413	$2.37-35.97\times10^4$
Aerobic nitrogen fixers	24—1962	$0.12-107.43\times10^3$
Anaerobic nitrogen fixers	2—307	15—2581
Nitrifying bacteria-I	1—25	50—3897
Nitrifying bacteria-II	2—140	106—3897
Denitrifying bacteria	1—25	2—2453
Ureolytic bacteria	6—2577	$0.15-46.68\times10^3$

Further, the nitrogen fixation rates in the water and sediment media were studied with reference to the processing patterns of cowdung and poultry manure in yard experiments. The heterotrophic bacterial counts (400—2000/ml) gradually reduced in the water medium, but increased in the sediment ($5-160\times10^3$/g) indicating organic accumulation. A higher nutrient flux to the water medium from poultry litter was evident in the levels of NH_4—N (0.005—2.00 mg/l), NO_2—N (traces—0.08 mg/l), NO_3—N (0.01—0.25 mg/l) and P_2O_5 (0.08—0.16 mg/l). The populations of aerobic nitrogen-fixing bacteria in the water medium were higher in poultry manure series (200—8000/ml) than in cowdung series (20—1400/ml). The corresponding ranges in the sediment were $0.20-60.00\times10^3$/g and $0.20-10.00\times10^3$/g. The nitrogen fixation presented a similar pattern (water, ARA, 0.09—120.26 and 0.21—30.41 n mol C_2H_4/l/h ; sediment 13.46—172.80 and 0.34—134.93 n mol C_2H_4/kg/h), with the mean levels in the two series being 2.24 and 1.04 mg N/m^3/d. The study showed the superiority of poultry manure with regard to nutrient release, nitrogen-fixing bacterial populations and nitrogen fixation rates for further application in fish ponds (Ayyappan *et al.*, 1990 c).

Bio-fertilization in fish ponds with heterocystous blue-green algae such as *Anabaena*, *Nostoc*, *Calothrix* and the Angiosperm, *Azolla* is being contemplated. In this direction, strains of blue-green algae are being isolated and evaluated for their growth and nitrogen-fixation potentials. The ranges of chlorophyll-a and acetylene reduction activity of three strains studied were 0.0001—15.595 μg/ml and 0.0002—2.5417 n mol C_2H_4/ml/h respectively over a period of six weeks. Mass culture and application of blue-green algae in fish ponds are planned. A culture system for *Azolla* sp. (*Azolla caroliniana* presently) has been laid out covering an area of over 1000 m^2 with a production capacity of about 100 kg/day and work is in progress at standardisation of its level and method of application to fish ponds.

Prospects and potentials

The studies at the Central Institute of Freshwater Aquaculture have provided data base on several new aspects of hydrobiology with regard to bacterial populations, bacterioplankton, photosynthetic pigments, community metabolism, organic decomposition, nitrogen fixation, etc. (Tripathi *et al*., 1988, 1989 b). In addition, applied research projects on intensive carp culture utilizing biotechnological tools (funded by Department of Biotechnology), use of biogas slurry in aquaculture (Department of Non-conventional energy sources), integrated fish farming (Indian Council of Agricultural Research) and use of domestic sewage and composted garbage in carp culture (Indian Council of Agricultural Research) are in operation at the Institute, where microbiological studies would provide a greater insight into the ecocycling and optimisation of production efficiencies.

Further investigations on different aspects such as microbial activity patterns, nutrient cycling and bioenergetics of fish ponds under different trophic conditions, microbial processing and recycling of organic matter, biofertilization and pond environmental modifications and environment—produce interactions with reference to microbial coenoses, etc. have been planned. The use of biofilters, bioreefs and microbial treatment of wastes are comparatively new to Indian aquaculture, which need to be standardised considering the increasing intensification of practices. Microbiological studies are of great importance both from the point of view of monitoring and maintaining a proper aquatic environment as well as optimum utilization of the available and added nutrient for fish production. It is clear from the foregoing account that aquatic microbiology has added a new dimension to Indian hydrobiological studies in the recent years.

REFERENCES

Adoni, A.D., (1975). *Studies on microbiology of Sagar lake.* Ph.D. Thesis, Univ. Sagar, Sagar, M.P.. 254 pp.

Ayyappan. S , (1987). *Investigations on the limnology and microbial ecology of a lentic habitat.* Ph.D. Thesis, Bangalore University, 326 pp.

Ayyappan, S and B.K. Pandey, 1989. Biological nitrogen fixation in freshwater fish ponds. *Proc. Natl. Sem. on* Forty years of Freshwater Aquaculture in India, 7-9 Nov. 1989, Bhubaneswar, Abstract No. SVE/4.

Ayyappan, S. and S.D. Tripathi, 1989. Bacterial populations in fish farming systems based on aquatic vegetation and biogas slurry. *Proc. 29th Annual Conf. Assoc. Microbiologists* of India, 9-11 Feb. 1989, Hisar.

Ayyappan, S , J. Olah, S.L. Raghavan, V.R.P. Sinha and C.S. Purushothaman, 1986. Macrophyte decomposition in two tropical lakes. *Arch Hydrobiol.*, 106 (2) : 219-231.

Ayyappan, S., P.K. Saha, C.S. Purushothaman and K.C. Pani, 1988. Decomposition of water hyacinth, *Eichhornia crassipes* (Mart) Solms. in response to nutrient enrichment. Proc. *First Indian Fisheries Forum, AFSIB,* 4-8 Dec. 1987, Mangalore, pp. 41-44.

Ayyappan, S , N.G.S. Rao, G.R.M. Rao, K. Janaki Ram, C.S. Purushothaman, P.K. Saha, K.C. Pani, H.K. Muduli, V.R.P. Sinha and S.D. Tripathi, 1990a. Production efficiencies of carp culture ponds under different management practices. *J. Aqua. Trop.*, 69-75.

Ayyappan, S., B.K. Pandey and S.D. Tripathi, 1990b. Nutritive quality of microbial communities as feed components in experimental carp culture, *Indian J. Exp. Biol.* (in press).

Ayyappan, S., P.K. Saha, B.K. Pandey and K.C Pani, 1990c. *Aquatic nitrogen fixation with reference to manure* processing patterns. Proc. Second Indian Fisheries Forum, AFSIB, 27-31 May, 1990, Mangalore, p. 25.

Ayyappan, S., C.S. Purushothaman, P.K. Saha and K.C. Pani. 1990 Heterotrophic bacterial populations and primary production in fish ponds under different management practices. (Accepted in Journal of Aquaculture).

Bagde, U,S. and A.K. Varma, 1982. Distribution and periodicity of total, faecal coliform bacteria in an aquatic ecosystem. *Int. J. Env. Stud.*, 19 : 215-220.

Das, S.M. and Srivastava, J.K. (1959). Quantitative studies on fresh-water plankton. III. Quantitative composition and seasonal fluctuations in plankton composition. *Proc. Nat. Acad. Sci India*, **29** : 174-189.

Dilworth, M.J., 1966. Acetylene reduction by nitrogen fixing preparations from *Clostridium pastacricuum.* Biochem. *Biophys. Acta*, 127 : 285-294.

Dugdale, R.C, V.A. Dugdale, J. Neess and J. Goering, 1959. Nitrogen fixation in lakes *Science*, 130 : 859-860.

El Samra, M J. and J. Olah, 1979. Significance of nitrogen fixation in fish ponds. *Aquaculture*, 18 : 367-372.

Ganapati, S.V., 1957. Limnological studies of the upland waters of Madras state. *Arch. Hydrobiol.*, 53 : 3-61.

George, M.G., 1966. Comparative plankton ecology of five fish tanks in Delhi, India. *Hydrobiol*, 27 : 81-108.

Gore, P.S., 1971. Some observations on the bacterial flora of the beach and beach seawater. *Indian J. Env. Hlth.*, **13** : 115-119.

Gulati, R D. and G. Wurtz-Schulz, 1980 Remarks on the present status of limnology in India based mainly on the Indian publications in *Hydrobiologia* and suggestions for further approach. *Hydrobiol.*, 72 : 211-222.

Horne, A.J. and A.B. Viner, 1971. Nitrogen fixation and its significance in tropical lake George, Uganda. *Nature*, 232 : 417-418.

Jana, B.B. and S.K. Roy, 1983. Estimates of microbial populations associated in the N cycle and their activity in water and sediments of fish farming ponds under mono and polyculture systems. *Int Res. ges. Hydrobiol*, 68 : 581-590.

Jana, B.B. and S.K. Roy, 1986. Seasonal and spatial distribution pattern of nitrogen fixing bacteria in fish ponds under different management systems. *Hydrobiol.*, 137 : 45-54.

Jana, B.B., U.K. De and R N. Das, 1980. Environmental factors affecting the seasonal changes of net plankton in two tropical fish ponds in India. *Schweiz Z. Hydrol.*, 42 (2) : 225-246.

Jana, B.B., G N. Patel. S.K. Roy and U.K. De, 1982. Heterotrophic bacteria, physico-chemical regime and stocking of fish in the tanks under three trophic conditions. *Limnological*, 14 (2) : 363-375.

Kuznetsov, S I., 1970. *The microflora of lakes and its geochemical activity.* Ed. C.H. Oppenheimer, Univ. Texas Press 503 pp.

Mathew, P.M., 1975 Limnology and productivity of Govindgarh Lake, Rewa, Madhya Pradesh. *J. Inland Fish. Soc.*, India, 8 : 16-24.

Michael, R.G., 1980. A historical resume of Indian limnology. *Hydrobiol.*, 72 : 15-20.

Moriarry, D.J.W. and R.S.V. Pullin (Ed.), 1987. *Detritus and microbial ecology in aquaculture.* ICLARM Conf. Proc., 420 pp.

Nair, S, 1979. Microbial characteristics of the Laccadives sea (Lakshadweep). *Indian J. Mar. Sci.*, 81 (4) : 227-231.

Olah, J., V.R.P. Sinha, S. Ayyappan, C S. Purushothaman and S. Radheyshyam, 1986. Community metabolism in tropical undrainable rural fish ponds. *Int. Revue ges. Hydrobiol.*, 71 (6) : 807-812.

Olah, J., V.R P. Sinha, S. Ayyappan, C.S. Purushothaman and S. Radheyshyam, 1987a. Sediment oxygen consumption in tropcial undrainable fish ponds. Int. *Revue gas. Hydrobiol,*, 72 (3) : 297 305.

Olah, J., V.R.P. Sinha, S. Ayyappan, C.S Purushothaman and S. Radheyshyam, 1987b. Detritus associated respiration during macrophyte decomposition. *Arch. Hydrobiol*, 111 (2) : 309-315.

Olah, J., S. Ayyappan and C S. Purushothaman, 1990. Processing and utilization of fermented water hyacinth, *Eichhornia crassipes* (Mert.) solms. in carp culture. *Aquaculture Hungarica*, 6.

Parimala, V. and S. Ayyappan, 1989. Population and distribution of heterotrophic bacterial flora in freshwater bodies of Bangalore. *Indian J. Env. Hlth.*, 31 (2) : 162-170.

Parimala, V and M.K.C. Sridhar, 1982. Some bacterial flora of beach sand at Madras, Sout i India. *J. Hlth. Sci*, 8 : 40-47.

Rodina, A.G., 1972. *Methods in Aquatic Microbiology.* Ed. Tr. Rev. R.R. Colwell and M.S. Zambruski, Univ. Park Press, Baltimore and Butterworth and Co , London, 481 pp.

Santhakumary, E., 1966. Some bacterial characteristics of the Cochin backwater. *Curr. Sci*, 35 : 10-11.

Saran, H.M and A.D. Adoni, 1985. Limnological studies on the seasonal variations in phytoplanktonic primary productivity in Sagar lake. *Bull. Bot. Soc.* Sagar, 32 : 71-76.

Schollhorn, R. and R.H. Burris, 1966. Study of intermediates of nitrogen fixation. *Fed. Proc.*, 25 : 710.

Sreenivasan, A., 1963. Primary production in three upland lakes of Madras State, India. *Curr. Sci.*, 32 : 130-131.

Sreenivasan, A., 1976. Limnological studies and primary production in temple pond ecosystems. *Hydrobiol.*, 48 (2) : 117-123.

Sumitra, V., 1969. *Limnological studies on some tropical ponds*. Ph D. Thesis, Madurai University, Madurai.

Tripathi, S.D., S. Ayyappan and S.D Gupta, 1988. Potential applications on biotechnology in Indian freshwater aquaculture. Natl. Sem. on *Application of biotechnology for Agriculture and rural development*, 12-15 Sept. 1988 NIRD, Hyderabad.

Tripathi, S.D , P.K. Aravindakshan, S. Ayyappan, Ranjit Singh, A N. Mohanty and H.K. Muduli, 1989a. Carp polyculture with cowdung and water hyacinth-based biogas slurry as manures. Natl. Sem. on *Forty years of Freshwater Aquaculture in India*, 7-9 Nov. 1989, Bhubaneswar, Abstract No. SIV/1.

Tripathi, S.D., S Patnaik, S. Ayyappan, P.K. Saha and C.S. Purushothaman, 1989b. Effects of mahua oilcake on the water quality and biota of fish ponds. *J. Aqua. Trop.* 4 : 29-36.

Valankar, N.K , (1950). Bacteriological survey of seawater from the coast of Madras city (Bay of Bengal). *Proc. Indian Acad Sci.*, 32 : 80-86.

Velankar, N.K., (1957). Bacteria isolated from seawater and marine mud of Mandapam (Gulf of Manner and Park bay). *Indian J. Fish*, 4 : 208.

11

HYDROBIOLOGICAL STUDIES IN RELATION TO WATER POLLUTION

K.P. Krishnamoorthi[1]
and
Mrs. Sarkar[2]

INTRODUCTION

Hydrobiology is not a discipline on its own account, but is the part of biology which is concerned with the life and life cycles of organisms in water. At the same time it is a special section of limnology, the science of inland waters. The objects studied by hydrobiology are animals, plants and bacteria so far as they live in water. Hydrobiology therefore covers nearly all the methods generally adopted in biology.

The primary objectives of hydrobiology work is not collection and determination of aquatic organisms alone, but to gain insight into relationship between organisms and environment. Fresh water in comparison with giant marine water occurs in small quantity as inland waters. Rivers and lakes are the most striking examples of this type, to some extent underground water also. However, organisms are affected by factors in its environment. The organisms, however, not only respond to the factors in the environment and they also mutually influence one another and form communities (biocoenoses). This animal biocoenoses developes in the standing natural water which is temporarily parallel to the higher vegetative pattern. The courses of the live of an animal is always bound up with plant in question and relationship is maintained.

With the advent of complicated discharges of wastewater of organic and inorganic *i.e.*, source from industrial and urban

1. Retired Scientist, NEERI. 2. Scientist, NEERI, Nagpur.

sectors the flora and fauna living in them are put to the large scale stress and this pehnomenon is now global in character. One of the most striking and relevant studies to be carried out in aquatic environment based on the impact of water pollution is impact assessment studies. This means the actual stress *i.e.* being exerted on the flora and fauna of the aquatic resources of waste discharges have to be studied critically to evaluate the impact assessment. Toxic discharges on the one hand tend to more or less wipe out flora and fauna water discharges like heavy metals and pesticides tend to results in bioaccumulation and biomagnification phenomenon.

Normally aquatic resources are associated with plankton, necton, periphyton and benthos and most of these are happened to be fish food.

Protection and conservation of water resources implies that its natural flora and fauna be protected. Impact of pollution, toxic or organic affect the living organisms which respond to the pollutants in several ways. The manifestation may be lethal or sublethal altered behaviour, morphology, physiology and life cycle. The ultimate impact, nevertheless, is damage to the ecosystem.

By careful study the aquatic communities based on the response of animals and plant communities to the quantitative effects of pollution, yardsticks can be arrived for monitoring water pollution levels. In other words, biomonitoring is a tool to evaluate water quality based on hydrobiological studies and they can be grouped as ecological response, physiological response and toxicological response. There has been a great thrust for developing water quality criteria based on quantitative enumeration of ecological data which are called "Biological Indices" of water quality. Several systems of biological classification exists. The fundamental concept underlying all these systems, however, being the same *i.e.* biota respond to pollution and the type of response is regulated by the quantity of pollution and time of exposure.

Several methods have been devised for the purpose of biological assessment of water quality. Three major groups of responses that are quantified for biological monitoring in waters are :

(1) Ecological Responses,
(2) Physiological Responses, and
(3) Toxicological Responses.

Under these major categories again there are several indices. The approach to be adopted and the group of organism selected for data acquisition and interpretation differ in each method. The different methodologies adopted are indicated in Table I. Broad classification under each category are listed. Thus at least 30 indices of water quality are available in water pollution biology.

In NEERI efforts have been concentrated on the ecological aspect of biomonitoring with special reference to water quality in rivers and lakes. Ecological responses of certain groups specially ciliated protoza have also been studied in laboratory simulating field conditions.

The major ecological units selected for these investigations are :

1. Benthic macroinvertebrates (Bottom fauna),
2. Ciliated protozoa,
3. Periphyton, and
4. Algae.

In this talk the results obtained during study of lakes and rivers with special reference to benthic fauna is presented briefly. Result of ciliated protozoa distribution and their relation to pollution is also discussed. Work on periphyton is under investigation, and only highlights of the trend of the result on periphyton is included in the talk.

Results

(*i*) *Studies on Benthic fauna.* Baseline data on distribution of bottom fauna in the rivers and lakes were collected on monthly basis. Three rivers of varying water quality were selected for the study. In these rivers eleven sampling stations

Table I. Biological Indices of Pollution—Various Systems

Ecological Methods
↓

(1) Saprobic Indices (quantitative)	(2) Biotic Indices (Semi quantitative)	(3) Diversity Indices (quantitative-very precise)	(4) Algal Indices
1. Saprobic system of Kolkwitz and Harsoon	1. Biotic index of Beck	1. Margalef's Index	1. Palmers Score
2. Liebman's Gute classes water quality	2. Trent Biotic Index	2. Menhinnicks Index	2. Nyggards Index
3. Saprobic index of Pantle and Buck	3. Empirical Biotic Index of Chutter	3. Shannon Weaver Index of Diversity	3. Chlorophyll estimation
4. Method of Zelinka and Marvan	4. Chandlers score system	4. Dominance Index I (Berger and Parker)	4. Autotrophic Index

(*Contd.*)

5. Saprobic Index of Knopp	5. Species Deficit of Kothe	5. Dominance Index II (MacNaughton)	5. Watanabe's Index
	6. Horasawa's Biological Index of Pollution	6. Simpson's Index	

Physiological Methods	Toxicological Methods
1. The Biomass Titre of Bringmann and Kuhn	1. Algal Assay
2. Knopps Supplementary Consumption	2. Fish Bioassay
3. Determination of Inhibition threshhold for biochemical activity	3. Assay with invertebrates
	4. Assay with ciliated protozoa
	5. Bioaccumulation.

Table II. Summary of Work Done in NEERI on Biological Indices

Ecological Indices			
Bottom fauna		Ciliated Protozoa	Periphyton
Logic Waters	Lentic Waters		
1. Species Richness Index	1. Faunal Associations	1. Species composition	1. Species composition
2. Density Distribution	2. Standing Crop	2. Frequency Distribution	2. Chlorophyll Content
3. Shannon Weaver Index	3. Biomass	3. Laboratory Model Studies on Succession of Species	3. Biomass
4. Margalef's Index			4. Autotrophic Index
5. Manhannicks Index			
6. Dominance Index I			
7. Dominance Index II			
8. Graphic and Pictorial Presentation			

were included. The static water or lakes included—an oligotrophic lake and a set of six experimental sewage fed fish ponds representing eutrophic condition.

Assessment of Riverine System. Three rivers *viz.*
Kanhan – clean water (potable water supply)
Pili —mildly polluted
Nag – highly polluted

were selected for the investigation. The degree of pollution was recorded mainly based on BOD values and nutrients *i.e.* phosphate and nitrates.

In the above three rivers benthic fauna were collected using Ekman dredge of size 6′×6′×′6′. The samples of bottom mud were screened through number 60 standard B.T.L. sieve and a jet of water was passed through the sample to wash off the debris. The macroinvertebrates retained in the sieve were back washed in white enamelled tray and the organisms were sorted and preserved. All these were counted and identified using standard taxonomic literature. Replicate sample in each station were taken for statistical validity. The benthic fauna species composition *i.e.* species list and the number of organisms in each species or in other words the differential count of benthic fauna were utilized to calculate seven indices of pollution. Of these S (species richness index), d_1 (Menhinnicks index 1964), d_2 (Margalef's index 1951) and I (Shannon Weaver Index 1963) are Diversity indices and N (Abundance), DI (Dominance Index of Berger and Parker 1970) and DII (Dominance index of Mac Naughton 1967) are indices of Dominance. It is seen that in case of Diversity Indices higher values signify clean water. Diversity index value decrease with pollution. On the other hand dominance values are lower in clean water and higher in polluted water. This pattern is evident in the results of our findings. The comparative ability of the indices to demarkate clean water zones, zone of sewage outfall and recovery zone in the river strech were statistically tested by analysis of variance. It was found statistically that Shannon Weaver Index is the best Index for water quality classification in rivers. Species richness index was found to be a very simple yet effective index in this regard.

Assessment of lake systems

Two static waters were evaluated and compared. These are

Ambazari lake (potable water source)

Sewage fed fish ponds (six ponds fed with effluent from stabilization pond with one pond serving as control).

For static waters it was found that the Diversity and Dominance indices did not yield very encouraging results. For lake system the trophic status of the water was determined. Benthic productivity was taken as a measure of the organic enrichment in water. Based on benthic fauna standing crop and biomass the trophy was compared. For this the populations were estimated as number of organisms/square meter and the biomass was expressed as dry weight in gms/square meters of bottom. This estimation was done gravimetrically. The observations indicate the oligotrophic condition of Ambazari lake and the highly eutrophic condition existing in the fish ponds.

Ciliated Protozoa as Indicators of Water Quality

Ciliated protozoa respond rapidly to changes in aquatic environment. Since they are primarily bacterial feeders they serve as important indicators of organic pollution. The saprobic valencies of ciliated protozoa are well documented.

Observations on ciliated prtozoa were divided into two phases :

(*a*) Field studies
(*b*) Laboratory studies.

Field studies. Data on species composition and frequency distribution of ciliated protozoa were obtained from benthos of river by a special technique. It was found that in clean water number of species is less and also percent frequency distribution is lower. Maximum number of species were recorded in mildly polluted mesosaprobic water. The percent frequency distribution is higher. However in polysaprobic water the species number decline because only tolerant species survive.

Laboratory Experiments on Autoecology. These experiments are models of self-purification in rivers. The degradation of organic matter during self-purification is simulated in the laboratory. The ciliate species and counts are recorded during various stages of stabilization of organic matter which is determined by estimating B.O.D. values in the experimental containers. Various organic substrates *viz.*, Peptone, Balamul (a baby food) and sterilized raw sewage were used as substrates for autoecological experiments. The results obtained during stabilization of sterilized raw sewage seeded with an inoculum of soil suspension containing naturally occurring ciliates. It is seen that both species composition and counts of ciliates gradually change with the stabilization of B.O.D. from 250 mg/l to 40 mg/l. The trend of result compare favourably with the direct observation made in the field conditions.

Studies on Periphyton and Algae

Periphyton includes the total assemblage of sessile organisms attached to natural or artificial substrates. Periphyton community as a whole has immense potential in biomonitoring of aquatic environment. Periphyton, specially the ciliated protozoa and algal component on artificially introduced substrates were denoted in a clean river, an oligotrophic lake and a set of eutrophic fish ponds. Comparative study on the species diversity, total cell count, biomass value and chlorophyll-a content of periphyton show a distinct gradient from oligotrophic, mesotrophic to eutrophic conditions. Also the distribution of diatoms, desmids, greens and blue greens in the periphytic flora show very clear demarkation in oligotrophic and eutrophic waters. Biomass and chlorophyll-a content in whole periphyton were found to be good indices for ascertaining trophic status of lakes.

Algae are very useful for eutrophication estimates. Quality and distribution of algal flora in rivers, lakes and ponds have been carried out systematically to evolve algal indices of pollution. Palmer's Index and Nygaard's Index have been studied for classifying some lakes for determining trophic status and fisheries potential. It is felt that even in lake plankton studies chlorophyll-a estimations can serve as an overall index of its trophic status.

In the present communication only ecological aspects of biomonitoring has been encompassed. However in the Institute extensive work on the toxicological aspects of biomonitoring is also being undertaken. This is done with special reference to heavy metal toxicity on aquatic food chain organisms, bioassay using fish and microassay using ciliates, periphyton, cladoceran and insect larvae. Algal assay using heavy metals and organochlorine pesticides is also an aspect of work which is being investigated. Lethal and sublethal responses and accumulation of metals and pesticides in algae is being estimated to assess the hazards of metals and pesticides to these organisms.

12

FRESHWATER MOLLUSCS OF INDIA

N. V. Subba Rao

Zoological Survey of India
New Alipore, Calcutta-700053

INTRODUCTION

The phylum Mollusca is a large assemblage of animals having diverse shapes, sizes, habits and occupying different habitats. In a general way, based on their habitats, molluscs can be distinguished into aquatic and land molluscs. All the seven classes of molluscs are represented in the aquatic ecosystem. Qualitatively and quantitatively Mollusca is one of the few common phyla that occur in the aquatic ecosystem. But the taxa with which hydrobiologists are more concerned are the rotifers, crustaceans, insects and fishes. The first two are well represented in the plankton, whereas the last two constitute important macrofauna. Although molluscs are common in the benthic communities, there are not many studies bringing out their role in the dynamics or aquatic ecosystem and their contribution to biomass production. The freshwater habitats are taxonomically impoverished in comparison to the marine habitats. In India, marine molluscs have received more attention because of their aesthetic and gastronomic appeals. In comparison freshwater molluscs are drab coloured and have attractive less attention. Notwithstanding their importance in the ecosystem freshwater molluscs are less explored and studied. These can be considered as natural treasures within our arm's reach. Freshwater molluscs have significant roles in the public and veterinary health. These have the potentialities that can be developed and harnessed for the benefit of man. An attempt is made in this paper to highlight the importance of freshwater molluscs and the need for more planned studies.

TAXONOMIC BACKGROUND

The ecological and biogeographic assessment of any fauna has to be based on sound taxonomic knowledge. A recent comprehensive treatment of the freshwater malacofauna (Subba Rao, 1989) provides the necessary backdrop for any analysis of the freshwater molluscs. Although there are still certain taxonomic confusions or uncertainities, a fair measure of accuracy and stabilit} has been achieved in presenting the data on taxonomic composition of these molluscs.

The freshwater molluscs comprise of two classes, namely Gastropoda and Bivalvia, which include a total of 21 families and 52 genera. The percentage of gastropods is higher (62%) than that of bivalves (38%). The families Iravadiidae, Stenothyridae and Assimineidae among gastropods, and Cultellidae and Solecurtidae among bivalves are euryhaline and occur primarily under estuarine conditions. The representatives of the three gastropod families have generally very small shells (length maximum of shell not more than 7 mm), a character which is comparable with a few species of the family Birthyniidae and Pomatiopsidae. Two families, namely Physidae (Gastropodo) and Margaritiferidae (Bivalvia) are distributed in Pakistan and Burma respectively, but nct in India. The former is however represented by fossil species in India. The seven families mentioned above together comprise 20 species, a small number compared to 284 species reported from India and adjacent countries. The rest of the families show species diversity ranging from one to 52. The largest number of species are recorded in the family Amblemidae (formerly Unionidae) with 52 species followed by Thiaridae (44 nos.). Planorbidae (22 nos.), Corbiculidae (21 nos.), Pisidiidae (18), Bithyniidae (17), Lymnaeidae (15), Neritidae (14) and Viviparidae (14). All the species are described on the morpho species concept and the number would definitely come down if the biological species concept is applied. The large number of species in bivalves is a reflection of their morphological variability rather than species with well-defined boundaries. Gastropods have successfully colonised a number of available ecological niches but their species diversity do not seem to be more than that of bivalves.

A total of 212 species are recorded from within the political boundaries of India (Table 1). When the ratio of average number of species for a family between the two classes is compared it is 10 : 10. It indicated that as far as the speciation is concerned both the classes exhibit same trend. Speciation is controlled by the adaptive nature of the group. While gastropods are abundant in the plains and hillstreams of western Ghat and North-eastern India, bivalves have successfully colonised even glacial lakes in the Himalayas. *Pisidium* is recorded at about 4600 m in the Himalayas.

HABITAT DISTRIBUTION OF MOLLUSCS

Freshwaters in general are classified into lentic or lotic types. The former includes ponds, tanks, lakes, ditches, paddy fields etc. while the latter category include flowing waters such as canals, streams and rivers. In terms of biomass lentic waters are more productive and molluscs play an important role in the dynamics of this ecosystem.

Distribution pattern of molluscs in relation to lentic and lotic waters is presented in the Table 1. Each family includes species that occur both in lentic and lotic habitats. But there are certain families which are restricted to or dominant in only one habitat. The families Viviparidae and Pilidae are generally restricted to lentic habitats with only a few exceptions. *Bellamya bengalensis* may be rarely seen in lotic habitats, while the genus *Turbinicola* of the family Pilidae occurs only in lotic habitats. The family Neritidae and subfamilies Cremnoconchinae (family Littorinidae) and Paludominae (family Thiaridae) are exclusively found in lotic habitats only. Their shells exhibit certain structural features in response to their life in flowing waters. Broadening of the aperture, flattening of shell whorls, and limpet like shape in some are certain adaptive changes seen in the representative of these families. Freshwater limpets of the family Ancylidae are typical of lotic habitats. Besides the flowing nature or otherwise of waters there are other factors like aquatic vegetation, water chemistry, nature of substratum etc. which influence the distribution and occurrence of freshwater molluscs. Very few studies were undertaken to investigate

Table 1. Habitat Distribution of Indian Freshwater Molluscs

Sl. No.	*Name of the family*	*No. of species*	*River & stream*	*Canals*	*Paddy fields*	*Ponds*	*Lakes*
1.	Neritidae	14	+	–	–	–	–
2.	Viviparidae	8	–	+	+	+	+
3.	Valvatidae	1	+	–	–	–	–
4.	Pilidae	7	–	–	+	+	+
5.	Littorinidae	3	+	–	–	–	–
6.	Pomatiopsidae	2	+	–	–	–	–
7.	Iravadiidae	4	+	–	–	–	–
8.	Bithyniidae	13	+	+	+	+	+
9.	Stenothyridae	10	+	–	–	–	–
10.	Thiaridae	31	+	–	–	+	+
11.	Lymnaeidae	11	–	+	–	+	+
12.	Ancylidae	5	+	–	–	–	–

(Contd.)

13.	Planorbidae	22	−	+	+	+	+
14.	Arcidae	2	+	−	−	−	−
15.	Unionidae	7	+	−	−	+	+
16.	Amblemidae	42	+	−	−	+	+
17.	Actheriidae	1	+	−	−	−	−
18.	Corbiculidae	14	+	−	−	−	+
19.	Pisidiidae	11	+	−	−	−	+
20.	Cultellidae	2	+	−	−	−	−
21.	Solecurtidae	2	+	−	−	−	−
	Total	212*					

*Species distributed within India.

the influence of these biotic and abiotic factors on the distribution of freshwater molluscs of India (Janaki Ram & Radhakrishna, 1984 ; Starmuhlner, 1982). More studies are needed to establish the importance of various environmental factors that produce a collective effect on the nature and distribution of freshwater molluscs.

Ecology of freshwater snails has been dealt in brief (Subba Rao & Mitra, 1989). The physiological and ecological requirements of freshwater snails are very broad and hence they have the capacity to overcome unfavourable conditions. They occur even in temporary water ponds with aquatic vegetation and can survive for some time by burying themselves under dead weeds or mud. Freshwater bivalves are absent in temporary ponds, since they are filter feeders, and have a pelagic larval form, called glochidium which parasitizes fish. However, it is extremely difficult to separate the effect of any one factor and conclude that it is critical.

Zoogeographical significance of the fauna

Many of the species are widely distributed in India and adjacent countries. However, quite a few species are highly localised being restricted to streams or lakes in north-east India, Western Ghats and a few to Himalaya mountains. Two species, namely *Neritina perottetiana* and *Cremnoconchus syhadrensis* belong to the two families (Neritidae and Littorinidae respectively), which are primarily marine. The occurrence of these two species in Western Ghats far away from the sea is interesting and perhaps can be explaind by the theory of transgression (Subba Rao, 1989). While for Littorinidae it is the only genus and species reported from freshwaters, the families Thiariidae and Neritidae include several eurghaline species and are more widely distributed in the Indo-Pacific. The stenohaline forms are found on the Asian mainland.

The subfamily Bellamyinae and the family Pilidae have Gondwanaland origin. The former is distributed in Africa, India and South-east Asia and the latter in tropical South America, South Africa, India, Sri Lanka and South-east Asia. The hill-stream genus *Turbinicola* (family Pilidae), which is an inhabitant of streams around Khandalla, Western Ghats resembles South American hill-stream form *Asolene*, suggesting convergent adaptation (Prashad, 1931).

The bivalve fauna include families, like Unionidae, Amblemidae and Pisidiidae, which are of great antiquity and which have cosmopolitan distribution. The family Aetheriidae, which include freshwater oysters, has a discontinuous distribution in South America, tropical Africa, Madagascar and India. The genus *Acostaea* is confined to South America and India. The distribution of the members of the family as a whole can only be attributed to the ancient geographical connections as proposed by Gondwana hypothesis.

The Importance of Malacological Studies

The level of our understanding of molluscs is such that we have not been able to comprehend the value of freshwater molluscs in the ecosystem and the anthropic significance of their studies. Their production rates and turnover values have not been determined. High biomass and high productive values were reported in two common species, namely *Bellamya bengalensis* and *Pila globosa* (Khan and Chaudhuri, 1984; Haniffa, 1978.)

In Medical and Veterinary Public Health

Freshwater molluscs, specially gastropods are important from the medical and veterinary public health point of view. About 100 species of freshwater gastropods are reported to act as intermediate hosts for the digonetic trematode parasites. Most of these were reported from African countries. But in India, only 14 species are implicated as intermediate hosts, of which only 10 species are considered significant (Subba Rao and Mitra, 1989). Eleven species of schistosomes, lung flukes and liver flukes and 33 species of amphistomes are reported to spend their lives in intermediate snail hosts (Srivastava, 1989; Mukherjee, 1989).

Some of the important trematodes and their snail hosts are given in the table 2. The distribution of the parasite as a whole

Table 2. Important Trematodes and their molluscan intermediate hosts

	Name of the parasite	*Host*	*Intermediate host*	*Family*	*Distribution*
	Schistosomes				
1.	*Schistosoma haematobium*	Human, Monkey	*Ferrissia tenuis*	Ancylidae	India (Gimvi, Maharashtra) Africa, Australia, Asia etc.
2.	*S. spindale*	Cattle, buffalo, bandicoot.	*Indoplanorbis exustus Lymnaea acuminata, L. luteola*	Planorbidae Lymnaeidae	India, Malaysia, Sumatra South Africa
3.	S. *indicum*	Horse, goat, sheep, dog, camel, donkey, cattle, zebra, antelope	,, ,, ,,	,,	India (Throughout)
4.	*S. incognitum*	Pig, dog	*Lymnaea luteola*	Lymnaeidae	India (U.P., Madras, Bihar, West Bengal)
5.	*S. nasale*	Cattle, buffalo, goat.	*Indoplanorbis exustus*	Planorbidae	India, Burma
6.	*Orientobilharzia dattai*	Buffalo	*Lymnaea luteola*	Lymnaeidae	India (Bareilly)

(*Contd.*)

7.	*Orientobilharzia turkestanicum*	Cattle, buffalo, sheep, goat, water buffalo, Horse donkey Mule, camel	*Lymnaea auricularia* *Lymnea acuminata rufescens*	Lymnaeidae	India (Kashmir) Pakistan; Russian Tur-kestan, Mangolia, Iran, China, Korea etc.
8.	*Fasciola gigantica*	Sheep, cattle other herbivores	*Lymnaea auri-culario, rufescens*	Lymnaeidae	Pakistan, India
		human	*L. acuminata*	,,	India
	Intestinal flukes				
9.	*Fasciolopsis buski*	Human, pig	*Hippeutis cantori*	Planorbidae	India (Assam)
			segmentina trochoideus	,,	,,
10.	*Echinostoma ilocanum*	Rat, human	1st Intermediate host *Gyraulus Prashadi*	,,	India
11.	*Gastrodiscoides hominis*	Pig, human	*Helicorbis coenosus*	,,	India (Assam)

Source : Srivastava (1989) and others.

and its intermediate host recorded in India only are mentioned. One of the significant aspect of these records are that all the important hosts belong to two basommatophoran families, namely Planorbidae and Lymnaeidae, which are widely distributed in India. Among Prosobranchs, members of the families Pilidae and Thiaridae were recorded to harbour larval trematodes. Certain species of these two families were found to carry natural infections of trematodes, which cause diseases in man. *Pila scutata* which is distributed from Andaman and Nicobar Islands to Malaysia is found to carry natural infection of rat lung worm *Angiostrongylus cantonensis* in Malaysia. The species *Thiara* (*Melanoides*) *granifera* and *T.* (*M.*) *tuberculata* are widely distributed in India, China and Formosa and in the later two countries they serve as intermediate hosts for the lung fluke *Paragonimus westermani* (Malck and Cheng, 1974). Thus the occurrence of some of the freshwater gastropods in India have dangerous portents for human health.

Choking of water distribution system

Some of the freshwater mollusc species which breed at a rapid rate and which aggregate in large numbers do cause problems in Public water supplies. Their presence in large numbers create choking in raw water source, the feeder-pipe lines and in the distribution system. At times these molluscs choke water meters and put them completely out of commission. They create tastes and odours to drinking water. The bivalves which have the tendency for natural aggregation into groups are commonly known to choke the various water distribution systems and other unit processes in a water treatment plant. The finger nail clams (Pisidudae) and Asiatic clam (Corbiculidae) are the two major pest groups in U.S.A. (Ingram 1956, 1959).

In India, mollusc nuisance in Public Water works had been reported in Calcutta, Nagpur, Pune and Trichinopoly (Krishnamoorthi and Rajagopalon, 1970). The juvenile stages of molluscs settle down in the pipe lines of distribution system and grow there reducing the diameter of the pipe. As a consequence the flow of water gets reduced ultimately leading to water shortage. A total of 14 species and 11 species were found to cause nuisance in the Palta and Baranagar Water Works of Calcutta and in different water bodies of Nagpur respectively. The following is the list of species recorded from these two places.

Table 3.

	Name of the species	Locality	
		Calcutta	*Nagpur*
	Gastropods		
1.	*Bellamya bengalensis*	×	×
2.	*B. dissimilis*	×	×
3.	*B. variate*	×	×
4.	*Neritina (Dostia) crepidularia*	×	—
5.	*Septaria caeralescens*	×	—
6.	*Thiara (Melanoides) tuberculatus*	×	×
7.	*Thiara (Tarebia) lineata*	×	×
8.	*T. (T.) scabra*	×	—
9.	*Lymnaea (Pseudosuccinea) acuminata*	×	×
10.	*L. (P.) luteola*	—	×
11.	*Indoplanorbis exustus*	×	×
12.	*Gyraulus* sp.	—	×
13.	*Digoniostoma* sp.	—	×
	Bivalves		
14.	*Modiolus striatulus*	×	—
15.	*Corbicula striatella*	×	—
16.	*Novaculina gangetica*	×	—
17.	*Saphula deltae*	×	—

Although the list is very big, all the species do not create problems. Most of them are smaller in size and are neither prolific breeders nor gregarious in nature. The common pond snail, *Bellamya* spp. has the potential to create problems. In Baranagar and Palta water works of Calcutta, the estuarine bivalve *Modiolus striatulus* and the two common freshwater pond snails *Bellamya bengalensis* and *B. dissimilis* occur abundantly and hence create problems.

Aquaculture Potentiol

Although freshwater molluscs have attracted more attention because of their harmful nature, these have the potential to be developed into an exploitable living resource. Molluscs are eaten in certain parts of the country and according to one survey there is a regular sale of shell-fish in certain markets. Since fish and others that supply animal protein are becoming dearer day by day common man will have to depend in shell-fish for the supply of animal protein. A list of edible molluscs and their selling prices in various markets are given in the table 4 (From Subba Rao and Day, 1989). The demand for shell fish may gradually increase and there is a need for taking up an anticipatory research programme to develop cultural techniques for some of the common molluscs. If developed properly the apple snail *pila globosa* can be turned out into a gourmet's delight and even earn foreign exchange for India.

In recent years there have been reports of the natural occurrence of freshwater pearls (Raut & Biswas, 1982; Subba Rao & Day, 1989). Some attempts were even made towards the production of cultured pearl in freshwater mussels (Janaki Ram, 1989). The initial success at culturing freshwater pearls is an encouraging aspect and freshwater molluscs deserve all the attention from hydrobiologists and aquaculturists.

Biological Indicators

In general microorganisms such as protozoans, rotifers etc. are considered as biological indicators of pollution. But macro-invertebrates are easily observable and should serve as better indicators. Hitherto no attempt has been made to undertake studies on the possible use of molluscs as pollution indicators. Certain species of molluscs occur in profuse numbers in slightly polluted waters and they can form good indicators of pollution, especially in the B-mesosaprobic zone (Abdulappa and Lakshminarayana, 1970).

CONCLUSION

Freshwater molluscs, are thus important because of their two broad implications of causing harm and also in contribu-

Table 4. A list of edible freshwater molluscs

Sl. No.	*Name of the species*	*Local Name*	*State in which used as food*	*Rate*
1.	Pila globosa (Swainson)	Genri—Bihar Samuk—W. Bengal	Bihar, West Bengal	Re. 1/=for 20 to 25 nos.
2.	Bellamya bengalensis (Lamarck)	Samuk/Gugli—Bihar, W. Bengal	Bihar, West Bengal	Re. 1/=to 2/= per kg.
3.	Brotia costula (Rafinesque)	Chenkal—Mizo Chikangbook—Riang	North Bengal (W. Bengal) Arunachal Pradesh, Mizoram	Re. 1/=per tin which include 15 to 20 average size specimens.
4.	Paludomus conica Gray	—do—	Arunachal Pradesh, Mizoram	Rs. 2/=per tin which include 40 to 50 average size specimens.
5.	Paludomus blanfordiana Nevill	—do—	—do—	—do—

(Contd.)

Table 4. (*Contd.*)

Sl. No.	*Name of the species*	*Local Name*	*State in which used as food*	*Rate*
*6.	Lamellidens marginalis (Lamarck)	Jhinuk—Bihar Katli—W. Bengal	Bihar, West Bengal	Re. 1/=for 20 nos. in lot.
7.	Lamellidens corrianus (Lea)	—do—	—do—	—do—
8.	Solenaia soleniformis (Benson)	Tuikep—Mizo Clampy—(Beta)—Riang	Mizoram	Not observed in market
9.	Trapezoideus exolescens (Gould)	Tuikep—Mizo Clampy—(Beta)—Riang	—do—	—do—
10.	Parreysia sikkimensis (Lea)	—do—	Arunachal Pradesh	—do—
11.	Parreysia caerulea (Lea)	—do—	—do—	—do—

* Species producing pearls.

ting to human welfare. Hydrobiologists should divert their attemption in elucidating the biological details of some common species and also in developing suitable techniques for the culture of the apple snail *Pila globosa* and freshwater mussel *Lamellidens marginalis*, the former to be developed into an edible shell-fish and the latter for its meat and also for producing pearls. When once the initial know-how is developed, commercial operations can be undertaken which will go a long way in adding to our rural economy. Freshwater molluscs are going to be the future living resources of our country.

REFERENCES

Abdulappa, M.K. and Lakshminarayana, J.S.S. 1970. Pathogenesis of helminths in the Mollusca and their relation to water quality *Proc. Symp. Mollusca, Mar. biolassn. India*, pt. III, 758-779.

Haniffa, M.L. 1978. Secondary productivity and energy flow in a tropical pond. *Hydrobiologia*, 59 : 49-65.

Ingram, W.M. 1956. Snails and clam infestations of drinking water supplies, JAWWA, *48* : 269-274.

Ingram, W.M. 1959. Asiatic clams as potential pests in Californian water supplies. JAWWA, 51 : 363-370.

Janaki Ram, K. 1989. Studies on culture pearl production from freshwater mussels. *Curr. Sci.*, *58* (8) : 474-476.

Janaki Ram, K. and Radhakrishna, Y. 1984. The distribution of freshwater mollusca in Guntur District (India) with a description of *Scaphula nagarjunai* sp.n. (Arcidae) *Hydrobiologia* 199 : 49-55.

Khan, R.A. and Chaudhuri, S. 1984. The population and production ecology of a freshwater snail *Bellamya bengalensis* (Lamarck) (Gastropoda : Viviparidae) in an artificial lake of Calcutta, India. *Bull. zool. Surv. India*, *5* (2 and 3) : 59-75.

Krishnomoorthi, K.P. and Rajagopalan, S. 1970. Survey of mollusc nuisance in some water supplies of Calcutta. *Proc. Symp. Mollusca. Mar. biol. assn. India*, pt. *III* : 746-754.

Malek, A.E. and Cheng, C.T. 1974. Medical and Economic Malacology. Academic Press, New York and London, ix+398 pp.

Mukherjee, R.P. 1989. Amphistomes of domesticated animals of India and their life histories. *In* : *Snails and Flukes*, (Mimeographed) : 103-109.

Prashad, B. 1931. Some noteworthy examples of parallel evolution in the molluscan faunas of South eastern Asia and South America. *Proc. Roy. Soc. Edinburgh*, *51*(1) : 42-53.

Raut, S.K. and Biswas, A. 1989. Pearls in freshwater mussels *Lamellidens marginalis* in India (Bivalvia : Unionidae). *Basteria,* 53 : 105-109.

Starmuhlner, F. 1982. Occurrence, distribution and geographical range of the freshwater gastropoda of the Andaman Islands. *Malacologia,* 22 (1-2) : 455-452.

Srivastava, C.C. 1989. Schistosomes and schistosomiasis with particular reference to India. In : *Snails and Flukes,* (Mimeographed) : 111-122.

Subba Rao, N.V. 1989. Handbook Freshwater Mollusca of India. Zoological Survey of India, xxiii+289 pp.

Subba Rao, N.V. and Dey, A. 1989. *Freshwater Molluscs in aquaculture.* In : Handbook Freshwater Molluscs of India, Zoological Survey of India, 225-232.

Subba Rao, N.V. and Mitra, S.C. 1989. Systematics and ecology of freshwater gastropod molluscs of parasitological importance. In : *Snails and Flukes* (Mimeographed) : 47-64.

13

IMPACT OF ASH EFFLUENTS ON THE DISTRIBUTION AND BIOMASS OF PERIPHYTON AROUND SARNI THERMAL POWER STATION

K.S. Unni
and
S.K. Khare
Deptt. of Botany,
Govt. P.G. College, Chhindwara, M.P.

INTRODUCTION

Sarni thermal power station situated in Betul District, Madhya Pradesh has nine generating units that produce 1150 mw of electricity. It burns 10,500 tons of coal per day and an estimated 3000-4000 tons of ash is discharged into its environment. Most of the bottom ash and a part of the fly ash from the electrostatic precipitators are sluiced through pipes into the ash bund. A canal carrying wastes (from the station) discharges into Tawa reservoir—that serves as the intake of water. The ash bund discharges, the bottom ash into the Tawa river downstream. We studied Tawa upstream, reservoir and Tawa downstream—(12 km downstream) at Banspur by establishing sampling sites for periphyton and water chemical parameters.

The fact that 90% of our electrical energy is generated by coal-fired stations emphasises the need for impact studies of the drainage systems that receives the bottom ash effluents. There have been very few or little detailed studies on the impact of ash effluents. both abroad and in India on the biota. The present detailed studies on the impact of ash effluents on periphyton distribution were undertaken since attached algae form an important group of indicator organisms that either tolerate or avoid pollution (Kamat 1981). The impact of thermal discharges from thermal power stations have been investigated by a number of workers in the field (Klarer and Hickman 1975, Moore 1978, Hickman 1974, Eloranta 1980 *a*, 1980 *b*, 1982, Simmons *et al.* 1974, Heim and Koopen 1979).

MATERIAL AND METHODS

Periphyton sampling from natural substratum only was possible. Following the methods suggested by Zafar two leaves of either *Potamogeton crispus* or *Vallisneria spiralis* was collected in small glass tubes along with surrounding water. These were thoroughly washed in water and made into 50 ml, preserved and later counted on 10 slides for quantitative estimation. The biomass was calculated from dimensions of the cell as suggested by Vollenweider (1974). The water samples were analysed for 20 physico-chemical parameters from all the above stations as per name of the method.

OBSERVATIONS

(*a*) Physico-chemical parameters

The mean (28.7±3.6°C) and range (22-36°C) of water temperature in the reservoir was greater than that of Tawa upstream and downstream. The upstream and downstream mean difference (Table-1) consisted of 2.5°C.. Although the water mass involved in circulation between the reservoir, upstream and hotwater discharge is much greater, it appears that mixing of hotwater does not take place and the thermal effluent heats up the reservoir water by 2.5°C. We did not observe the water temperature of the reservoir and downstream simultaneously and this aspect require further detailed studies.

The mean conductivity at the upstream station was greater than that of the reservoir, which may be attributed to washing of flyash and its deposition upstream. The catchment of the river is not populated and there was no addition of sewage or organic waste into the river. The turbidity was also greater at upstream. Phenolphthalein alkalinity increased in the reservoir whereas M.O. alkalinity recorded a reverse trend. Calcium and magnesium recorded only negligible difference between the upstream and reservoir. Magnesium however, increased 2 times at downstream. D.O. exceeded 9.0 mg l^{-1} at all stations. Nitrate nitrogen and total phosphates were greater at downstream in comparison to upstream.

(*b*) Distribution of periphyton

In all 25 species of periphyton were recorded out of which 14 species belonged to Bascillariophyceae. Chlorophyceae (Fig. 1) was represented by only 10 species and Cyanophyceae by a single species *Merismopedia*. Bascillariophyceae contributed 81 to 99% in density followed by 65-100% in the reservoir and 72 to 100% at Tawa downstream. From 2753 organisms cm^{-2} (Fig. 1) at Tawa downstream in February the maximum of 7600 cm^{-2} was recorded in May. The growth being poor in 1986 the total organisms did not exceed 1355/cm^{-2}. In the reservoir from 1289/cm^{-2} (February) the maximum was recorded in May (4928 cm^{-2}). At Tawa downstream 8325 and 7170 organisms cm^{-2} were recorded in November and January. The density of periphyton declined to 1200/cm^{-2} in summer.

Amongst Bascillariophyceae (Fig. 2 and 3) *Diatoma elongata*, *Cymbella prostrata*, *Navicula rhyncocephala* and *Synedra ulna* were regularly found during all sampling occassions. *Gomphonema olivaceum* and *Gomphoneis* sp. showed a reduction downstream while the greater density of *Cocconeis placeutula* and *Synedra ulna* indicated their tolerence towards ash pollution.

The biomass of periphyton (Fig. 4) ranged between 1.6 to 181.3 $\mu g\ cm^{-2}$ at different stations. The maximum was found in May 1985 and the minimum in April 1986. The contribution by *Navicula rhyncocephala* towards biomass was maximum and ranged between 4.2 to 84.1 $\mu g\ cm^{-2}$ in different months. In between the stations the biomass did not show any trend but the correlation (p=? r=?) between periphyton and some physico-chemical parameters were significant.

Table 1. Selected Physico-chemical factors and their fluctuation in the source water (Tawa upstream) reservoir and Tawa downstream mean and range (in brackets).

S. No.	*Factor*	*Tawa upstream*	*Reservoir*	*Tawa downstream*
1.	Depth Unit	1.0—3 m	28.97 m	1.0—3 m
2.	Water temperature °C	28.7±3.69 (22.4—34.8)	29.1±4.0 (22—36)	26.2±3.0 (17.0—32.2)
3.	Conductivity μ mhos cm^{-1}	208±67 (151—404)	191±39 (151—303)	405±92 (272—646)
4.	Turbidity NTU	28.5±21 (10—80)	24.1±14 (20—60)	87.2±48 (20—170)
5.	Sulphate mg l^{-1}	23.0±10 (6.4—51.2)	21.7±13 (6.4—51.2)	66.0±48 (9.6—185.3)
6.	Sodium ppm	14.1±7.0 (4—28)	12.5±6.8 (1—25)	29.4±4.4 (21—35)
7.	Potassium Unit	3.7±1.8 (1.6—8.3)	3.4±1.7 (1.6—7.2)	10.1±5.9 (3.2—16.0)

(Contd.)

8. pH	8.33±0.12 (7.7–8.5)	8.2±.2 (7.79—1)	8.62±.23 (7.9–9.6)
9. Total hardness ppm	55.0±5.13+ (40–60)	66.0±9.4 (32–76)	102±4.4 (90–140)
10. Nitrate nitrogen mg l^{-1}	2.2±2.2 (.07–8.8)	1.7±2.0 (.03–6.6)	1.89±1.7 (.01–5.0)
11. Total phosphate μg l^{-1}	84±53.5 (10–190)	162±181 (10–520)	118±86 (30–250)
12. DO Unit	9.8±2.6 (7.6–15.2)	9.2±2.4 (6.3–16.8)	10.0±2.4 (7.2–14.4)

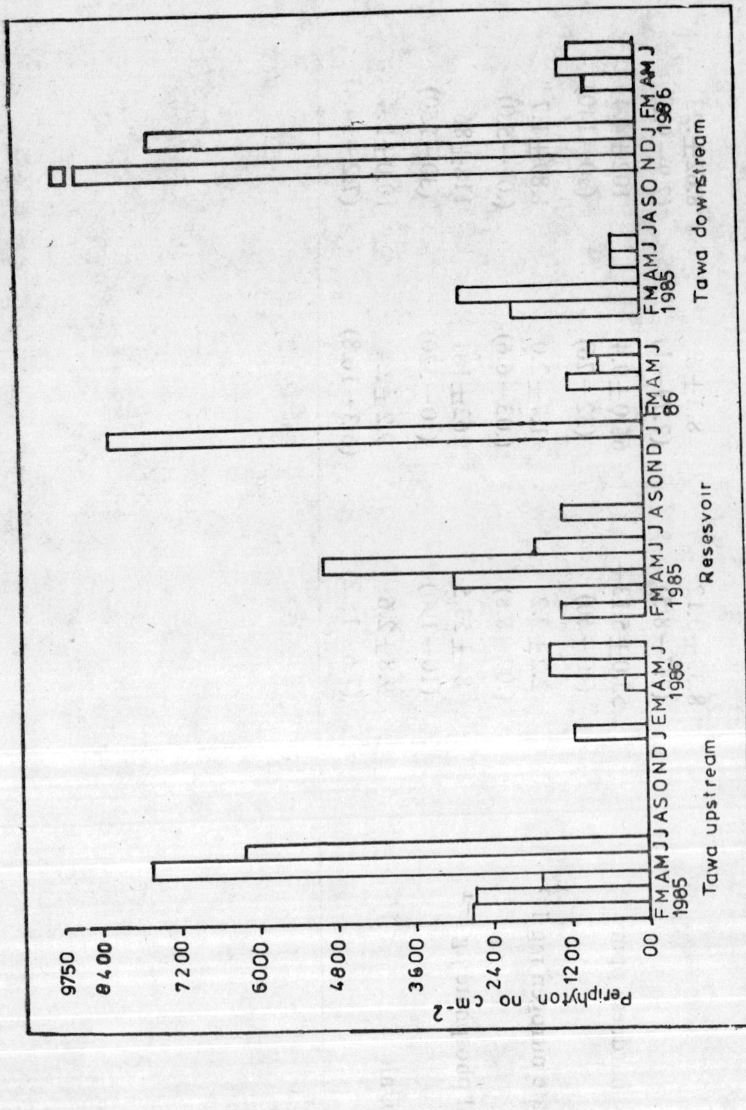

Fig. 1. Monthly distribution of Periphyton at three sampling stations round Sarni Thermal Power

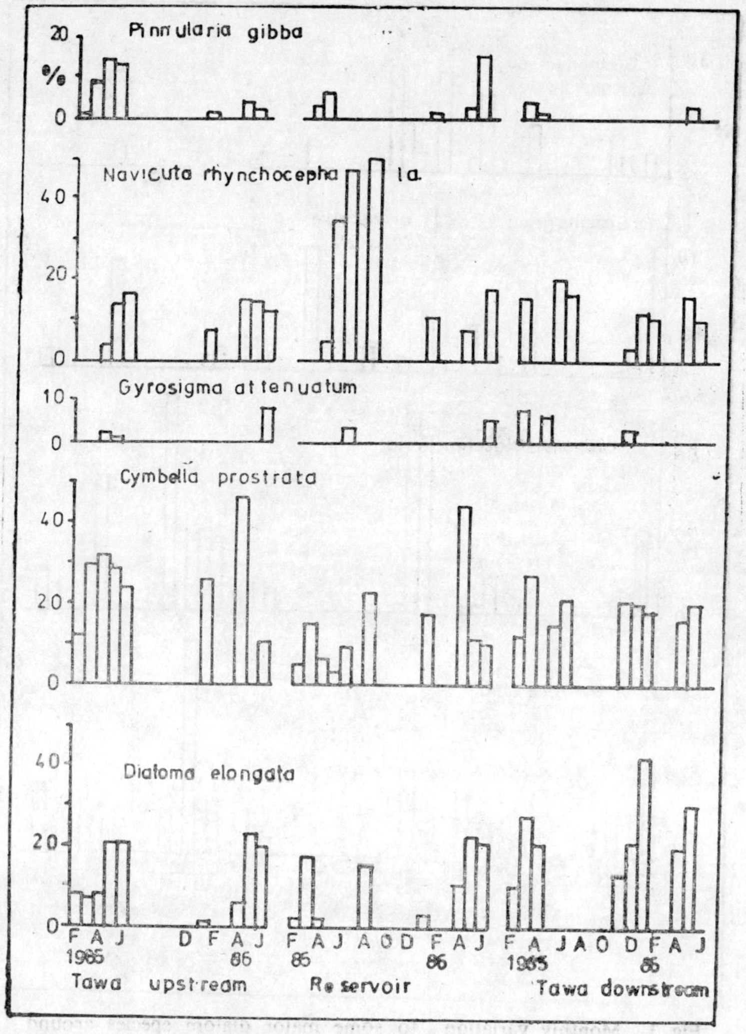

Fig. 2. Distribution of major diatoms and their monthly variation at different sampling stations.

The total number of periphyton and turbidity showed a significant correlation ($p=?$ $r=?$). However correlation between the turbidity and distribution of *Navicula rhyncocephala* was negative.

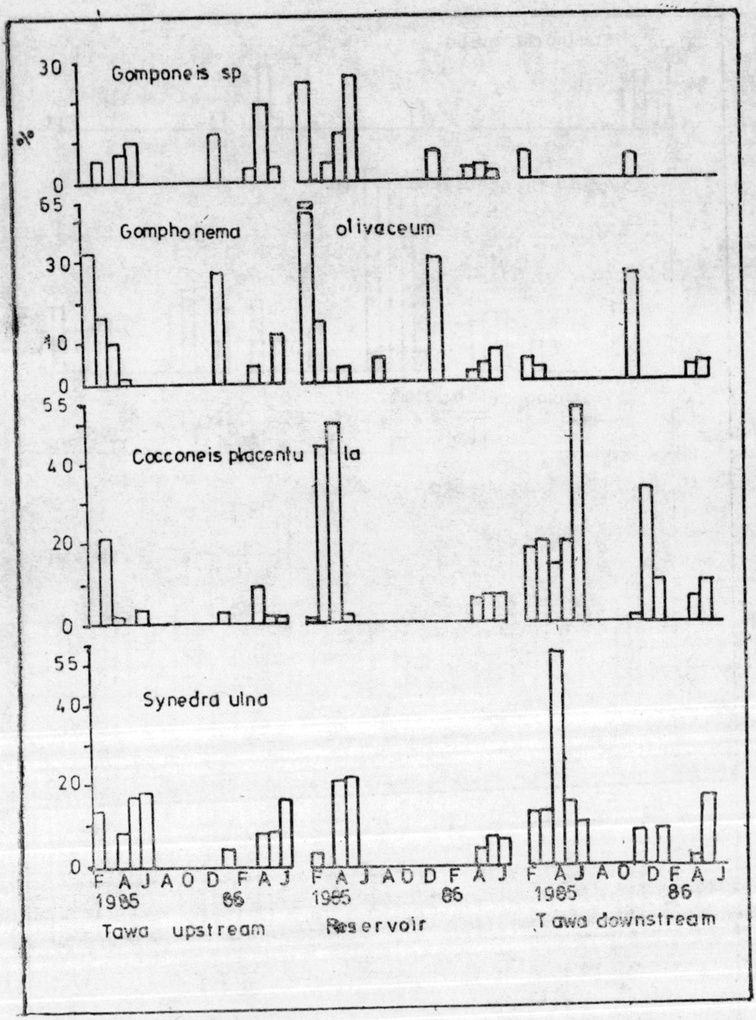

Fig. 3. Monthly variation to some major diatom species around Sarni Thermal Power stations.

DISCUSSION

During the 1985 sampling period the water level in the upstream and reservoir (during the summer) was lower than in 1986 and the water temperature in 1985 was lower, at both

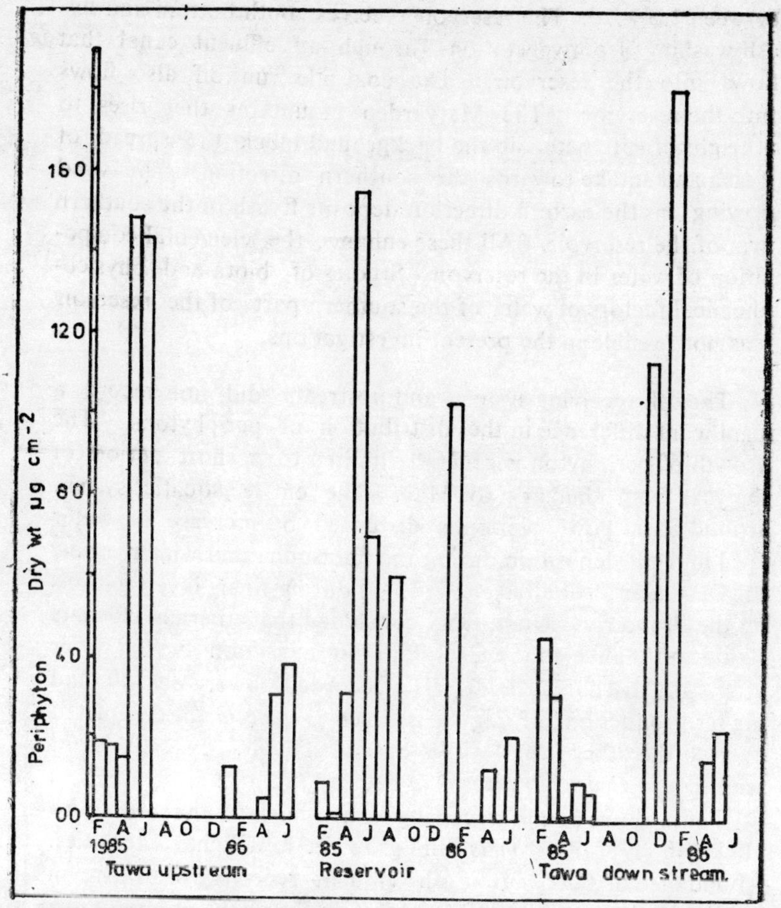

Fig. 4. Monthly variation of total biomass of periphytons around Sarni Thermal Power Stations.

upstream and downstream stations. The water level was manipulated by raising the height of the dam in 1986. But this did not decrease the temperature of water. Reservoir, Tawa downstream and upstream stations are stressed due to heated effluents which enhanced the reservoir water temperature (Table 1) and the ash effluents impacted the distribution of biota during the course of the river downstream. Elemental concentration at Tawa downstream increased by a factor ranging between 1.5—2.0. The reservoir receives both bottom and fly-ash washings from vegetation through an effluent canal that flows into the reservoir. The coal pile run off also flows into the reservior. The Matyardeo mountains that rises to a height of 950 meters in the background block the spread of flyash and smoke towards the southern direction. The wind fiowing in the eastern direction deposits flyash in the southern arm of the reservoir. All these enhance the elemental composition of water in the reservoir. Studies of biota and physico-chemical factors of water of the southern part of the reservoir was not feasible in the present investigations.

The ash receiving systems and upstream did not record a significant difference in the distribution of periphytons. The growth of periphyton is generally limited to a short period of the year from January to May. The entire aquatic system around Sarni power station is disturbed by increase in water level and silt deposition during the monsoon and winter rains, till December. Squires *et al.* (1979) during their investigations on the Provo river Utah, USA concluded that thermal effluents significantly affected the algal flora in a section of the river after the discharge point. He recorded lower density and higher production of algae. *Cocconeis placentula* dominated over all the other forms at the thermal effluent. In the present studies also *Cocconeis placentula* occurred in maximum numbers at Tawa downstream receiving ash effluents. According to Hickman (1982) *Cocconeis* can grow at low light intensities. Abundance of *Cocconeis* at ash effluent receiving downstream station having turbid waters agree with the above observation.

Philipose (1960) and Kamat (1965) reported that diatoms dominate in alkaline waters at 8.0 pH. According to Kamat (1981) diatoms grow best at 20-22°C and the number may vary between $10\text{-}10^5/1$. He observed that organic rich waters show association of *Cocconeis placentula*, *Eunotia*, *Gomphonema*, *Navicula radiosa*, *Pinnularia*, *Stauroneis* and *Tabellaria* sp. In the present observation association of diatoms of *Navicula rhyncocephala*, *Diatoma* and *Cymbella* indicate that the waters are poor in organic matter supporting the view of Kamat. The BOD (Mean) ranged between 2.4 ± 0.8 to 3.2 ± 1.6 at Tawa upstream and downstream, respectively indicating the organically poor waters.

Evanson *et al.* (1981) studied the effect of physico-chemical factors on the attached diatoms in the Unitah basin using slides, recorded 22 species at the high and low end of the physico-chemical gradient. During the present work only 18 species out of total 25 species were recorded at the ash affected station. A number of members of Chlorophyceae appears to be intolerent to ash effluent waters. These include 2 species *Pediastrum,* 2 species of *Scenedesmus Ankistrodesmus* sp. and *Closterium* sp. These members are present in the reservoir and the limiting factor appears to be the increased turbidity caused by the bottom ash.

ACKNOWLEDGEMENT

The senior author is grateful to the Deptt. of Environment, Govt. of India for financial assistance to Project No. 20.5.82 MAB/RE, and to Principal Govt. P.G. College, Chhindwara for facilities.

REFERENCES

Eloranta, P (1980 a). Winter phytoplankton in a pond warmed by a thermal power station *Ann. Bot. Fenn. 17* : 264-275.

Eloranta, P. (1980 b). Annual succession of phytoplankton in an ice free pond in central Finland. *Acta Hydrobiol. 22* : 421-438.

Eloranta, P. (1981). Yearly succession of phytoplankton in a ice free pond in Central Finland. *Schweiz* Z. *Hydrol.* 43/1 Birkhauser Verlaug Basel.

Eloranta, P. (1982). Periphyton growth and diatom community structure in a cooling water pond. *Hydrobiologia 96* : 253-265.

Evanson, W.E Rushforth, S.R. Brotherson, J.D. and Fungladda, N. (1981). The effects of selected physical and chemical factors on attached diatoms in the Unitah Basin of Utah, U.S.A. *Hydrobiologia 83* : 325-330.

He in, M.K and Koopen, J.D (1979). Effects of thermally elevated discharges on the structure and composition of esturine periphyton diatom assemblages *Esturine. Coastal. Marine.* Sci. 9, 385.

Hickman, M. (1974). Effects of the discharges of thermal effluent from a power station on lake Wabamun, Alberta Canada-the epipelic and epipsammic algal communities. *Hydrobiologia 45* : 199-216.

Hickman, M. (1982). The removal of a heated water discharge from a lake and the effect upon an epiphytic algal community. *Hydrobiologia 87* : 21-32.

Kamat, N.D. (1965). Ecological notes on the algae of Kolhapur *J. Biol. Sci. 8* : 47-54.

Kamat, N.D. (1981). Diatoms and diatom populations indicating water quality and pollution. WHO workshop on Biol. indicators and indices of *Environmental Pollution* Osm. Univ. Hyderabad India.

Klarer, D.K. and Hickman (1975). The effect of a thermal effluent upon the standing crop of an epiphytic algal community *Int. Rev. Ces. Hydrobiol 60* : 17-62.

Moore, L.F. (1978). Attached algae at thermal generating stations. The effect of temperature on *cladophora*. *Vesh. Intl. ver. Limnol* (Gur). *20* : 17-27.

Philipose M.T. (1960). Freshwater phytoplankton of inland fisheries. In : *Symposium on algology*, ed. Kachroo P. ICAR. New Delhi 272-291.

Simmons, G.M. Armitage, B.J. and White J.C. (1974). An ecological evaluation of heated water discharge on phytoplankton blooms in potomaci River. *Hydrobiologia 45* : 441-465.

Squares, L.E. Rushforth, R. and Brotherson J.D. (1979). Algal response to a thermal effluent study of power station on the Provo River Utah, *Hydrobiologia 63* : 17-32.

Unni, K.S. (1985). Comparative limnology of several reservoirs in Central India. *Int. Revue. gcs Hydrobiol. 70* : 6, 845-856.

Vollenweider, R,A (1974). A manual on methods for measuring primary production in aquatic environments. IBP. handbook no. 12. Blackwell Scientific Publication Oxford, London.

14

IMPACT OF DISTILLERY EFFLUENTS ON THE RIVER WAINGANGA PART—III

Anil Chauhan

SRF, ICAR Project, School of Studies in Zoology,
Vikram University, Ujjain (M.P.)

Abstract—The study has been carried out during January 1987 to December 1988 to assess the impact of distillery effluents on River Wainganga. The distillery discharged hot (48°C) effluents, contrary temperature was lowest at station D and showing some endothermal reactions. Turbidity most severely affected dissolved oxygen concentration ($p<0.001$; $r=-0.64$) and calcium ($p\leq 0.001$; $r=-0.57$) after the closure of distillery. Beside the oxygen demand and temperature, other inorganic reductance ammonia nitrogen, nitrite+nitrate nitrogen cause D.O. depletion in the system. BOD was always above the prescribed limit of ISI and reached upto 320 ppm in the Wainganga. The calcium content of Wainganga varied directly with the bicarbonate content ($p<0.001$). Nitrification in the river was significant at station W_3. BOD and COD values caused by plankton were higher than distillery waste after the closure of distillery.

INTRODUCTION

Industrial waste water constituted broad and varied quantities of gasesous, liquid and solid pollutants. Untreated or partially treated release of wastes create serious health hazards and gradually demolishing the ecological equilibrium, is assuming disastrous impact of pollution all through the world. Third world countries, like India more prominent to undergo overall degradation of water, atmosphere and soil pollution. In India, Central and 14 state Water Pollution Control Boards (1981) identified 1,700 large and medium water polluting industries, about half of them were in urban areas.

The quantity of outfall and chemical composition is keep confidential so that people around are not provoked but recent awaken of environmental awareness has created problems, and forced the industries to abide by standards. Existing concepts of standards, of outfall are not strictly followed even in big industrial concerns owned by public and private sectors. Hence collection of data is very essential in all reported cases.

Limited investigations has been carried out on the impact of distillery effluents on rivers water quality. These included David and Ray (1966), Ajmal *et al.* (1985) and Shankar *et al.* (1986). The studies of Chauhan (1989, 1990) reflacted broad and wide spectrum of characteristics, recovery of distillery and refinary wastes and its effect on River Wainganga. The present communication gives further information of impact of distillery effluents on the River Wainganga, during distillery operation period and after the closed of the distillery.

STUDY AREA

Wainganga River is a small fast flowing stream at Seoni that dwindles in summer. Wainganga originates at an altitudes of 450 metre above m.s. and flows down to 300 metre into the plains. After flowing more than 18 km., it is impounded at Lakhanwara to form a minor reservoir which serve water to Seoni town.

The present distillery was located in the outskirts of Seoni town discharging its untreated effluents into the tributary of the river. The tributary after covering a distance of about 10 km. confluences with the river just below the dam. The distillery was shifted else where in September 1987 therefore the whole communication categorised into Distillery Operation Period (January 1986 to September 1987) and Distillery Closed Period (October 1987 to December 1988).

Sampling Stations

Sampling stations were established during the course of are shown study in the Figure-1.

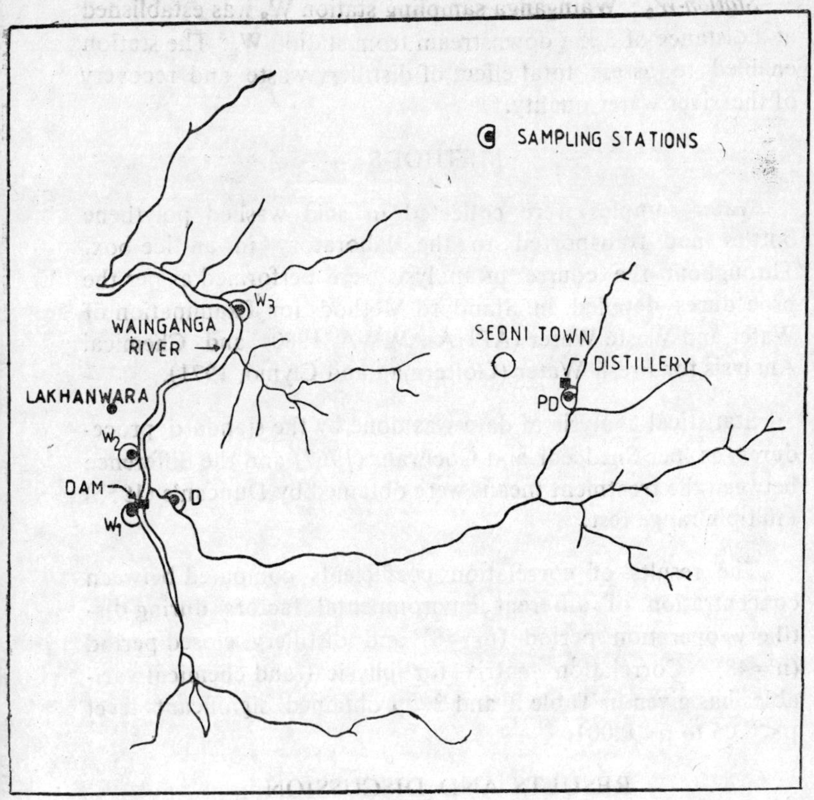

Fig. 1. Map of River Wainganga showing various sampling stations.

Station-W_1 : The first sampling station on Wainganga was kept at sluice gate of the dam, to understand the original river water quality before it receive any effluents.

Station-PD : Pure distillery effluents just after the discharged into the distributary of the river.

Station-D : Second sampling station on distillery waste having tributary, before its outfall to the river. This station was situated about 10 km. away from station PD.

Station-W_2 : Second station on Wainganga River, immediate after the discharged of distillery effluents. This station was situated at a distance of 0.5 km. from upstream station.

Station-W_3 : Wainganga sampling station W_3 was established at a distance of 3 km downstream from station W_2. The station enabled to assess total effect of distillery waste and recovery of the river water quality.

METHODS

Water samples were collected in acid washed polythene bottles and transported to the laboratory in an ice-box. Throughout the course of analysis were performed as per the procedures detailed in Standard Methods for Examination of Water and Waste Water (APHA-AWWA, 1980) and Chemical Analysis for Fresh Water (Golterman and Clymo, 1971).

Statistical analysis of data was done by the standard procedures as per Snedecor and Cochran (1967) and the difference between the treatment means were obtained by Duncan's (1955) multiple range test.

The results of correlation coefficients computed between concentration of different environmental factors during distillery operation period (n=48) and distillery closed period (n=48). Correlation matrix for physical and chemical variables has given in Table 2 and 3. I obtained significant level $p<0.05$ to $p<0.001$.

RESULTS AND DISCUSSION

Pure Distillery Effluents

The effluents was always highly acidic and contained considerable amount of other pollutants (Figure-2). The values of biochemical oxygen demand, chemical oxygen demand and suspended solids were recorded at 21,910 mgl^{-1}, 91,225 mgl^{-1} and 30,644 mgl^{-1} respectively. Likewise, the concentration of chloride, nitrogen, potassium and hardness were also high.

Physics

1. *Colour*

Daily discharged of crude residues from Seoni distillery carries huge quantities of suspended matter which gives dark brown colour to the effluents. Colour is frequently more

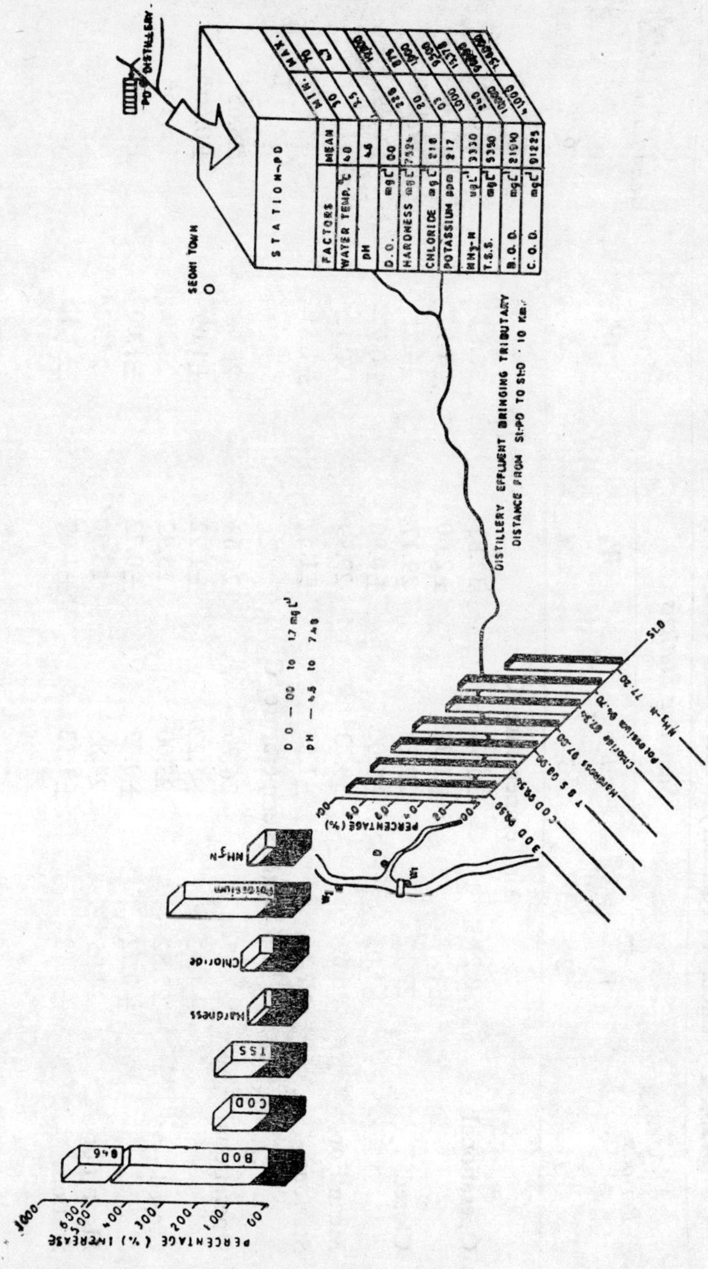

Fig. 2. Characteristics of pure distillery effluents, percentage reduction and its impact on River Wainganga.

Table 1. Effects of Distillery Waste Water on Different Physical and Chemical Factors.

Periods	Sampling Stations				Mean for Periods
	W_1	D	W_2	W_3	
1	*2*	*3*	*4*	*5*	*6*
		Air Temperature °C			
Operational	30.21 ±1.78	30.75 ±1.94	30.82 ±2.00	30.82 ±1.92	30.70^{A} ±0.96
Closed	29.17 ±1.03	29.71 ±1.00	29.17 ±0.90	29.21 ±0.97	29.31^{A} ±0.49
Mean For Stations	26.69^{A} ±1.46	30.23^{A} ±1.55	29.99^{A} ±1.57	30.13^{A} ±1.55	
		Water Temperature °C			
Operational	27.07 ±1.23	26.70 ±1.22	26.68 ±1.22	26.90 ±1.09	26.65^{A} ±0.60
Closed	22.82 ±0.72	23.00 ±0.72	23.92 ±0.72	23.29 ±1.00	23.26^{B} ±0.41
Mean For Stations	24.95^{A} ±1.18	24.85^{A} ±1.13	25.30^{A} ±1.09	24.73^{A} ±1.12	

(Contd.)

Table 1. (*Contd.*)

	Sampling Stations				Mean for periods
	W_1	D	W_2	W_3	
7	8	9	10	11	12
		Carbonate Alkalinity ppm.			
Operational	4.00	3.50	2.83	0.00	2.58^B
	±1.51	±2.06	±1.67	±0.00	±0.79
Closed	4.63	7.75	5.67	8.25	6.58^A
	±1.59	±2.30	±1.33	±1.89	±0.93
Mean for	4.32^A	5.63^A	4.25^A	4.13^A	
Stations	±1.56	±2.27	±1.57	±1.79	
		Bicarbonate Alkalinity ppm.			
Operational	120.83	202.08	139.92	179.17	150.04^A
	±19.89	±31.32	±17.33	±17.98	±12.04
Closed	133.75	197.42	138.08	138.92	150.54^A
	±8.10	±16.79	±9.22	±8.41	±6.84
Mean For	127.29^B	199.75^A	139.00^B	135.13^B	
Stations	±15.30	±25.14	±13.88	±14.05	

Table 1. (*Contd.*)

Periods	*Sampling Stations*				*Mean for Periods*
	W_1	D	W_2	W_3	
1	*2*	*3*	*4*	*5*	*6*
		Conductivity umhos cm^{-1}			
Operational	243.00 ±14.07	693.25 ±27.03	319.33 ±23.09	354.67 ±52.00	400.31^{A} ±29.76
Closed	284.42 ±30.99	457.00 ±26.27	329.83 ±38.69	292.50 ±24.90	340.94^{B} ±18.31
Mean For Stations	263.71^{B} ±24.80	573.13^{A} ±43.28	324.58^{B} ±31.90	319.08^{B} ±41.48	
		Turbidity NTU			
Operational	86.25 ±17.45	12.50 ±3.93	75.83 ±16.56	32.50 ±11.54	51.77^{A} ±8.04
Closed	25.83 ±7.12	23.75 ±13.18	24.17 ±8.70	20.83 ±8.49	23.65^{B} ±4.93
Mean For Stations	56.04^{A} ±15.93	18.13^{B} ±10.26	50.00^{A} ±15.18	26.67^{A} ±10.26	

(*Contd.*)

Table 1. (*Contd.*)

	Sampling Stations				*Mean for Periods*
	W_1	D	W_2	W_3	
7	*8*	*9*	*10*	*11*	*12*
			Total Alkalinity ppm.		
Operational	124.83 ±20.09	205.58 ±30.71	142.67 ±17.78	182.25 ±17.98	152.60[A] ±12.02
Closed	138.42 ±8.25	205.17 ±17.92	143.95 ±10.10	145.25 ±9.92	158.15[A] ±7.22
Mean For Stations	131.63[B] ±15.48	205.38[A] ±25.14	143.21[B] ±14.46	141.29[B] ±14.50	
			Calcium ppm.		
Operational	64.67 ±5.05	142.50 ±19.69	73.13 ±8.26	98.00 ±27.22	99.52[A] ±9.88
Closed	87.17 ±8.04	125.25 ±11.36	84.50 ±7.66	85.75 ±5.77	95.67[A] ±4.89
Mean For Stations	75.92[B] ±7.46	133.88[A] ±16.26	78.83[B] ±8.13	101.75[B] ±20.21	

Table 1. *(Contd.)*

Periods	*Sampling Stations* W_1	D	W_2	W_3	*Mean for Periods*
1	*2*	*3*	*4*	*5*	*6*
		Total Suspened Solids ppm.			
Operational	136.08 ±46.94	367.58 ±113.29	281.58 ±59.30	195.33 ±46.31	245.15^{A} ±38.18
Closed	76.83 ±8.90	167.33 ±30.19	87.67 ±9.42	136.75 ±13.10	117.15^{B} ±10.31
Mean For Stations	106.46^{B} ±34.85	267.46^{A} ±87.80	184.63^{AB} ±51.15	166.04^{AB} ±35.07	
		Dissolved Oxygen ppm.			
Operational	7.32 ±0.54	1.68 ±0.66	5.14 ±0.55	5.12 ±0.86	4.81^{B} ±0.44
Closed	8.49 ±0.74	8.92 ±1.17	10.58 ±1.11	9.82 ±0.77	9.45^{A} ±0.48
Mean For Stations	7.90^{A} ±0.67	5.30^{B} ±1.41	7.86^{A} ±1.13	7.47^{A} ±1.06	

(Contd.)

Table 1. (Contd.)

Periods	Sampling Stations				Mean for Periods
	W_1	D	W_2	W_3	
7	8	9	10	11	12
		Magnesium ppm.			
Operational	53.92 ±8.93	73.17 ±18.98	63.00 ±9.40	73.67 ±7.61	59.31^A ±6.22
Closed	40.75 ±5.94	81.50 ±7.65	51.83 ±6.94	55.75 ±4.73	57.46^A ±3.86
Mean For Stations	46.33^B ±7.82	77.33^A ±14.52	57.42^{AB} ±8.42	51.46^B ±6.46	
		Total Hardness ppm.			
Operational	118.58 ±7.29	215.83 ±27.32	144.50 ±10.06	180.00 ±26.55	160.90^A ±11.26
Closed	127.92 ±6.51	206.75 ±11.95	136.33 ±8.22	134.83 ±10.29	151.46^A ±6.62
Mean For Stations	123.25^B ±7.04	211.29^A ±21.12	140.42^B ±9.26	149.75^B ±20.59	

(Contd.)

Table 1. *(Contd.)*

Periods	*Sampling Stations* W_1	D	W_2	W_3	*Mean for Periods*
1	*2*	*3*	*4*	*5*	*6*
Bio-chemical Oxygen Demand ppm.					
Operational	2.63 ±0.44	91.08 ±24.41	44.00 ±11.24	26.00 ±4.97	40.93^A ±8.29
Closed	2.81 ±0.41	8.93 ±1.93	4.33 ±0.74	5.03 ±0.10	5.28^B ±0.68
Mean For Stations	2.72^C ±0.42	50.01^A ±20.99	24.16^B ±9.81	15.52^{BC} ±4.70	
Chemical Oxygen Demand ppm.					
Operational	30.17 ±3.40	422.58 ±94.33	102.83 ±13.48	89.41 ±113.94	161.25^A ±32.72
Closed	23.67 ±4.25	50.50 ±6.55	40.00 ±7.25	39.50 ±5.33	30.42^B ±3.28
Mean For Stations	26.92^B ±3.97	236.54^A ±85.76	71.42^B ±41.12	64.46^B ±12.78	

(Contd.)

Table 1. (*Contd.*)

Periods	*Sampling Stations*				*Mean for Periods*
	W_1	D	W_2	W_3	
7	*8*	*9*	*10*	*11*	*12*
		Sodium ppm.			
Operational	5.08	12.08	7.00	11.00	7.94^{A}
	±1.73	±2.39	±2.07	±2.27	±1.13
Closed	2.50	6.29	4.21	1.43	3.63^{B}
	±1.51	±1.67	±1.64	±0.96	±0.78
Mean For Stations	3.79^{B}	9.19^{A}	5.60^{AB}	4.52^{B}	
	±1.66	±2.23	±1.91	±1.96	
		Potassium ppm.			
Operational	0.33	10.97	4.36	8.77	$\pm 5.46^{A}$
	±0.11	±3.58	±1.26	±1.35	±1.15
Closed	0.00	4.08	0.63	0.35	1.27^{B}
	±0.00	±1.87	±0.30	±0.18	±0.53
Mean For Stations	0.17^{B}	7.53^{A}	2.50^{B}	3.26^{B}	
	±0.09	±3.02	±1.06	±1.28	

(*Contd.*)

Table 1. (*Contd.*)

Periods	*Sampling Stations*				*Mean for Periods*
	W_1	D	W_2	W_3	
1	*2*	*3*	*4*	*5*	*6*
		pH			
Operational	8.09 ±.10	7.47 ±0.16	7.85 ±0.12	7.61 ±0.11	7.76^A ±0.07
Closed	7.67 ±0.07	7.57 ±0.08	7.67 ±0.05	7.79 ±0.10	7.68^A ±0.04
Mean For Stations	7.88^A ±0.11	7.52^B ±0.13	7.76^A ±0.09	7.71^A ±0.11	
		Free CO_2 ppm.			
Operational	4.75 ±2.82	8.08 ±2.06	5.17 ±2.58	7.42 ±3.10	6.26^A ±1.35
Closed	1.60 ±0.79	1.82 ±1.26	1.14 ±0.65	0.68 ±0.48	1.31^B ±0.43
Mean For Stations	3.18^A ±2.12	4.95^A ±1.93	3.15^A ±1.97	3.86^A ±2.40	

(*Contd.*)

Table 1. *(Contd.)*

Periods	*Sampling Stations*				*Mean for Periods*
	W_1	*D*	W_2	W_3	
7	*8*	*9*	*10*	*11*	*12*
		PO_4—P ppm.			
Operational	0.106 ±0.029	0.103 ±0.017	0.065 ±0.007	0.080 ±0.003	0.077[A] ±0.009
Closed	0.050 ±0.009	0.032 ±0.009	0.040 ±0.010	0.021 ±0.011	0.036[B] ±0.005
Mean For Stations	0.078[A] ±0.023	0.067[A] ±0.017	0.052[AB] ±0.009	0.028[B] ±0.008	
		NH_3—N ppm.			
Operational	0.071 ±0.011	0.990 ±0.168	0.615 ±0.095	0.462 ±0.064	0.534[A] ±0.069
Closed	0.039 ±0.009	0.165 ±0.075	0.083 ±0.023	0.037 0.009	0.081[B] ±0.021
Mean For Stations	0.055[C] ±0.001	0.578[A] ±0.176	0.349[B] ±0.103	0.249[B] ±0.076	

(Contd.)

Table 1. *(Contd.)*

Periods	*Sampling Stations*				*Mean for Periods*
	W_1	*D*	W_2	W_3	
1	*2*	*3*	*4*	*5*	*6*
		Chloride ppm.			
Operational	10.72 ±0.75	24.28 ±1.95	14.00 ±1.55	22.58 ±1.19	16.23^{A} ±1.02
Closed	10.67 ±1.75	19.75 ±2.04	12.67 ±1.85	14.83 ±2.38	14.48^{A} ±1.12
Mean For Stations	10.70^{A} ±1.34	22.02^{A} ±2.10	13.33^{BC} ±1.72	15.38^{B} ±1.89	

Means bearing a common superscript columnwise for periods and rowwise for stations are not ($p<0.05$) significantly different.

Periods	*Sampling Stations*				*Mean for Periods*
	W_1	D	W_2	W_3	
7	8	9	10	11	12
	NO_3 and NO_2-N ppm.				
Operational	0.044	0.540	0.362	0.459	0.317^A
	±0.008	±0.056	±0.057	±0.053	±0.035
Closed	0.020	0.113	0.101	0.076	0.077^B
	±0.004	±0.043	±0.028	±0.024	±0.015
Mean For	0.032^C	0.327^A	0.232^B	0.198^{BC}	
Stations	±0.007	±0.080	±0.058	±0.054	

Means bearing a common superscript columwise for periods and rowwise for stations are not ($p<0.05$) significantly different.

Table 2. Correlation Matrix of Distillery Operation Period.

1	*Air Temp.* 2	*Wat. Temp.* 3	*pH* 4	*Con.* 5	*DO* 6	*Free CO_2* 7	*Carb alk* 8	*Bic. alk.* 9	*Tot. alk.* 10	*Ca* 11	*Mg* 12	*Tot. har.* 13	*Chl.* 14
Air temp.	1	0.10	0.03	0.03	0.02	0.03	0.16	—.10	—.09	—.05	0.03	—.01	—.04
Wat. temp.		1	0.01	0.09	—.09	—.08	0.35*	—.03	—.02	—.09	0.11	0.00	—.00
pH			1	—.01	0.11	0.03	0.07	—.00	0.00	—.05	0.08	0.00	—.87‡
Contd.				1	—.41@	0.02	0.11	0.17	0.18	0.31	0.17	0.28	—.09
D.O.					1	—.10	—.14	—.22	—.22	—.24	—.13	—.23	—.02
Free CO_2						1	—.15	0.30	0.28	0.10	0.06	0.08	—.08
Carb alk.							1	0.14	0.15	—.12	0.04	0.00	—.00
Bic. alk.								1	0.47@	0.10	0.16	0.14	—.10
Tot. alk.									1	0.09	0.15	0.14	—.10
Ca										1	—.04	0.35*	—.08
Mg											1	0.21	—.07
Tot. har.												1	—.11
Chl.													1
PO_4													
SO_4													
Tur.													
Na													

	PO_4	SO_4	Tur.	Na	K	Amn. nit.	Nit nit.	TS	TDS	TSS	BOD	COD
	15	16	17	18	19	20	21	22	23	24	25	26
Air. temp.	—.03	0.18	—.01	—.00	—.16	0.08	0.02	0.16	0.16	0.10	—.12	—.21
Wat. temp.	—.02	0.17	0.06	—.13	—.07	0.09	—.00	0.11	0.15	0.03	0.11	—.07
pH	—.19	—.14	—.24	0.08	—.07	—.01	0.04	00.2	—.01	0.05	0.00	—.06
Contd.	0.05	—.02	—.39*	0.23	0.37*	0.37*	0.38‡	0.13	0.05	0.18	0.50*	0.50*
D.O.	—.07	—.03	0.24	—.16	—.40*	—.43@	—.36*	—.18	—.07	—.27	—.44@	—.47@
Free CO_2	—.03	—.05	—.20	0.29	0.06	0.22	0.22	0.02	0.11	—.08	0.06	0.13
Carb alk.	0.02	0.04	0.09	—.16	—0.6	—.04	0.03	—.14	—.12	—.12	0.02	—.00
Bic. alk.	—.00	—.10	—.20	0.11	0.06	0.13	0.30	—.11	—.09	—.08	0.33*	0.33*
Tot. alk.	0.00	—.10	—.19	0.01	0.05	0.13	0.30	—.11	—.10	—.07	0.34*	0.33*
Ca	0.02	—.08	—.33*	0.34*	0.35*	0.15	0.19	0.05	0.01	0.09	0.13	0.26
Mg	—.04	0.03	—.11	0.02	—.04	0.11	0.18	0.00	0.01	—.00	0.50@	0.32*
Tot. har.	—.00	—.05	—.34*	0.30	0.28	0.18	0.27	0.03	0.00	0.06	0.38*	0.40*
Chl.	0.17	0.18	0.26	—.07	—.05	—.07	—.08	—.02	—.00	—.03	—.05	—.02
PO_4	1	0.00	0.20	0.04	—.03	0.09	0.00	0.05	0.01	0.08	—.01	0.02
SO_4		1	0.33*	—.30	—.06	—.07	—.06	—.05	0.01	—.12	—.08	—.08
Tur.			1	—.17	—.18	—.21	—.26	—.20	—.17	—.17	—.19	—.24
Na				1	0.19	0.16	0.13	0.36*	0.31	0.31	0.52@	0.17

(Contd.)

Table 2. (Contd.)

1	*Air Temp.* 2	*Wat. Temp.* 3	*pH* 4	*Con.* 5	*DO* 6	*Free CO_2* 7	*Carb alk.* 8	*Bic. alk.* 9	*Tot. alk.* 10	*Ca* 11	*Mg* 12	*Tot. har.* 13	*Chl.* 14
K													
Amn. nit.													
Nit. nit.													
TS													
TDS													
TSS													
BOD													
COD													

$n=48$
*=p 0.05
@=p 0.01
‡=p 0.01

	PO_4	SO_4	Tur.	Na	K	Amn. nit.	Nit. nit.	TS	TDS	TSS	BOD	COD
	15	16	17	18	19	20	21	22	23	24	25	26
K					1	0.17	0.19	0.11	0.12	0.08	0.25	0.55‡
Amn. nit.						1	0.42@	0.39*	0.27	0.41*	0.37*	0.28
Nit. nit.							1	0.07	0.06	0.07	0.47@	0.41@
TS								1	0.68‡	0.64‡	0.15	0.19
TDS									1	0.20	—.01	0.06
TSS										1	0.27	0.26
BOD											1	0.64‡
COD												1

$n=48$
*$=p$ 0.05
@$=p$ 0.01
‡$=p$ 0.001

Table 3. Correlation Matrix of Distillery Closed Period.

1	*Air Temp.* 2	*Wat. Temp.* 3	*pH* 4	*Con.* 5	*Do* 6	*Free CO_2* 7	*Carb alk* 8	*Bic. alk.* 9	*Tot. alk.* 10	*Ca* 11	*Mg* 12	*Tot. har.* 13	*Chl.* 14
Air. temp.	1	0.55‡	0.06	0.27	0.05	—.09	0.41@	0.24	0.29	0.22	—.07	0.07	—0.2
Wat. temp.		1	—.04	0.1	—.13	—.01	0.23	—.03	0.02	—.29	—.07	—.32*	—.32*
pH			1	0.3*	0.41@	—.00	—.14	—.16	—.15	0.12	0.07	0.11	0.22
Contd.				1	0.23	0.18	—.06	0.33*	0.31*	0.54@	0.22	0.53*	0.54*
D.O.					1	0.1	—.16	—.13	—.14	0.38	—.17	—'17	0.25
Free CO_2						1	—.45@	0.08	0.01	—.08	—.01	—.06	0.25
Carb alk.							1	0.33*	0.45@	0.08	0.06	0.06	—.23
Bic. alk.								1	0.98‡	0.45@	0.32	0.56@	0.09
Tot. alk.									1	0.45@	0.31*	0.55‡	0.09
Ca										1	—.00	0.75‡	0.32*
Mg											1	0.6‡	0.44@
Tot. har.												1	0.51@
Chl.													1
PO_4													
SO_4													

(*Contd.*)

Table 3. *(Contd.)*

	PO_4	SO_4	Tur.	Na	K	Amn. nit.	Nit. nit.	TS	TDS	TSS	BOD	COD
	15	16	17	18	19	20	21	22	23	24	25	26
Air. temp.	—.43@	0.05	—.13	—.04	—.06	—.02	—.01	—.06	—0.1	0.08	—.08	—.41@
Wat. temp.	—.13*	—.02	0.32*	—.05	0.00	—.07	—.15	0.2	0.24	0.10	—.29	—.61‡
pH	0.00	—0.1	—.25	0.02	.00	—.15	—.07	—.00	—.00	0.00	—.15	—.08
Contd.	—.17	0.05	—.35*	0.3*	0.37*	—.01	0.03	0.21	0.03	0.47@	0.42@	0.16
D.O.	—0.1	0.39*	—.64‡	0.35*	—.12	—.01	—.14	—.02	0.03	—.11	0.02	0.03
Free CO_2	0.12	—.03	—0.7	0.14	0.02	—.09	0.36*	0.29	0.22	0.27	0.3	0.05
Carb alk.	—.29	0.01	0.06	—.13	—.01	0.34*	0.17	—.11	—.12	—.00	—.28	—.36*
Bic. alk.	—.31	—.19	—.05	0.03	0.06	0.53@	0.41@	0.02	—.01	0.11	0.29	0.05
Tot. alk.	—.35	—.17	—.05	0.00	0.06	0.53@	0.41@	0.00	—.04	0.12	0.23	0.00
Ca	—.41*	0.12	—.57‡	0.26	0.15	0.12	0.1	—.04	—.15	0.22	0.38*	0.27
Mg	0.14	0.02	0.31	0.23	0.35‡	0.3*	0.15	0.14	0.09	0.15	0.25	0.28
Tot. har.	—.18	0.12	—.22	0.35	0.33*	0.28	0.19	0.08	—.03	0.28	0.45@	0.4@
Chl.	0.2	0.16	—.22	0.35*	0.33*	0.28	0.19	0.08	—.03	.28	0.45@	0.4@
PO_4	1	—.01	0.31*	0.21	—.04	0.02	—.13	0.00	0.09	—.19	—.14	0·07
SO_4		1	—.21	0.31*	0.05	.01	0.04	0.13	0.13	0.01	0.04	0.06

(Contd.)

Table 3. (*Contd.*)

	Air Temp.	Wat. Temp.	pH	Con.	DO	Free CO_2	Carb alk.	Bic alk.	Tot. alk	Ca	Mg	Tot. har.	Chl.
Tur.													
Na													
K													
Amn. nit.													
Nit. nit.													
TS													
TDS													
TSS													
BOD													
COD													

n=48
*=p 0.05
@=p 0.01
‡=p 0.001

Table 3. *(Contd.)*

	PO_4	SO_4	*Tur.*	*Na*	*K*	*Amn. nit.*	*Nit. nit.*	*TS*	*TDS*	*TSS*	*BOD*	*COD*
Tur.			1	—.16	—.15	—.00	—.09	0.04	0.14	—.17	—.27	—.35*
Na				1	0.14	0.31*	0.11	0.02	0.03	—.00	0.15	0.29
K					1	0.25	0.19	0.27	0.05	0.6‡	0.25	0.2
Amn. nit.						1	0.36*	—.02	0.04	—.13	—.02	0.1
Nit. nit.							1	0.07	—.04	0.31*	0.22	0.06
TS								1	0.91‡	0.5@	0.31*	—.01
TDS									1	0.13	0.17	—.01
TSS										1	0.35*	0.15
BOD											1	0·57‡
COD												1

$n = 48$
$* = p$ 0.05
$@ = p$ 0.01
$‡ = p$ 0.001

objectionable to the public because it is so obvious and have easthatic impact (Chauhan, 1990). Addition of this distillery effluents given black brown appearance to the Wainganga making it undesirable for various routine purposes.

2. *Odour*

Odour nuisance is a problem for Seoni people because of microbile decomposition of organic matter, accumulated along the side of the stream having distillery effluents. This stream encircled Seoni town 4 to 6 km. before confluencing to the river (Figure-1). Settling tank near the distillery were low quality and creating direct hazards for people, whose houses are located along the side of polluted stream and around the distillery. People further handicapped by seasonal discomfort from noxious odour.

3. *Temperature*

The hot temperature (48°C) of distillery effluents enabled to cool down after flowing 10 km. from station PD to station D. The lowest water temperature at station D (Table-1) revealed that the distillery effluents may have some endothermal reactions. The mixing of the effluents to the Wainganga was lowering water temperature of the river 0.39°C at station W_2 then station W_1. It shows an increase towards recovery station W_3 but not maintained even after 3 km.

4. *Turbidity*

The turbidity values in the Wainganga were higher during flood and disturbance, in comparison to upstream station. Distillery effluents encountered lower values of turbidity favoured the dilution at river downstream. Turbidity most severely affected the dissolved oxygen concentration ($p<0.001$; $r=0.64$) and calcium ($p<0.001$; $r=0.57$) after close of distillery.

5. *Suspended Solids*

The distillery effluents mainly contributed suspended solids to the Wainganga (Chauhan, 1989), and it increased more than 200%. The mean total suspended solids reduced to about

30% at recovery station W_3 due to natural purification of the river as per reported by Sastry *et al.* (1972) in River Khan. Relationship of TSS with other factors shown in Table 2 and 3, clearly indicating that TSS were one of the important elimination factor of trouts in Wainganga.

Chemistry

1. *Dissolved Oxygen*

Dissolved oxygen changed inversely proportional to temperature ($p<0.05$) and BOD ($p<0.05$) (Chauhan, 1991). Beside the oxygen demand and temperature in the system other inorganic reductance ammonia nitrogen, nitrate+nitrite nitrogen cause dissolve oxygen depletion in Wainganga showing resemblance to the study of Venkateswarlu (1969), Ramanathan and Periswamy (1987). The relationship evident in Figure-3, that dissolved oxygen tended to rise and fall inversely with NH_3—N, NO_2 and NO_3—N and BOD in Wainganga River.

2. *BOD and COD*

High BOD and COD in the River Wainganga was impact of distillery spent wash containing high percentage of organic matter and other inorganic constituents. The tolerence limit for discharge of industrial effluents have been prescribed by ISI, 2490 : 1974, that BOD should not exceed 30 mgl^{-1} into inland waters. But the present values of BOD at stations D, W_2 and W_3 crossed several the limit. The maximum values recorded were 320 ppm in the Wainganga.

3. *pH*

The pH varied between 3.5—8.97 at all sampling stations. Highly acidic distillery effluents reached at the Wainganga appears to have neutralized the pH. Still, the values of pH at station W_2 were lower then W_1, indicating prolonged pollutional condition, the pH at station D was slight acidic to alkaline.

4. *Free CO_2*

Mishra *et at* (1990) pointed out that high free CO_2 concentration associated with low oxygen content indicates greater

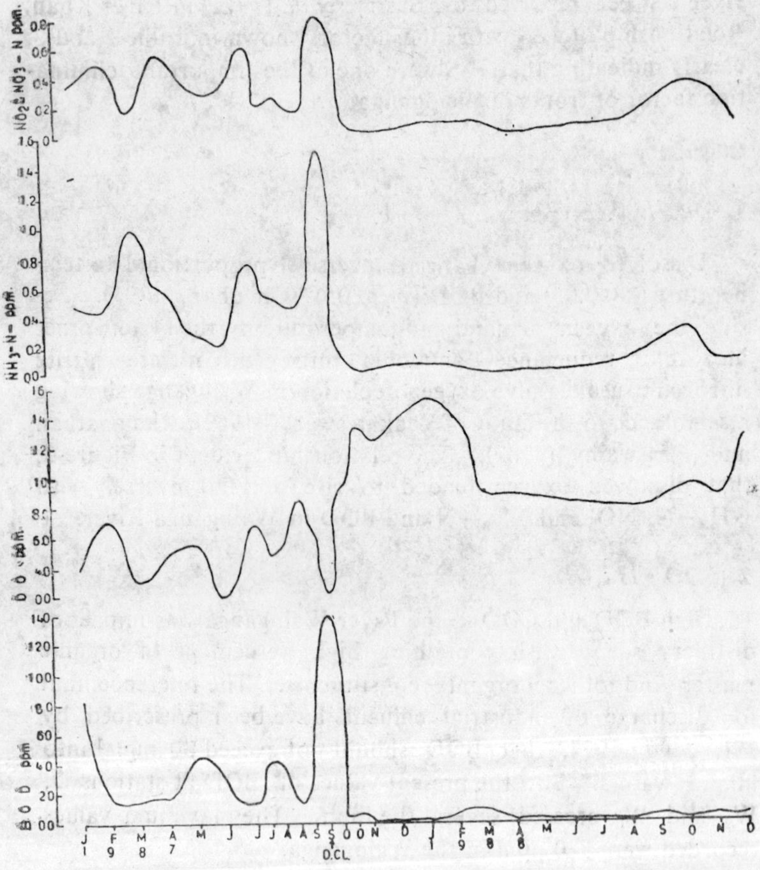

Fig. 3. **Relationship of nitrate and nitrite nitrogen, ammonia nitrogen, DO and BOD.**

pollution, the values of free CO_2 should not exceed 10 mgl^{-1} in fresh water bodies. In the Wainganga free CO_2 values ranged from 00 to 42 ppm. The station W_3 represented by free CO_2 all through distillery operation period from January 1987 to October 1987. The peak of free CO_2 in summer revealed rapid decomposition of organic matter through the activities of micro-organisms. Free CO_2 in Wainganga changed inversely proportional to carbonate alkalinity ($p<0.01$, $r=-0.45$); and it was subsequent by carbonate alkalinity from February 1988 onwards.

5. *Alkalinity*

The alkalinity present in Wainganga River was mainly due to bicarbonates and little amount of carbonates (Chauhan, 1989). Bicarbonate alkalinity varied between 60 to 240 ppm. at downstream station W_2. Saxena *et al* (1966) reported that the alkalinity and hardness correspondingly increased in Ganga with pH on contrary pH declined in Wainganga with increase in alkalinity and hardness.

6. *Hardness*

Calcium content of Wainganga was varied directly with bicarbonate content ($p<0.001$) possibly due to the conversion of the insoluble carbonate into soluble calcium bicarbonate in the presence of carbon dioxide ($p<0.001$, $r=0.89$). This is in keeping with the results reported by Pearsall (1930), Zafar (1964), Prasad and Saxena (1980). It was sighted that the bicarbonates inversely correlated with ammonia nitrogen ($p<$ 0.001, $r=-0.53$) during the present study.

7. *Potassium and Sodium*

Potassium and sodium were much greater in distillery effluents at station D, but mean values showed decline at station W_2. Both these elements increased (Table-1) towards recovery station. Potassium in the Wainganga shows strong coefficient of correlation to suspended solids ($p<0.001$; $r-0.6$).

8. *Chloride*

In the present data it is obvious that the increase in chloride content was accompanied by an increase in ammonia and organic matter. Chloride at station W_3 apparent to increase with conductivity ($p<0.001$; $r=0.54$) and potassium ($p<$ 0.001 ; $r=0.82$) due to reduction in flow. Chloride inversely related to DO contents ($p<0.001$; $r-0.87$) at this station.

9. *Phosphate*

Phosphate level in Wainganga varied between 00—4.0 ppm. at upstream but did not show any undesirable growth of planktons as reported by Nat. Tech. Adv. Com. (1968). The high phosphate content in the Wainganga might be derived from agricultural fields because of distillery effluents showed lower values. The present data numarically not supporting the findings of Beeton (1969), however, correlation coefficient showed positive relationship ($p<0.01$) of phosphate with sodium ($r=0.73$), potassium ($r=0.72$) and chloride ($r=0.79$).

10. *Nitrogen*

Distillery effluents added higher amount of nitrogen in to Wainganga (Chauhan, 1989, 1990). In the Wainganga nitrification is considered significant ($p<0.01$) only at recovery station W_3, even though high concentration of ammonia existed at station W_2. At station W_3, nitrification is obvious because of the increase in nitrate+nitrite nitrogen concentration. Depletion oxygen level ($p<0.001$) associated with nitrification at this station. Nitrate and nitrite nitrogen and dissolved oxygen gone more or less hand in hand and inversely proportionate to the ammonia content of the aquatic environment from January to September 1987 (Figure-3). But during closure of the distillery NO_2 and NO_3—N tend to rise and fall together with ammonia. The present conditions at Wainganga are in concordant with James and Evison (1979) in that nutrient increase in water will not always result in an increase standing crops, other factors also limit the production.

Distillery closed

Distillery closed period further classified in three months, seven months and twelve months to assess gradual recovery in physical and chemical factors. The process of self-purification involves break down of organic substances through the activities of micro-organisms (Mercer, 1967). The BOD5 loading of the river remain somewhat high for about 0.5 km., but despite this DO frequently reached a point of super-saturation during distillery operation period. But after the close of distillery high BOD and COD values in January and February, 1988 caused by plankton rather than by distillery waste. Elements concentration fluctuated during distillery closure, and were higher till the last observation indicating that the ionic activities remained ineffective by internal changes of the system.

CONCLUSION

Continuous distillery waste release shall render the atmosphere unsuitable for the people of Seoni. Already decline in aquatic environmental quality has been evident in the present study. After the closure of the distillery the quality of the Wainganga water did improve but not to wholesome as desired. Indicators of various pollution were the net result of undesirable system of controls and regulated releases.

ACKNOWLEDGEMENT

The author is thankful to Dr. K.S. Unni for their valuable guidance, and to the Department of Environment, Government of India for providing financial assistance.

REFERENCES

APHA-AWWA (1980) : *Standards methods for examination of water and waste water.* IS ed. ; Amer. Publ. Hlth. Assoc. Inc. New York.

Ajamal, M., Razi Uddin and Khan Ashan Ullah (1985) : Physical-chemical aspects of pollution in Kali Nadi *IAWPC. Tech. Annual XII*, pp. 106-114.

Beeton, A.M. (1969) : Changes in the environment and biota of the great Lakes. *Nat. Acad. Sci.* Washington, D.C. pp. 150-187.

Chauhan, A. (1989) : Impact of distillery effluents on River Wainganga and self-purification. M.S. accs. by *Env. Hlth.* NEER I, Nagpur.

Chauhan, A. (1990) : Distillery and refinary wastes : Characteristics, recovery and its impact on aquatic environment with reference to periphyton. *IAEM Annual* NEER I, Nagpur.

Chauhan, A. (1991) : Impact of distillery effluents on river Wainganga at Seoni, M.P. Thesis Approved for the Ph.D. degree of Dr. Harisingh Gour Vishwavidyalaya Sagar.

David, A. and Ray, D. (1966) : Studies on the pollution of the River Daha (N. Bihar) by sugar and distillery wastes *Env H'th*, Vol. III, pp. 6-35.

Duncan, D.B. (1955) : Multiple range and multiple 'F' test. *Biometrics* 11, 1-42.

Golterman, H L. and Clymo, R.S. (1971) : *Methods for chemical analysis for fresh waters.* IBP handbook No 8 Blackwell Sci Publ. Oxford and Edinbugh.

James, A. and Evison, L. (1979) : *Biological indicators of water quality.* Ed. John Willey and Sons Ltd.

Mishra, P.C., Dash, M.C. and Kar, G.K. (1990) : Pollution studies in River IB—Physico-chemical characteristics. *River Pollution in India*. ed. by R.K. Trivedy. pp. 32-39.

Mercer, D (1967) : The effects abstractions and discharges on river water quality. In river management (Ed. ISAAC, P.C G.), Maclaren and Sons Ltd. London, pp. 153.

Natation Technical Advisory Committee (1968) : Water quality criteria. Report to the secretary of the indicator. Fed. Water Poll. Controll Administration (US Env. Prot. Agency) Washington D.C. pp. 234.

Prasad, B N. and Saxena, M. (1980) : Ecological study of blue green algae in River Gomati. *Ind. J. Env. Hlth Vol. 22 No. 22*, pp. 151-168.

Pearsall, W.H. (ed) (1930) : Biological servey of the River Whafe. *J. Ecol.* 17. pp. 278-305.

Ramanathan, P.S. and Periswamy, R.M.S. (1987) : An overview of DO in water Part-1. *J. Ind. Poll. Unt*. 3(1), pp. 19-25

Shankar, V., Sangu, R.P.S. and Joshi, G.C. (1986) : Impact of distillery effluents on the water quality and ecosystems of River Reh. in Doon Valley. *Poll. Res. Vol. 5* (3 and 4), pp. 137-142.

Saxena, K L., Chakrabarty, R.N., Khan, A.Q., Chattopadhya, S.N. and Chandra Harish (1966) : Pollution studies of the River Ganges near Kanpur *Env. Hlth*. Vol. 8, pp. 270-285.

Sastry, C A., Khare, G.K. and Rao, A V. (1972) : Water Pollution Problems in Madhya Pradesh *Ind. J. Env. Hlth. Vol 14, No. 4*, pp 297-309.

Snedecor, G.W. and Cochran, W.G. (1967) : *Statistical Methods* 6th ed. Oxford and IBH Publishing Co. Calcutta.

Venkateswarlu, V. (1969 a) : An ecological study of the algae of the Moosi River, Hyderabad (India) with special reference to Water poll. Part-1 Physico-chemical complexes. *Hydrobiolo* , *33*. 1, pp. 117-43.

Zafar, A.R. (1964) : On the ecology of algae in certain fish ponds of Hyderabad India II Distribution of Unicellular and colonial forms. *Hydrobiologia* 24(4). pp. 556-566.

15

TREND INPUTS IN AQUATIC ECOSYSTEM STUDIES

P.S. Dubey and M. Mishra
School of Studies in Botany
Vikram University, Ujjain (M.P.) India

INTRODUCTION

Since the inception of Water Pollution Control Act 1972 and subsequent establishment of Pollution Control Boards, a lot of water has flown in Indian rivers. The analysis of water for determining pH, colour, temperature, DO, BOD, COD or the load of chloride, phosphate sulphate etc. initially was the right approach but should the same slate be rewritten or reorientation involving new concepts processes and understandings be incorporated in any assessment, specially in the Indian reference. Practically even small water bodies have been explored but the parameters remained unchanged, while at international levels significant changes with broader vision emerged which are of relevance of application even in the biomonitaring (Dubey, 1991).

The paper is an attempt to visualise what could/should be these newer components for correct appraisal of the aquatic ecosystems *vis a vis* pollutants and their behaviour. The toxicity testing water quality analysis should become predictive tools for the net effect of chemicals, effluents, pesticides, organic compounds or heavy metals.

FUNCTIONS AND PROCESSES IN ECOSYSTEM COMPONENTS

The aquatic environment is living and even its nonliving systems are active in contrast to the exclusive air and soil environments. The water itself and the soil bed, both interact and response mutually at certain places and in phases depending on the temperature regimes, light and dark conditions and physicochemical and biological characteristics of both, the compounds and of course will depend on the bed characteristics. Since even metal or compound or element has its specific or group specific entry mechanism at every stage the fate in the ecosystem thus relies on these factors at each stage of transfer, obviously transfer rates at each step should be estimated. For more perfection transport studies could be coupled with transfer rates on the entire food chain including micro consumer subsystems.

Obviously, the fundamental transport and transformation process models are to be evolved, evaluated and identified. This however is not simple since physical, chemical and biological processes are to be counted, assembled and integrated.

(*a*) **Physical Processes**

Hydrolysis, Transport, Sorption, Adsorption, Volatilisation, Settling Suspension, Resuspension, Sediment mechanisms, Diffusive exchange, Boundry layer transport.

(*b*) **Chemical Processes**

Physicochemical properties and regimes, Oxidation-Reduction Molecular mass water solubility, Vapour pressure, Octanol/Water partition coefficients, Abiotic degradation, Chemical Hydrolysis, Indirect and direct photolysis, UV radiation induced effects, Partitioning between environment media, Metal speciationand chelation, Soil enzymes activities and Free redical exchange.

(*c*) **Biological Processes**

Biosorption, Bioaccumulation, Microbial transformations, Algal transformation, Food chain transfers of original compounds/metabolities and many chemical processes like Hydrolysis, Cleavage, Deamination etc. operated through biological systems.

Infact it all amount to simple approach to complex models where a total transfer from Technosphere to Ecosphere is to be assessed.

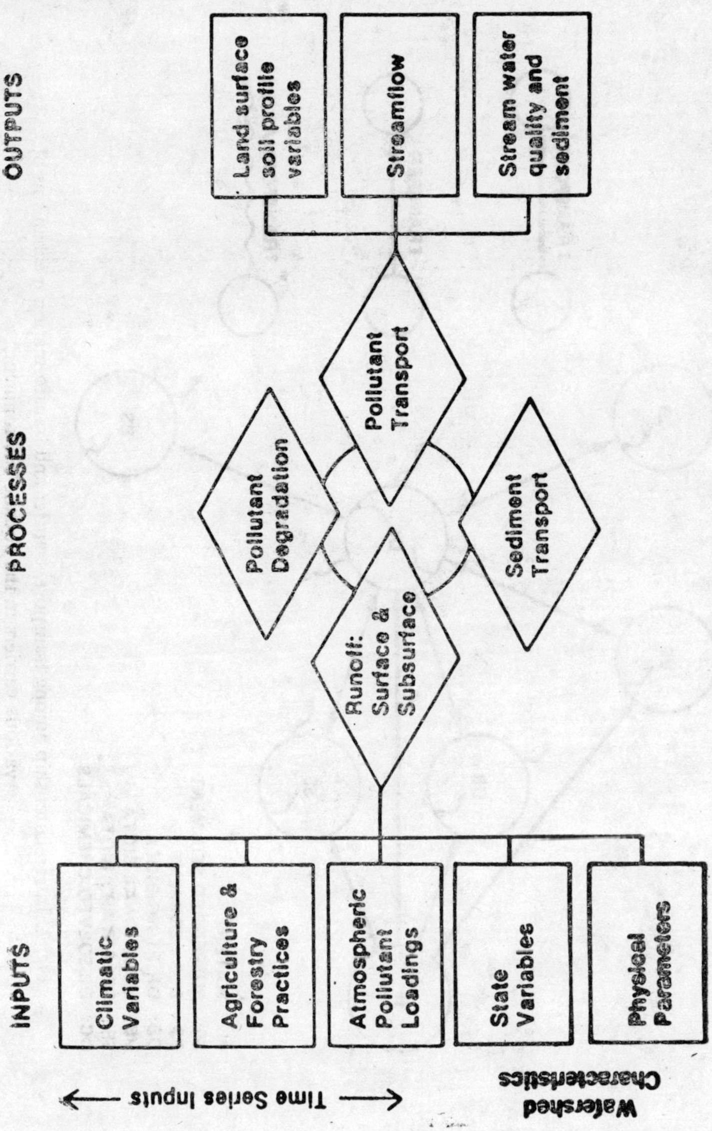

Fig. 1. Model input/output.

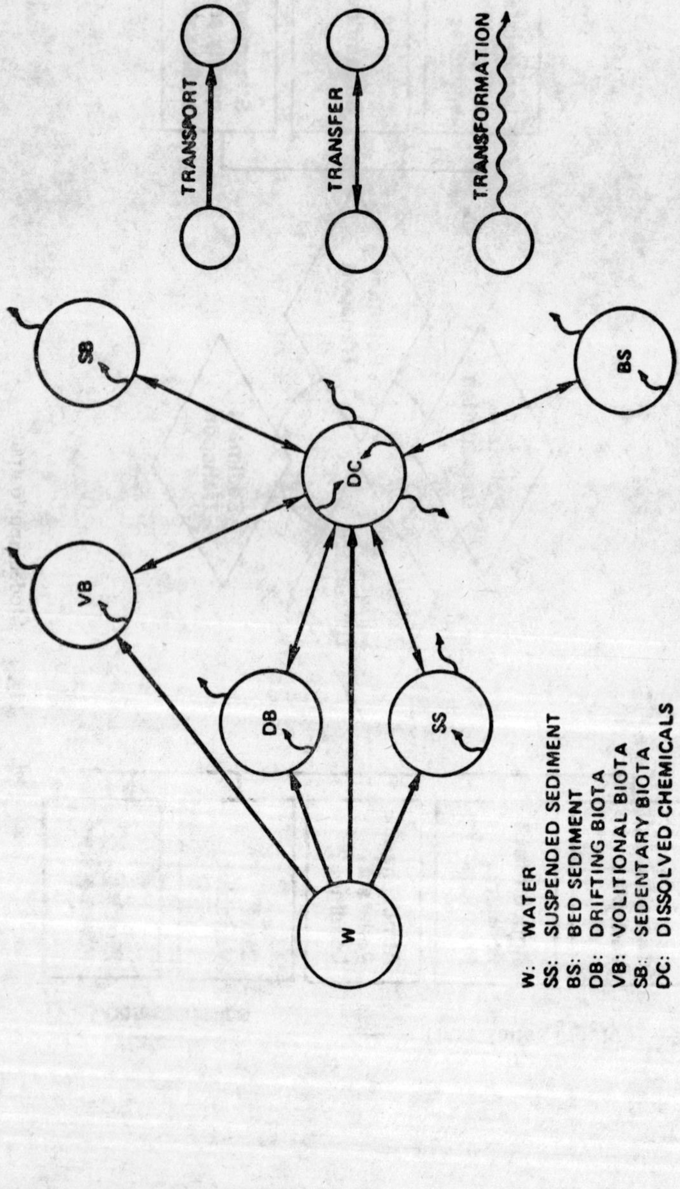

Fig. 2. Interrelationship among transport, transfer and transformation mechanisms of various carriers in the aquatic environment.

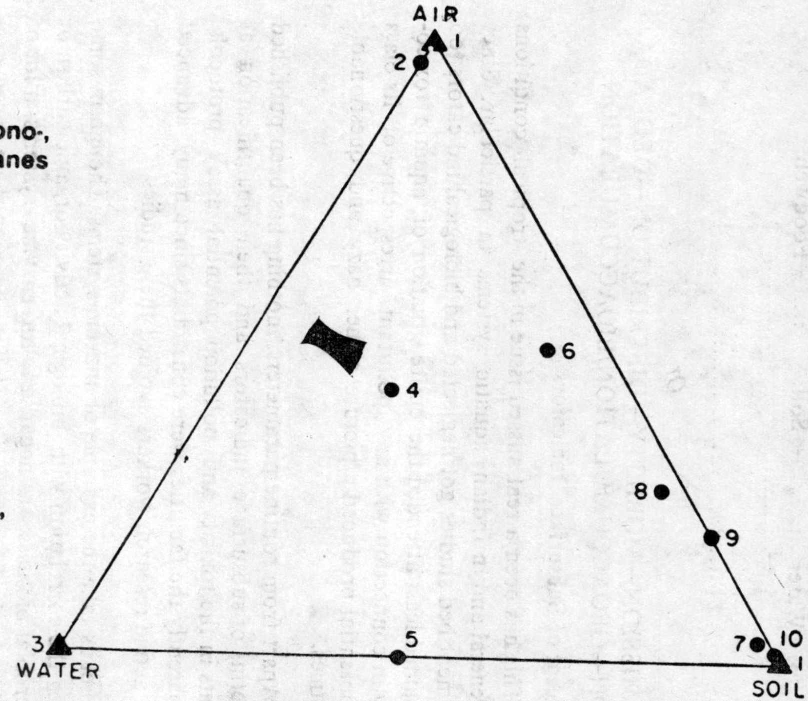

1. Benzene, toluene, and r ono-, di- and trichlorinated ethanes and ethenes
2. *a*-Endosulfan
3. Endosulfan sulfate
4. Phenol
5. Benzidine
6. Dimethylphthalate
7. Diethylphthalate
8. 2,4-Dichlorophenol
9. PCB-1254
10. TCDD
11. Benzo[a]pyrene, chrysene, *bis*-(2-ethylhexyl)phthalate, DDT, DDD, DDE

Fig. 3. Equilibrium distribution of several priority pollutants between environmental media (percentages).

Water...........→Soil...........→Foodweb
↑ ←...........   ←........... ↑
|.................................... |

Or

EMMISSION–MOBILITY—DISTRIBUTION—DEGRADATION—GEOACCUMULATION/BIOACCUMULATION.

Concept of Subsurface Indicators

This has been a real missed issue in the tropical conditions in general and in Indian aquatic systems in particular. Since sediment bed studies got neglected and biological indicators too remained least attended the entire situation of aquatic toxicology in connection with any pollutant, irrespective of its class or potential produced a picture rather hazy and questionable at times.

Apart from routine parameters, nothing has been published in terms of subsurface indicators and their quantification as inputs in the toxicity and pollution potential assay protocols. Apparently the facilities were constraints since many advanced and active research workers avoided these studies.

Let us take the example of enzyme systems. Though in some cases the correlation with biological activity organic carbon or microbial biomass was negative with enzyme systems, a lot of evidence exists for a positive relation between these parameters and soil dehydrogenase, cellulase, and urease activities and even catalase and glycoside hydrolases. Some times individual enzymes emerge as reliable subsurface indicators which include oxidases like Glucose oxidase, Aminoacid oxidase, Urate oxidase etc. The specific enzyme concept is now a reality since even shodanese a transferase enzyme is of importance in sediments activity. Since preciseness can not be ignored, the enzyme system study itself got ignored, ignoring the fact that waters and solids both bear enzymes and substrate both.

In addition to the specific enzyme studies, the macromolecules level too can sharpes the testing and assessments precisely. Such class of compounds consists of proteins, RNA, DNA and even the components of bacterial cell walls (Muramic acid).

Recently, presence of metabolic intermediates in water and soil systems like ATP and GTP microbial metabolic products like luminol and even excretary products of higher animals *i.e.*, fecal sterols (specially coprostanol) because of the persistance in domestic sewage, too serve the purpose of bioindication with a potential of getting quantified satisfactorily. The respirometric measurements of CO_2 output or O_2 uptake on the other hand are also advantageous and reliable indicators (Mahendru and Dubey, 1987 ; Mishra, 1991 ; Mishra and Dubey, 1992).

The tool of Microcosm Technology

The Unit World Models are, synonymous to microcosm technology, now applied successfully in toxicant fate and behaviour studies. Renamed as Laboratory Aquatic Microecosystem Models, substantial use of these UK, USA, Japan, France etc. in studying the toxicant behaviour has made them quite reliable tools because they are seminatural, versatile, reproducible, convenient and economical. The authors too introduced these in India for the first time in the research of Chaturvedi (1990) and modified to produce a terrestrial-aquatic model system in the work of Kumawat (1990) but before this author received a full training at USDA, BARC, Beltsville (USA), the most established centre of laboratory microecosystem modelling. The EPA infact developed a full programme on these systems while one of the author (Dubey, 1984) reviewed its concept, origin, development, recent status working design, type and pros and cons of these systems, the so called "Unit World Models" or infact to be renamed as "Sliced Ecosystems" or "Spare Part Ecosystem Models" than "Sliced Earth Models."

The question of indices and Bioindicators

Nothing now has been added to our knowledge of various indices to establish diversities of species, their distribution etc. or to predict pollution or water quality (Saprobity index, Palmer).

But a comparison made by our teammet (Trivedi, 1981) further stresses a careful use of these in the tropical aquatic systems. However this all does not quantify the pollutant load

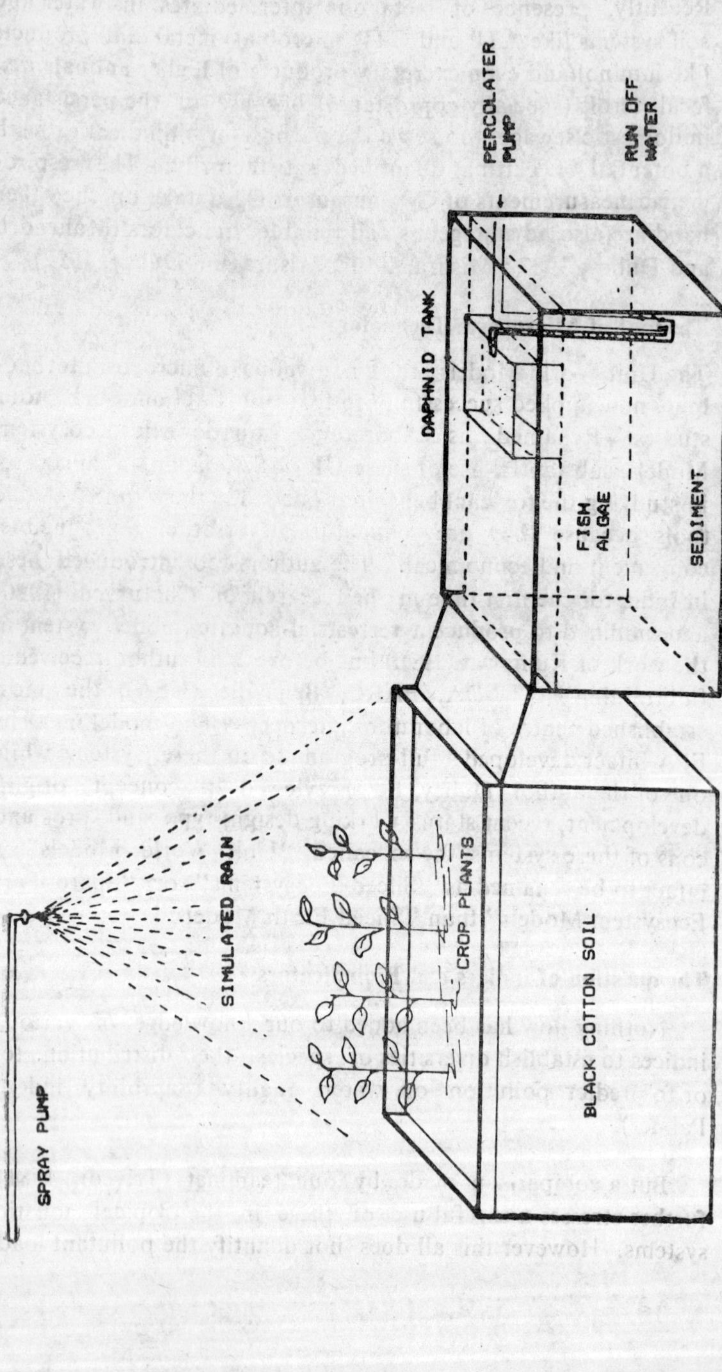

Fig. 4. Kumawat Dubey Model combining Terrestrial (Agro) and Aquatic Microecosystems.

vis a via parameter values. Recently, Shrotriya and Dubey, (1990) realised the problem and after a painstaking research established a Correlative Occurrence Index connecting quantitative load of pollutant to quantities of the bioindicators, in addition to the specificity to the pollutant type. The method and methamatical calculations involve the pollutant load monitoring as well as quantitative values of frequency, density and abundance, on monthly basis throughout the year and calculating value and computation of an index with which by knowing the quantity of the organisms (mostly algal forms and microbes were employed) the pollutant load can be predicted with minor variation in the range. But the need to establish a more reliable diversity index suitable to tropics still remains like the change in the concept of BOD_5 at 20°C to BOD_3 at 27°C for tropics.

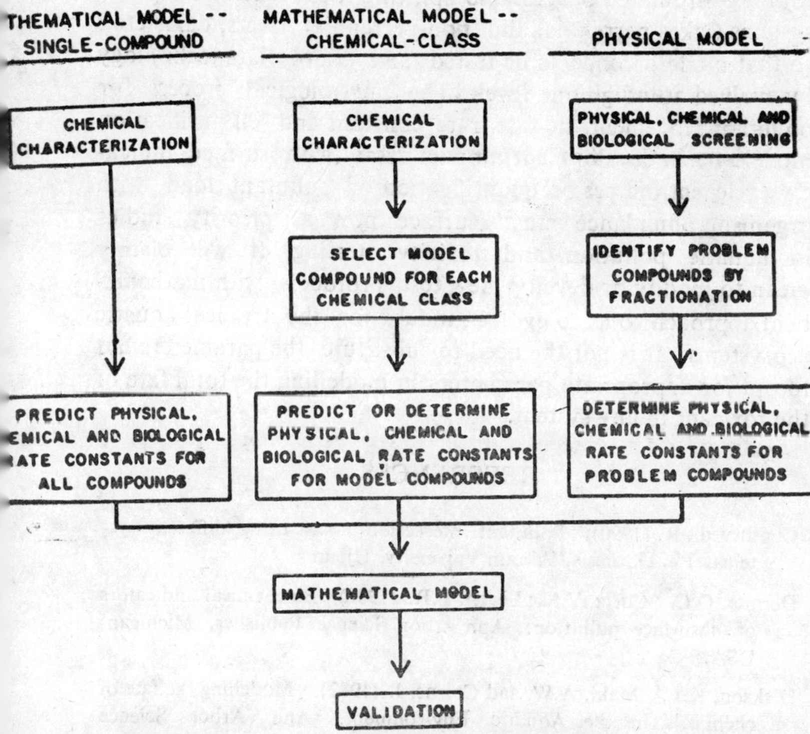

Fig. 5. Alternative modelling approaches for complex mixtures.

The need for change

Since last twenty to thirty years we are still depending on the simple traditional water quality parameters, we stand at a repeatitive stage of research with no gains with practically putting some parameters as objectives to all the aquatic systems or water bodies receiving effluents, runoffs, sewage etc. At times when the river system flow characteristics for the year is known in a splitted monthly design and so also the standard quality and quantity of the effluent in an industry with a fixed standard process, after calculating the employing above parameters and procedures sufficient quantum of labour, chemical and time could be saved. Even the pollutant load characteristics too needs a reorientation as done in US now-a-days.

Infact we will have to develop some centres of research, evolve coordinated programme splitting it to different aquatic systems (lakes perrennial and non-perrennial rivers, dams etc.) so that modern concepts be tested and appropriate technology by evolved at indiginous level. The physiological process for pollutant movement, its fate, rate constant and self purification model system development, impact assay with surface indicators, relevent indices or quantification of pollutant load with organism abundance must surface now as priority studies in aquatic pollution and toxicity testing. It will plainly mean to modify or develop new test protocols with mathematical approach so as to evolve models for the tropical aquatic ecosystems. It is not the need to substitute the parameters but to opt for appropriate parameters in modelling the total fate of the toxicant in any system.

REFERENCES

Chaturvedi, R. (1990). Pollutant interaction assay using different systems. Ph. D. thesis, Vikram Vniversity, Ujjain.

Dermer, O.C. ; Curtz, V.S. ; Leach, F.R. (1980). Biochemical indicators of subsurface pollution. Ann Arbor Science Publisher, Michigan, USA.

Dickson, K.L. ; Maki, A.W. aad Cairns, J. (1982). Modelling the fate of chemicals in the Aquatic Environment. Ann Arbor Science publishers, Michigan, USA.

Dubey, P.S. 1984). Laboratory Microecosystem Models—New tools to assess the toxicity behaviour in the ecosystems. In effect of Pesticides on Aquatic Fauna, Eds. S. S. Kulshreshtra, V. Kumar, M.C. Bhatnagar. 175-184 Acad of Env. Biol , Muzzaffernagar, India.

Dubey, P.S. (1991). Biomonitaring the environment—Theory, Practice and Problems. Jr. of Env. biol. (AEB Decade Commonneration Issue of Bioindicators) 10 : 233-241.

Kumawat, D.M. (1990). Plant response against air pollution stress with special reference to heavy metals and sulphur dioxidc. Ph. D. thesis, Vikram University, Ujjain.

Mahendru, P. and Dubey, P.S. (1987). Herbicide Toxicity and Incorporation response in Soil. Toxicity Assessment 1 (2) : 50-55.

Mishra, M. (1991). Evaluation of toxicity of a few pesticides and activity of antidote. Ph. D. thesis, Vikram Univ. Ujjain.

Mishra, M. and Dubey, P.S. (1992). Soil Respiration—A monitoring bioindicator under soil stress. Accepted in Proceedings of the BARC Symposium (National Symposium on Environmental Stress) 3-5 Feb. 92. BARC, Bombay.

Palmer, C.M. (1980). Algae and Water Pollution. Castle House Publications Ltd.

Shrotriya, L. and Dubey, P.S. (1990). Specificity of organisms to pollutants in a tropical area. Ed. Nalin Shastri, 23-28 Aashish Publishing Co., New Delhi.

Trivedi, R.C. (1981). Use of Diversity Index in Evaluation of Water quality. Proceedings of the workshop on "Biological Indicators and Indices of Environmental Pollution." Central Pollution Control Board, New Delhi, 175-188.

16

ACUTE TOXICOLOGICAL STUDIES OF LEAD TO CERTAIN TELEOSTS

Rao, K.S., Anil K. Singh and B.S Rao
School of Studies in Zoology,
Vikram University, Ujjain, 456 010

INTRODUCTION

The toxicity of lead in water, like that of other heavy metals, is affected by pH heardness, organic materials, and the presence of other metals. The aqueous solubility of lead ranges from 500 μg/l in soft-water to 3 μg/l in hard water. Lead enters the aquatic environment through precipitation, lead dust fallout, erosion and leaching or soil, municipal and industrial waste discharges, and the runoff of fallout deposits from streets and other surfaces. Extrapolations from recent studies (EPHA, 1972 ; University of Illinois, 1972) indicate that nationally as much as 5,000 tons of lead per year may be added to the aquatic environment as a result of urban runoff.

Mediterranean and Pacific surface waters contain up to 0.20 and 0.35 mg/l of lead, respectively (NAS, 1972), which is about 10 times the estimated pre-industrial lead content of marine waters. The lead content of rivers and lakes also has increased in recent years (NAS, 1972). It may be inferred from available data that the mean natural lead content of the world's lakes and rivers ranges from 1 to 10 μg/l (Livingstone, 1963) : the lead content of rural U.S. soils is 10 to 15 μg/g (Chow and Patterson, 1962) ; and the usual range of lead in soil concentrations is 2 to 200 ppm, exclusive of areas near lead or deposits (Motto, *et. al.*, 1970), although many urban soil concentrations are much higher.

MATERIALS AND METHODS

The experimental fishes *viz.*, *Lebistes reticulatus*, *Sarotherodon mossambicus*, *Channa striatus*, *Mystus tengara* and *Cyprinus carpio* were collected from the vicinity of Ujjain area and adapted to laboratory conditions for about 15 days.

The physico-chemical condition of stored tap water was analysed before the experimentation by following the method of APHA (1985).

For determination of acute toxicity values 10 fishes were exposed in each concentration. The Lc_{50} values were obtained by Litchfield and Wilcoxon (1949) graphical interpolation. There is no simple formula to translate the various Lc_{50} values into permissible concentrations considered safe for long periods of exposure. Although a chemical may not be directly toxic to given fish ; its cummulative action may eventually eliminate the species indirectly. Inspite of these limitations in the estimation of long term least effective concentration (LEC) from the acute toxicity assays some workers have developed certain formulate to translate the experimental data into numerical and usable values. Thus the Chio River Sanitation Commission (1955) has recommended and orbitrary factor of 1/10 to the 48 hrs. TLM values Hart *et. al.* (1945) suggested the following formula for finding out the presumably harmless concentration for toxicant.

$$C=\frac{48 \text{ hrs } Lc_{50}\times 0.3}{S^2}$$

C=Presumably harmless concentration

$$S=\frac{24 \text{ hrs } Lc_{50}}{48 \text{ Hrs } Lc_{50}}$$

In the present study least-effect-concentration was calculated by formula suggested by Hart *et. al.* (1945) as given above.

RESULTS AND DISCUSSION

In the present study five fishes *viz.*, *Lebistes reticulatus*, *Sarotherodon mossambicus*, *Channa striatus*, *Mystus tengara* and *Cyprinus carpio* were selected to report the acute toxicity of lead nitrate and lead chloride.

Table 1. The acute toxicity (24, 48, 96 hrs. TL_{50} values) of Lead to several species of fish in water of various water qualities.

Species	*Size (g/fish)*	*Compound*	*Exposure time (hr)*	*Exposure type*	*Tempera-ture (°C)*	*Concen-tration (mg/l)*	*pH*	*Alkalinity*	*Hordness*	*Reference*
(1)	(2)	(3)	(4)	(5)	(6)	(7)	(8)	(9)	(10)	(11)
Rainbow trout (*Salmo gairdneri*)	31.5 g/fish	Pb $(NO_3)_2$	24	FT*	14.7	3 75 Pb	7.15-7.5	41-42	43-45	Benoit and Holcombe
Rainbow trout	24.0 g/fish 132 mm/ fish	,,	96	S**	7	1.38 free Pb	6.89-8.78	7-21 228 in control	300.0	Davies and Everhart (1973).
Rainbow trout	,,	,,	,,	,,	,,	471 Pb	,,	,,	,,	,,
Rainbow trout	5.9 g/fish 85.9 mm/ fish	,,	,,	S	13.8-14.2	542 Pb	6.23-7.17	6-25 228 in control	385	,,
Brook trout (*Salvelinus fontinalis*)	102 g/fish	,,	,,	FT	12	4.5 Pb	6.9-7.1	43	45	Benoit and Holcombe
Brook trout	40 g/fish	,,	,,	,,	14	5.8 Pb	7.0-7.4	40-42	43-45	,,
Coho Salmon (*Oncorhynchus kisutch*)	4 wks post hatch	$PbCl_2$	,,	S	10	0.8 Pb	—	—	17-26	Chapman
Coho salmon	5 wks post hatch	,,	,,	,,	,,	,,	—	—	,,	,,
Coho salmon	6 wks post hatch	,,	,,	,,	,,	0.52 Pb	—	—	,,	,,
Coho salmon	7 wks post hatch	,,	,,	,,	,,	,,	—	—	,,	,,

(*Contd.*)

(1)	(2)	(3)	(4)	(5)	(6)	(7)	(8)	(9)	(10)	(11)
Sticklebacks (*Gasterosteus aculeatus*)	—	—	96	S	—	0.1 Pb	7.0-8.5	—	—	Hawksley (1967)
Mosquitofish (*Gawbusia affinis*)	Adult female	$Pb(NO_3)_2$	24	S	22·24	240 Pb $(NO_3)_2$ (150 Pb)	7.7-8.3	100	—	Wallen *et.al*. (1957)
Mosquito fish	,,	,,	48	,,	,,	,,	,,	,,	—	,,
Mosquito fish	,,	,,	96	,,	,,	,,	,,	,,	—	,,
Mosquito fish	,,	Lead oxide	24	S	18-20	56,000 PbO	7.1-7.2	100	—	,,
Mosquito fish	,,	,,	48	,,	,,	,,	,,	,,	—	,,
Mosquito fish	,,	,,	96	,,	,,	,,	,,	,,	—	,,
Gold fish (*Carassius auratus*)	1.2 g/fish 1.5-2.5 in/fish	$PbCl_2$	24	S	25	45.4 Pb	7.5	18	20	Pickering and Henderson (1965)
Gold fish	,,	$PbCl_2$	48	,,	,,	31,5 Pb	,,	,,	,,	,,
Gold fish	,,	,,	96	,,	,,	,,	,,	,,	,,	,,
Guppies (*Poecilia reticulatus*)	0.1-0.2 g/fish 0.75-1 in/fish	,,	24	,,	,,	24.5 Pb	,,	,,	,,	,,
Guppies	,,	,,	48	,,	,,	,,	,,	,,	,,	,,
Guppies	,,	,,	96	,,	,,	20 6 Pb	,,	,,	,,	,,
Blue gills (*Lepomis macrochinus*)	1,2g/fish 1 5-2 .5 in/fish	,,	24	S	25	25.9 Pb	7.5	18	20	,,
Blue gills	,,	,,	48	,,	,,	24.5 Pb	,,	,,	,,	,,
Blue gills	,,	,,	96	,,	,,	23.8 Pb	,,	,,	,,	,,

(Contd.)

Table 1. (*Coutd.*)

(1)	(2)	(3)	(4)	(5)	(6)	(7)	(8)	(9)	(10)	(11)
Blue gills	58 g/fish 7 cm/fish	$(C_2H_5)_4$ Pb	24	S	20	2.0 Pb	6.9-7.5	35-81	84-163	Turnbull *et. al.* (1954)
Blue gills	,,	,,	48	,,	,,	1.4 Pb	,,	,,	,,	,,
Blue gills	1-2 g/fish 1.5-2.5 in/fish	Pb Cl_2	24	S	25	482 Pb	8.2	300	360	Pickering and Henderson (1965)
Blue gills	,,	,,	48	,,	,,	468 Pb	,,	,,	,,	,,
Blue gills	,,	,,	96	,,	,,	442 Pb	,,	,,	,,	,,
Fatheads (*Pimephales promelas*)	,,	,,	24	,,	,,	8.10-11.5 Pb	7.5	18	20	,,
Fatheads (*Pimephales promelas*)	1-2 g/fish 1.5-2.5 in fish	Pb Cl_2	48	S	25	5.99-11.5	7.5	18	20	,,
Fatheads	,,	,,	96	,,	,,	5.58-7.33 Pb	,,	,,	,,	,,
Fatheads	—	,,	48	FT	—	1.1 Pb	7.4	18	20	Henderson
Fatheads	—	,,	96	FT	—	0.97 Pb	,,	,,	,,	,,
Fatheads	—	,,	96	S	—	5.6 Pb	,,	,,	,,	,,
Fatheads	—	,,	96	—	—	2.4 Pb	,,	,,	,,	Tarzwell and Henderson (1960)
Fatheads	1-2 g/fish 1.5-2.5 in/fish	Pb $(C_2H_3O_2)_2\ 3H_2O$	24	S	25	14.6 Pb	7.5	18	20	Pickering and Henderson (1965)
Fatheads	,,	,,	48	,,	,,	10.4 Pb	,,	,,	,,	,,
Fatheads	,,	,,	96	,,	25	7.48 Pb	,,	,,	,,	,,

(*Contd.*)

(1)	*(2)*	*(3)*	*(4)*	*(5)*	*(6)*	*(7)*	*(8)*	*(9)*	*(10)*	*(11)*
Fatheads	,,	$Pb\ Cl_2$	24	,,	,,	482 Pb	8.2	300	360	,,
Fatheads	,,	,,	48	,,	,,	,,	,,	,,	,,	,,
Fatheads	,,	,,	96	,,	,,	,,	,,	,,	,,	,,
Fatheads	,,	,,	96	,,	,,	75 Pb	,,	360	400	Tarzwell and Henderson (1960)
Puntius ticto	4.1 cm/fish 1.6 g/fish	$Pb\ (NO_3)_2$	96	,,	,,	3.0Pb	7-5	,,	293	Saksena and Bhatnagar (1981)
Lebistes reticulatus	Male	$Pb(NO_3)_2$	96	,,	20-26	12.5	7.8-8.6	88-124	88-118	Gadkari and Marathe (1983)
Lebistes reticulatus	Female	,,	,,	,,	,,	15.5	,,	,,	,,	,,
Lebistes reticulatus	Fry	,,	,,	,,	,,	1.23	,,	,,	,,	,,
Sarotherodon-mossambicus	Hatching	,,	,,	,,	24-27.2	10	7.5-8.6	240-290	185-195	Bhalerao (1986)
	Fry	,,	,,	—	,,	45	,,	,,	,,	,,
	Fingerling	,,	,,	—	,,	115	,,	,,	,,	,,
	Adult	,,	,,	—	,,	230	,,	,,	,,	,,

Table 2. The acute toxicity (96 hrs. Lc_{50} values) of Lead to some teleost fishes (Rao and Singh, 1988 unpublished data from School of Studies in Zoology, Vikram University, Ujjain).

Species (1)	*Size (g/fish)* (2)	*Compound* (3)	*Exposure time (hr)* (4)	*Concentration (mg/1)* (5)	*pH* (6)	*Temperature (°C)* (7)	*Alkalinity (mg/1)* (8)	*Hardness (mg/1)* (9)	*Reference* (10)
Lebistes reticulatus	0.15±0.01 g/fish. 1.7±0.03 cm/fish	$Pb(NO_3)_2$	96	10.1	7.3-7.4	22-23	158-164	150-160	Rao and Singh, 1988
		Pb Cl_2	,,	8.2	,,	,,	,,	,,	,,
Sarotherodon mossambicus	15±2g/fish 11±0.1 cm/fish	$Pb(NO_3)_2$	,,	198.6	,,	,,	,,	,,	,,
		Pb Cl_2	,,	164.5	,,	,,	,,	,,	,,

(*Contd.*)

(1)	*(2)*	*(3)*	*(4)*	*(5)*	*(6)*	*(7)*	*(8)*	*(9)*	*(10)*
Channa striatus	12±1 g/fish 10±2 cm/fish	$Pb(NO_3)_2$	,,	225.2	,,	,,	,,	,,	,,
		Pb Cl_2	,,	206.1	,,	,,	,,	,,	,,
Mystus tengara	5±1 g/fish 8 cm/fish	$Pb(NO_3)_2$	,,	215.3	,,	,,	,,	,,	,,
		Pb Cl_2	,,	196.4	,,	,,	,,	,,	,,
Cyprinus corpio	7±2 g/fish 10 cm/fish	Pb $(NO_3)_2$	,,	178.1	,,	,,	,,	,,	,,
		Pb Cl_2	,,	150.6	,,	,,	,,	,,	,,

A. Lead nitrate toxicity levels.

It was observed that the Lc_{50} and Lc_{100} values for *L. reticulatus* are 10.1 mg/l. and 17.2 mg/l. For *S. mossambicus* 198.6 mg/l and 234.7 mg/l. For *C. striatus* 225.2 mg/l and 290.5 mg/l. For *M. tengara* 215.3 mg/l and 242.4 mg/l and For *C. carpio* 178.1 and 210.5 mg/l of $Pb(NO_3)_2$ at 96 hours respectively.

B. Lead Chloride toxicity levels.

In the present study acute toxicity patterns of $PbCl_2$ were observed that Lc_{50} and Lc_{100} for *L. raticulatus* 8.2 mg and 12.6 mg/l. S. mossambicus 164.5 mg/l and 200 mg/l, *C. striatus* 206.1 mg/l and 235.6 mg/l, *M. tengara* 196.4 mg/l and 214.2, C. carpio 150.6 mg/l and 175.6 mg/l at 96 hours respectively.

DISCUSSION

As far as is known, lead has no beneficial or desirable nutritional effects. Lead is a toxic metal that tends to accumulate in the tissues of man and other animals. Although seldom seen in the adult population, irreversible damage to the brain is a frequent result of lead intoxication in children. Such lead intoxication most commonly results from ingestion of lead containing paint still found in older homes. The major toxic effects of lead include anemia, neurological dysfunction, and renal impairment. The most common symptoms of lead poisoning are anemia, serve intestinal cramps, paralysis of nerves (particularly of the arms and legs), loss of appetite, and fatique, the symptoms usually develop slowly. High levels of exposure produce severe neurologic damage, often manifested by encephalopathy and convulsion, such cases frequently are fatal. Lead is strongly suspected of producing subtle effects (*i.e.*, effects due to low level or long term exposures insufficient to produce overt symptoms) such as impaired neurologic and motor development and renal damage in children (EPA, 1973). Subclinical lead effects are distinct from those of residual damage following lead intoxication.

Studies dealing with the effects of Pb in aquatic environment are rare. For the fathead minnow, *Pimephales promelas*, Pickering and Henderson (1966) have determined the 96-hour

TLM values for lead (chloride) to be 5.6 to 73 mg/l in soft-water (20 to 45 mg/l as $CaCO_3$). Brown (1968) reported a 96-hour Lc_{50} of 1 mg/l for rainbow trout, *Salmo gairdneri*, in soft water (50 mg/l as $CaCO_3$). The 96-hours Lc_{50} value for lead in hard water was 482 mg/l for fathead minnows (Pickering and Henderson, 1966).

Preliminary information on 2 to 3 month exposures of rainbow *Salmo gairdneri*, and brook trout, *Salvelinius fontinalis*, indicated detrimental effects at 0.10 mg/l of lead in soft water (20-45 mg/l as $CaCO_3$) (NAS, 1974). Growth of guppies, *Lebistes reticulatus*, was affected by 1.24 mg/l of lead (Grandall and Goodnight, 1962): Jones (1939) and Hawksley (1967) found chronic or sublethal effects on three spine stickelbacks, *Gasterosteus aculeatus*, from lead concentrations of 0.1 and 0.3 mg/l ; the conditioned behaviour of goldfish, *Carassius auratus*, in a light-dark shuttlebox was adversely affected by 0.07 mg/l of lead in soft-water (50 ppm $CaCO_3$) added to de-ionized tap water (Weir and Hine, 1970).

A concentration of 30 mg/l lead in a 3-week exposure produced 16 percent reproductive impairment in *Daphnia magna* in water with a hardness of 45 ml/l $CaCO_3$ (Biesinger and Christenson, 1972). The 96 hour Lc_{50} for rainbow trout, *Salmo-gairdneri*, in hard water (alkalinity 243 mg/l) for total lead was 471 mg/l and the highest mean continuous flow concentration that did not have an adverse effect on survival, growth, and reproduction was 0.12 and 0.36 mg/l (Davies and Everhart, 1973). For dissolved lead, the 96-hour Lc_{50} was 1.38 mg/l and the no effect level was 18 to 32 mg/l. Total and free lead were considered to be the same in soft water. The 18-day LC_{50} in soft water (alkalinity 26.4 mg/l) was 140 ug/l and the highest mean continuous flow concentration that did not have an adverse effect on survival, growth, and reproduction was 6.0 to 11.9 ug/l. The no-effect concentrations were determined on the occurrence. In general, the salmonids are most sensitive to lead in soft water, but the influence of pH and other factors on the solubility and form of the lead preclude the recommendation of a freshwater criteria based on acute toxicities alone.

There are few data from which to draw conclusions about the relationship of safe levels for a given species in hard and soft water. The differences in lethals between very soft and very hard water are 100 to 500 times because of insolubility and precipitation. The single source of information for chronic and acute effects in hard and soft water is for rainbow trout, *Salmo gairdneri*, (Davies and Everhart, 1973) and shows these significant factors :

Hard water data were reported in terms of both total and free lead. The static acute hard water bioassay (mean hardness, 353.5 mg/l ; methyl orange alkalinity, 243.1 mg/l, and pH 8.02) demonstrated 96-hours TL_{50} (50 percent tolerance limit) concentration of 471 mg/l total lead and 1.38 mg/l free lead.

In the chronic hard warter tests, lead attributed mortalities occured at levels of 3240 ug/l total lead and 64 ug/l free lead, while the maximum acceptable toxicant concentration, based on the occurrence of black tails, were found to be 120 ug/l to 560 ug/l total lead and 18 ug/l to 32 ug/l free lead.

Soft water data were reported in terms of free lead. The flow through acute soft water bioassay (mean hardness 27.2 mg/l ; methyl orange alkalinity 26.4 mg/l ; and pH 6.88) demonstrated an 18-day TL_{50} (50 percent tolerance limit) of 140 ug/l free lead.

In the chronic soft water tests, lead attributed mortalities occurred in fly and fingerling fish at a concentration of 95.2 ug/l, while 30 year-old broad fish produced viable eggs and fry when exposed to any but the highest (27.9 ug/l) concentration. The maximum acceptable toxicant concentration, based on the occurrence of black tails, was determined to be between 11.9 and 6.0 ug/l.

A criterion involving an application factor of 0.01 multiplied by the acute 96-hour Lc_{50} value expressed as dissolved lead is used to estimate the safe concentration for various fish species. Based upon the existing data, as well as upon the sensitivity of various species of fish to other metals, it is highly probable that salmon trout, minnows, and catfish will be especially sensitive to lead as compared to bass, sunfish and perch. Therefore, tests for acute toxicity should be performed on the more sensitive species when establishing standards for lead. This approach requires the experimental determination of Lc_{50} values before a criterion can be determined, but the extreme effect of various water characteristics on lead solubility and toxicity warrants this additional efforts.

There is no question that some marine organisms can concentrate the lead present in seawater. Wilder (1952) reported lobster daying in 6 to 20 days when held in lead lined tanks. Calabrese, *et. al* (1973) found a 48 hours Lc_{50} of 1,730 ug/l and a 48-hour LC_{50} of 2,450 ug/l for oyster. *Crassostrea virginica* eggs. The remarkable ability of the eastern oyster, *Crassostrea virginica*, to cencentrate lead was demonstrated (Pringle *et. al.* 1968) by exposing them to flowing seawater containing lead concentrations of 25 ug/l, 50 ug/l, and 200 ug/l, after 49 days, the total accumulation of lead amounted to 17,35,75 and 200 ppm (wet weight), respectively, and those oyster exposed to the two highest lead levels, upon gross examination, showed considerable atrophy and diffusion of the gonadal tissue, adema, and less distinction of hepatopancreas and mantle edge. Saksena and Bhatnagar (1981) determined toxic level as 96 hours Lc_{50} for *Puntius ticto* to lead about 3 mg/l and tolerance initial concentration used for lead 0.3 mg/l.

Lloyed *et. al.* (1962) observed that Pb toxicity was increased in low O_2 level. Hoglund and Hordig (1969) studied the "Reaction of Young salmonoids to sudden changes of pH, CO_2 tension and O_2 content. He recorded that a decrease in pH, combination with an increase in CO_2 tension provoked changes in the locomotor behaviour and ventilation among fish *Salmo Salar* while an extremely adverse PH$-$pCO_2 condition in combination with low p.o. depletion down to 1.3 mg/l provoked hyperventilation, but did not initiate any changes in locomotor activity. According to Maclead Pessah (1973) the toxicity tolerance of fishes decreases with an increase in temperature and fish become less tolerant in tropical water condition.

REFERENCES

APHA, (1985), *Standard methods for the examination of water and waste water.* 16th ed. 1193.

Bencit D. and G. Holcombe. Unpublished date available from water Quality laboratory, Duluth, Minn.

Berry, R.A. 1924. The manurial properties of lead nitrate, *Jour. Agric. Sci. Londan,* 14 : 58.

Biesinger, K.E. and G.M. Christensen, 1972. Effects of various metals on survival, growth, reproduction and metabolism of *Dophnia magna. Jour. Fish. Res. Bd.* of Canada, 29 : 1691.

Bhalerao, V. (1986). *Toxicity of Lead* and *Cadmium on four life history stages of Sarotherodon mossambicus.* M. Phil thesis, Vikram University, Ujjain.

Brewer, R.F. (1966) Lead Page 213 in H.D. Chapman, ed. *Diagnostic criteria for plants* and *soils.* University of California Division of Agric. Sci. Berkeley.

Brown, V. M. (1968). The calculation of the acute toxicity of mixtures of poisons to rainbow trout. *Water Res.* 2. 723.

Chapman, G Unpublished data available from National Water Quality Laboratory, Duluth, Minu.

Chow, T.J., and C.C. patterson, 1962. The occurrence and significance of lead isotopes in pelagic sediments. *Geochim. Cosmochin. Acta* 26 : 263.

Crandall, C.A. and C. J. Goodnight, 1962. Effects of sublethal concentrations of several toxicants on growth of the common guppy, *Lebistes reticulatus. Limnol. Oceanog.* 1 : 233.

Davies, P.H. and W.H. Everhart (1972). Effects of chemical variations in aquatic environments : Vol. III. Lead toxicity to rainbow trout and testing application factor concept. *Ecol. Res. Series Report.* EPA-R3-73-011, D. S. *Environment Protection Agency.*

Environmental Protection Agency (1972). Water pollution aspects of streel surface contaminants. EPA-R2-72-081. U.S. *Environmental Protection Agency, Washington,* D.C.

Environmental Protection Agency (1973). EPA's position on the health implications of airborne Lead. U.S. *Environmental protection Agency, Washington,* D.C.

Gadkari, A.S. and V.B. Marathe (1983). Toxicity of cadmium and lead to a fish and a snail from Two different Habitats. *IAWPC Tech. Annual* X. 141-148.

Hart, W.B., P. Doudoroff and J. Greenback, (1945). *The evaluation of the toxicity of industrial waste chemicals and other substances to freshwater fishes* (Philadelphia. The atlantic Refining Co.), 317.

Hawksley, R.A. (1967). Advanced water pollution analysis by a water laboratory. *Analyzer*, 8 : 13.

Jones, J.R.E. (1939). The Relation between the electrolytic solution pressures of the metals and their toxicity to the stickleback (*Gasterosteus aculeatus L.*) *Jour. Exp. Biol.* 16 : 425.

King, B.G. (1971). Maximum daily intake of lead without excessive body lead burden in children. *Amer. Jour. Dist. Children*, 122 : 337.

Kopp. J.F. and R.C. Kroner (1969). Trace metals in water of the United States Federal Water Pollution Control Administration, U.S. Department of the interior, Cincination, Chio.

Lin. Fu, J.S. (1973). Vulnerability of children to lead exposure and toxicity. *New Eng. Jour. Med.* 259 : 1229.

Livingstone, D.A (1963). Chemical composition of rivers and lakes pp. Cl-C64. Geological Survey Professional Paper 440-G. *In M.* Fleischer, ed. *Data of geochemistry*, 6th ed. U.S. *Government Printing Office, Washington, D.C.*

National Academy of Sciences Committee on Biologic Effects of Atmospheric Pollutants (1972). *Lead : airborne lead in perspective. The National Academy of Sciences*, Washington, D.C.

National Academy of Sciences, National Academy of Engineering (1974) *Water Quality criteria 1972*. U.S. Government Printing Office, Washington, D.C.

Pickering, G H. and C. Henderson (1966). The acute toxicity of some heavy metals to different species of wormwater fishes. *Int. Jour. Air-water Poll.* 10 : 453.

Ryan, J P. (1971). *Minerals Yearbook*, 1971. U.S. Department of the Interior, Washington, D.C.

Saksena, D.N. and B. Bhatnagar (I981). Toxicity of some heavy metals to a freshwater Teleost. *Buntius ticto* (Ham.) *Bioresearch* 5 (1) : 113-116.

Tarzwell, C.M. and C. Henderson, (1960). Toxicity of less common metals to fishes. Industrial Wastes, 5 : 12.

University of Illinois (1972). *Environmental pollution by lead other metals* (NSFRANN Grant 61-31605). Progress report May Oct. 31. 1972. Chapter 6 Univ. of Illinois at Urbana-Champaign.

Weir, P.A. and C.H. Hine. (1970). Effect of various metals on behaviour of conditioned goldfish. *Arch. Environ. Health.* 20 : 45.

Wilder, D.C. (1952), The relative toxicity of cercain metals to lobsters. *Jour. Fish Res. Bd Can.* 8 : 486.

Llyod, R. and D.W.M. Herbert (1962). The effect of Environment on the toxicity of poisons to fish. *Jour. Inst. Pnblic Health* Engg. 61 : 132-145.

Mcleon, J.C. and S. person (1973). Temperature effects on accumulation toxicity and metabolic rate in rainbow trout (*Salmo gairdneri*). *J. Fish Rs. Bd. Can.* 30 :485-492.

17

PESTICIDE STRESS AND FISH REPRODUCTION

S.K. Kulshrestha
Department of Zoology
Motilal Vigyan Mahavidyalaya
Bhopal—462 008

INTRODUCTION

The pesticides are toxic compounds designed to destroy pests of agriculture and forests. Pesticides once sprayed in the field find their way to water resources through wind, run-off during monsoon or leaching. The contamination of aquatic environment by pesticides causes severe effects on fish life including its life history. The pesticides are known to cause behavioural changes in fish, impairments in incubation and post-embryonic development of fishes affecting chances of their survival. Successful survival potential of fishes depends upon the suitability of eggs and fry which occupy a lower level in aquatic food chain as compared to adults occupying higher levels. The eggs and sac-fry are vulnerable to changes in their physical environment. The pesticides may affect fish population in paddy-cum-fish culture.

Although the effects of pesticides on gonads and life history of fishes are well documented in western world, a few studies are available on Indian fishes. The present review is based on recent studies in India on the effects of pesticide stress on the reproduction of Indian and exotic fish species.

Effects of Pesticides on Gonads

The effect of pesticides on gonads has been studied on *Channa striatus* (Arora and Kulshrestha, (1984 ; Kulshrestha and Arora, 1984), *Heteropneustes fossilis* (Harilal and Sahai, 1987), *Sorathrodon mossambicus* (Pandey and Shukla, 1980, 1982 ; Shukla *et. al.*, 1984), *Cyprinus carpio* (Kaur and Virk, 1983), *Channa punctatus* (Saxena and Garg, 1978 ; Mani and Saxena, 1985 ; Saxena and Mani, 1987), *Rasbora daniconius* (Singh and Sahai, 1985 ; 1986 ; Rastogi and Kulshrestha, 1990), *Punctius ticto* (Sahai, 1989), *Anabas testudineus* (Sahai and Harilal, 1990), and *Lebistes reticulatus* (Sehgal and Pandey, 1984). These fishes were exposed to sublethal doses of pesticides such as Carbaryl, Carbofuran, Endosulfan, BHC, Lindane, Fenitrothion, Malathion and Methyl parathion. The effects of pesticides were dependent upon the doses, duration of exposure and the nature of pesticides. In most of the studies reduction in gonadosomatic index was observed on pesticide exposure.

On testes the pesticide stress induced degenerative lesions, general reduction in spermatogenesis at various stages ; shrinkage and necrosis of seminiferous tubules, spermatogonia and labule boundary ; involution, atrophy and necrosis of interstitial cells ; rupture of peritoneal lining of testes ; thickening of ducts and necrosis and fibrosis in connective tissue (Table 1).

In ovaries the pesticides stress induced various effects such as arrestation of ovarian recrudescence and vitellogenesis ; prevention of appearance of mature eggs, higher percentage of atretic oocytes, reduction in the number of oocytes, deformities in different stages of oocytes, degeneration of oocyte nuclei, lowering of the ovarian weight, development of interfollicular spaces, broken ovigerous lamellae, increase in the connective tissue of tunica albuginea, necrosis, fibrosis and dilation of blood vessels (Table 2).

Effect of Pesticides on Early Development

The effect of pesticides on early development, hatching, sac-fry, early juveniles and fingerlings of fishes has been studied recently. The fishes studied include major carps, *Cyprinus carpio*, *Rasbora daniconius*, *Hypophthalmichthys moltrix*, *Puntius conchonius* and two estuarine fishes *Caranx sp.* and *Opisthopterus*

Table 1. Summary of Effects PF Pesticides on Testes

Author (s)	*Fishes*	*Pesticide, Doses and Duration of exposure*	*Major Effects*
Arora and Kulshrestha (1984)	*Channa striatus*	Carbaryl 10,20 ppm Endosulfan 0.00075 and 0.001 ppm duration upto 30 days	Rupture of peritoneal lining and wall of seminiferous tubules ; failure of spermatogenesis ; necrosis of interstitial cells ; fibrosis in connective tissue ; reduction in gonadosomatic index (GSI) ; endosulfan more toxic.
Harilal and Sahai (1987)	*Hetero-pneustes-fossilis*	Carbaryl 20 ppm, upto 30 days	Broken peritoneal lining ; shrinkage of seminiferous tubules ; reduction in the number of spermatocytes.
Pandey and Shukla (1980)	*Tilapia-mossambica*	BHC 2, 4 ppm, 10 days	Vacuolated cells ; necrosis in seminiferous tubules ; atrophy of interstitial cells ; thickening of sperm duct and lowering of GSI.
Pandey and Shukla (1982)	*Sorathradon-mossambicus*	Malathion 2, 4 ppm, 10 days	Vacuolation ; distortion of tubules ; arrestation of spermatogenesis at spermatid stage : necrosis of interstitial cells ; thickening of ducts and reduction in GSI.

(*Contd.*)

Sahai and Harilal (1990)	*Anabas-testudineus*	Carbaryl 5 mg/L, 30 days	Arrestation of spermatogenesis ; degenerative changes in seminiferous tubules, connective tissue septa, germinal epithelium, interstitial cells and blood vessels.
Saxena and Mani (1987)	*Channa-Punctatus*	Fenitrothion 1.5 ppm ; Carbofuran 5 ppm ; duration upto 120 days	Delay information of spermatids ; arrestation of sperm formation ; necrosis of spermatids, spermatogonia and spermatocytes ; formation of collagenous capsules around necrotic germ cells ; involution of interstitial cells.
Sehgal and Pandey (1984)	*Lebistes-reticulatus*	BHC, 1 ppm, upto 30 days	Degenerative lesions in the testis ; impairment of spermatogenesis ; appearance of atretic cysts ; increase in the height of Sertoli cells and epithelial cells of sperm duct ; fall in GSI.

Table 2. Summary of Effects of Pesticides on Ovaries

Author (s)	*Fishes*	*Pesticides, doses and duration of exposure*	*Major Effects*
Harilal and Sahai (1987)	*Heteropneustes-fossolis*	Carbaryl. 20 ppm, 30 days	Broken ovigerous lamellae ; degeneration of oocyte nuclei ; necrosis of interstitial cells ; cytoplasmic clumping.
Kaur and Virk (1983)	*Cyprinus carpio*	BHC Bppm, Sumithion 1.5 ppm, 120 days	Arrestation of ovarian recrudescence ; decline in GSI ; BHC more deleterious.
Kulshrestha and Arora (1984)	*Channa striatus*	Carbaryl 10, 20 ppm ; Endosulfan 0.00075 and 0.001 ppm ; duration 30 days	Reduction in the number of oocytes ; reduction in size and deformity in different stages of oocytes ; dilation of blood vessels ; reduction in GSI ; endosulfan more toxic.
Mani and Saxena (1985)	*Channa punctatus*	Carbofuran 5 ppm, Fenitrothion 1.5 ppm, duration 120 days	Lowering of ovarian weight ; retardation of growth of different oocytes ; fenitrothion more toxic than carbofuran.

(*Contd.*)

Rastogi and Kulshrestha (1990)	*Rasboradani-conius*	Endosulfan 0.001 mg/L Carbofuran 0.1 mg/L Methyl Parathion 0.1 mg/L 75 days	Damage to peritoneal lining ; necrosis ; fibrosis ; dilation of blood vessels ; rupture of ovigerous lamellae ; reduction of GSI ; methyl parathion more toxic than carbofuran and endosulfan.
Sahai (1990)	*Puntius ticto*	BHC 0.14 ppm, Endosulfan 0.17 ppm, Lindane 0.16 ppm, Malathion 4 ppm ; 30 days	Nuclear deformities ; lesions ; clumping of cytoplasm ; vacuolation and degeneration of stroma ; degeneration and abnormalities in different stages of oocytes.
Saxena and Garg (1978)	*Channa punctatus*	Carbaryl 2 ppm, Fenitrothion 1.5 ppm, 150 days	Arrestation of ovarian recrudescence to different degrees ; carbaryl more damaging.
Shukla *et. al.*, (1984)	*Sarothradon-mossambicus*	Malathion 2, 4 ppm, 20 days	GSI reduced, ovary diameter reduced ; higher number of atretic oocytes.
Singh and Sahai (1985, 1986)	*Rasbora-daniconius*	Malathion 6 ppm, 30 days BHC 0.14 ppm, 15 days	Arrestation of vitellogenesis ; decrease in ovarian weight ; degenerative changes in oocytes, cytoplasm and nuclei.

tardoore. The pesticides studied include DDT, BHC, Chlorodane, Meta-systox, Carbaryl (Sevin), Diazinian, Fenitrothion, Malathion, Phosphamidon, Dimothoate, Stomp, Butachlor, Rogor, Carbofuran and Endosulfan (Table 3).

Malhotra and Sharma (1986) found that hatching duration is compressed at lower concentration and extended at higher concentrations of pesticides ; the pesticides also affected hatching success and egg mortality. Advancement in hatching time on pesticide exposure is well reported. Kulshrestha and Arora (1985, 1986) observed the effects of Carbofuran, Dimethoate and DDT on early development of *Cyprinus carpio* on egg mortality, hatching and early juveniles. The highest egg mortality was observed in Carbofuran and earliest hatching in Dimethoate. They also observed delayed development, increase in mortality rate and change in the behaviour of early juveniles. Abnormalities in fish fry such as displacement of yolk sac, vertebral column deformities, reduction in fin size, curving of the body and inclination of the caudal region have also been reported (Kaur and Toor, 1977 ; Venugopalan and Rao, 1979 ; Kulshrestha and Arora, 1985). On vertebral column and caudal fin 15 days exposure of sublethal doses of Carbofuran, Rogor and DDT produced anomalies such as deformities in fin, notochord, sclerotome, fibrous sheaths, arches, ribs, zygapophyses and centra (Kulshrestha, 1987).

Kulshrestha (1987) studied the effects of certain pesticides on the early development of *Cirrhinus mrigala, Rasbora daniconius, Catla catla* and *Hypophthalmichthya moltrix*. On fertilized eggs of *Cirrhinus mrigala* the exposure of sublethal doses of Metasystox, Endosulfan and Carbofuran produced anomalies such as delay in cleavage, unusual shape of blastodisc, deformation of blastomeres, irregular cleavage, irregular yolk formation, bent body axis and malformation of eye capsule with minor variations in three pesticides. In *Rasbora daniconius* sublethal doses of Metasystox and Carbofuran produced enhancement of hatching and anomalies in cleavage, irregularities in yolk sac formation and malformations in organogenesis ; the effects of DDT were more pronounced than those of metasystox. In *Catla catla* and *H. moltrix* early hatching and many structural deformities were produced by pesticide exposure.

Table 3. Summary of the Effects of Pesticides on Fish Life History

Author (s)	*Fishes*	*Pesticides, Doses and Duration*	*Major Effects*
Bansal *et al.*, (1980)	Indian major carps	Chlorodane, Meta-Systox and Sevin	The values for 96 hr LC_{50} and 30 days MATC were found to be different in early juveniles of each species and in all pesticides.
Kaur and Toor (1977)	*Cyprinus carpio*	Diazinion—eggs 0.008/ fry 3.11 Fenitrothion 0.062, 2.30 ppm Carbaryl 1.4 and 10.36 ppm Malathion 4.57 ppm Phosphomidon 111.2 and 163.4 ppm	The doses were LC_{50} values for eggs and fry ; Decrease in the survival of developing eggs and hatchings : reduction in hatchability and percentage of hatching ; deformities such as stunted growth of the body beyond yolk sac, upward and inward curving of the tail, deformed head region and vertebral column.
Kaur and Cheeme (1985)	*Cyprinus carpio*	BHC 0.5-35.0 ppm Dimethoate 0.01-50 ppm Stomp 0.05-55 ppm Butachlor 0.01-12.5 ppm	Safe concentrations for eggs and larvae determined ; mortality and abnormalities in newly hatched larvae ; premature hatching, deformities ; larvae more sensitive than eggs.

(*Contd.*)

Table 3. *(Contd.)*

Author (s)	*Fishes*	*Pesticides, doses and duration*	*Major Effects*
Kulshrestha and Arora (1985) 1986)	*Cyprinus carpio*	Carbofuran 0.1, 0.2 ppm Rogor 1, 5 ppm DDT 0.1, 0.5 ppm Duration—upto 3 days	Maximum egg mortality in carbofuran followed by DDT and Dimethoate ; advancement of hatching time ; the early juveniles exhibited highest mortality in carbofuran ; deformities such as upward curving of body, reduction of body length, enlargement of yolk sac and abnormalities in fine and vertebral column.
Kulshrestha *et. al.*, (1986)	*Labeo rohita, Catla catla, Cirrhinus mrigale*	DDT 6.3 to 6.8 ppm Meta-Systox 6.3 to 9.1 ppm Dimethoate 10.1 to 10.5 ppm Carbofuran 4.7 to 5.1 ppm duration upto 30 days	Variations in the toxicity of pesticides on the fingerlings of the three species ; carbofuran most toxic followed by DDT, Meta-Systox and Dimethoate.
Kulshrestha (1987)	*Cyprinus carpio*	Carbofuran 1 ppm, 2 ppm, Rogor 1, 5 ppm DDT 0.1, 0.2 ppm Fries-upto 15 days	Vertebral column deformities, degeneration of tail region, delay in the formation of vertebrae and caudal fin.

(Contd.)

Kulshrestha (1987)	*Cirrhinus mrigala*	Meta-Systox 5 ppm Carbofuran 1.5 ppm Endosulfan 0.0001 ppm Eggs-duration 24 hrs.	Delay in cleavage, irregular cleavage unusual shape of blastodisc, deformation of blastomeres and yolk formation malformation in eye capsule ; advancement of hatching time.
Kulshrestha (1987)	*Rasbora daniconius*	DDT 0.1 ppm Meta-Systox 0.5 ppm, 24 hrs	Anomalities in cleavage, yolk formation and organogenesis.
Kulshrestha (1987)	*Catla catla*	Carbofuran 0.5 ppm Endosulfan 0.001 ppm 24 hrs.	Delay in cleavage, irregular cleavage, deformities in blastomeres, bent body axis malformations in eye capsule and yolk sac.
-do-	*Hypophthalmichthysmoltrix*	Endosulfan 0.001 ppm Carbofuran 0.05 ppm Meth. Parathion 0.001 ppm 24 hrs.	
Malhotra and Sharma (1985)	*Puntius conchonius*	Malathion 0.0015-0.0125 ppm upto 96 hrs.	Hatching duration variable in different concentrations ; compression of hatching time ; decrease in hatching percentage ; abnormalities in caudal region, vertebral column, yolk sac, pericardial sac and head region.

(Contd.)

Table 3. (*Contd.*)

Author (s)	*Fishes*	*Pesticides, dose and duration*	*Major Effects*
Venugopalan and Rao (1979)	*Caranx sp.* *Opisthopterus tardoore*	Lindane 0.048 to 0.2 ppm upto 48 hrs.	Advancement of hatching time, stress on incubation and hatchability ; sustained embryonic condition in postembryonic stages ; abnormalities in body and tail region ; mortality.

Effects of Pesticides on Late Development

The effects of pesticides on fingerlings have not attracted sufficient attention in India. Bansal *et. al.*, (1980) studied the effects of Chlorodane, Metasystox and Sevin on juveniles of four major carps and emphasized predition of long term toxicity by subacute screening of pesticides. Kulshrestha *et. al.*, (1986) exposed fingerlings of three Indian major carps to selected doses of DDT, Dimethoate, Metasystox and Carbofuran and observed values for LC_{50} for 96 hours, maximum acceptable tolerant concentrations (MATC) and application factor (AF) on 30 days exposure. Carbofuran produced maximum toxicity followed by DDT, Metasystox and Dimethoate. Detailed studies on growth and reproduction of adult fish may be useful in specific circumstances when the population is multispecies and study relates to ecological significance. Use of early life history stages for experimental toxicity tests would make inexpensive system for toxicological assessments.

Conclusion and Recommendations

Considering the hazards associated with the use of pesticides on fishes vulnerable during breeding season, and developmental stages, it is necessary to evolve a system of monitoring pesticides in aquatic environment for implementing remedial measures. Following areas need further studies :

1. Development of new criteria for pesticides in aquatic ecosystems.
2. Mechanism of immunological and biochemical responses on chronic exposure to pesticides.
3. Effects of new formulations of pesticides on fish life history.
4. Monitoring of pesticide residues in water, sediments, phytoplankton, zooplankton, benthos and fish.
5. Pesticides persistence and bio-magnification.
6. Biological control of mosquitoes and pests.

A lot of information on scientific data is being generated in India but there is no central agency for evaluation of overall hazards of chemicals, management of toxicological formulations, risk assessment policy formations and implementation of guidelines for safe usage. There is need to establish a Central Environmental Protection Agency with branches at state level to monitor environmental hazards.

REFERENCES

Arora, N. and Kulshrestha, S.K. (1984). Comparison of toxic Effects of two pesticides on the testes of a fresh water tele st *Channa striatus* BL. *Acta hydrochim* et. *Hydrobio* ., 12 (4). 435-441.

Bansal, S.K., Verma, S.R., Gupta, A.K. and Dalela, R.C. (1980). Predicting long term toxicity by subacute screening of pesticides with larvae and early juveniles of four species of major carps. *Ecotoxicology* and *Env. Safety* (4), 224-231.

Harilal, K.S. and Sahai, Y.N. (1987). Effect of Carbaryl on the gonads of *Heteropneustes fossilis* (Bloch). In Environment and Ecotoxicology, 277-281. Academy of Environmental Biology, Muzaffarnagar.

Kaur, K. and Cheema, G. (1985). Acute toxicity of some pesticides to the eggs and larvae of the common carp *Cyprinus carpio*. *Zoologica Orientalis*, 2, 52-58.

Kaur, K. and Toor, H.S. (1977). Toxicity of pesticides to embryonic stages of *Cyprinus carpio*. *Ind. J. Exp. Biol.*, 15, 193-196.

Kaur, K. and Virk, S. (1983). Effect of BHC and Sumithion on ovarian recrudescence in common carp *Cyprinus carpio*. *Indian J. Ecol.*, 10(2), 182-187.

Kulshrestha, S.K. (1987). Studies on the effects of some pesticides on life history of some fishes. Technical Report, M.P. Council of Science and Technology, Bhopal.

Kulshrestha, S.K. and Arora, N. (1984). Impairments induced by sublethal doses of two pesticides on the ovaries of a fresh water teleost *Channa striatus* Bl. Toxicology Letters, 20, 93-98.

Kulshrestha, S.K. and Arora, N. (1985). Effect of Carbofuran, Rogar and DDT on the development of *Cyprinus carpio*. Part II. Early Juveniles. *Int. J. Acad. Ichthyol* (Proc. V. AISI), 6, 1-4.

Kulshrestha, S.K. and Arora, N. (1986). Effect of Carbofuran, Rogar and DDT as the development of *Cyprinus carpio*, Part I. Egg mortality and hatching. *J. Environ. Biol.* 7(2), 13-119.

Kulshrestha, S.K., Arora, N. and Sharma, S. (1986). Toxicity of four pesticides on the fingerlings of Indian major carps, *Labeo rohita*, *Catla catla* and *Cirrhinus mrigala*. *Ecotoxicology* and *Environ*. Safety, 12(2), 114-20.

Malhotra, Y.R. and Sharma, K.K. (1986). Effect of Malathion on some aspects of early developmental stages of *Puntius conchonius*. Ham. Proc. Natl. Symp. Fish and Envi. 1-4 Eds. Joshi, B.D., Chopra, A.K. and Bhatt, D. Zoology Department, G.K. University, Hardwar.

Mani, K. and Saxena, P.K. (1985). Effect of safe concentrations of some pesticides on ovarian recrudescence in the freshwater teleost *Channa punctatus* (Bl.) : A Quantitative study. *Ecotoxicology and Env. Safety*, 9, 141-249.

Pandey, A.K. and Shukla, L. (1980). Effect of BHC on the testicular histology in *Tilapia mossambica*. *GEOBIOS*, 7, 251-253.

Pandey, A.K. and Shukla, L. (1982). Effect of Malathion on testicular histophysiology of *Soratrodon mossambicus*. Nat. Acad. Sci. Letters 5(4), 141-142.

Rastogi, A. and Kulshrestha, S.K. (1990). Effect of sublethal doses of three pesticides on the ovary of a carp minnow *Rasbora daniconius*. *Bull Environ. Contam. Toxicol*., 45(5). 742-747.

Sahai, S. (1989). Toxicological effects of some pesticides on the ovaries of *Puntius ticto*. In Trends in Pollution and Toxicology. 21-29. Ed. Sashikant Jagmander Book Agency, New Delhi.

Sahai, Y.N. and Harilal, K.S. (1990). Carbaryl induced histopathological changes in the testis of a fresh water teleast *Anabas testudineus* In Environ. Pollut. and Health Hazards, 121-126.

Saxena, P.K. and Mani, K. (1987). Effect of safe concentrations of some pesticides on testicular recrudescence in the freshwater teleost *Channa punctatus* (Bl.). A Morphological study. *Ecotoxicology* and *Env. Safety*, 14, 56-63.

Saxena, P.K. and Garg, M. (1978). Effect of insecticidal pollution on ovarian recrudescence in *Channa punctatus*. *Ind. J. Exp. Biol*., 6, 689-691.

Sehgal, R. and Pandey, A. (1984). Effects of two different pollutants on the testis of a viviparous teleost *Labistis reticulatus* (Peters). Uttar Pradesh J. Zool., 4(1), 105-111.

Shukla, L., Shrivastava, A., Merwani, P. and Pandey, A.K. (1984) Effect of sublethal Malathion on ovarian histophysiology in *S. mossambicus*. Comp. Physiol. Ecol., 9(1), 13-17.

Singh, S. and Sahai, S. (1985). Effect of Malathion on the immature ovary of *Rasbora daniconius*. A histopathological study. *Int. J. Acad. Ichthyol.* (Proc. V. AISI), 6, 11-15.

Singh, S. and Sahai, S. (1986). Toxic effects of BHC as the immature ovary of *Rasbora daniconius* Proc. Nat. Acad. Sci. India, 56(B), II, 108-110.

Venugopalan, V.K. and Rao, P.S.B. (1979). Pesticides induced impairments in incubation and post embryonic development of planktonic eggs of estuarine fish of Porto Novo waters. *Proc. Symp. Environ. Biol.*; 397-408. Eds. Verma, S.R., Tyagi, A.K. and Bansal, A.K. Acadamy of Environ. Biol. Muzaffarnagar.

18

INSECTICIDE INDUCED CHANGES IN EGG YOLK PROTEIN PROFILES OF LEBISTES RETICULATUS

Anil K. Singh and K.S. Rao
School of Studies in Zoology,
Vikram University, Ujjain, 456 010 (India)

Abstract—The gel electrophoretic separation of yolk proteins in an ovo-viviparous teleost *Lebistes reticulatus* showed definite patterns and revealed four fractions of different Rf values. However, variations were observed in the yolk protein patterns under 0.025 ppm, 0.020 ppm and 0.0125 ppm malathiona organo-phosphate insecticide exposures. These fractions showed significant variations in their Rf values in these fishes.

INTRODUCTION

A number of investigators have attempted to isolate lipovitellin and phosvitin like yolk proteins from ovaries of a variety of teleost species such as—the herring (Barman *et. al.* 1964), trout (Ando, 1965; Campbell and Idler, 1980; Ito *et. al.*, 1966; Wallacc *et. al.*, 1966), Pecific Salmon (Markert and Vanstone, 1968), Atlantic Salmon (Idler *et. al.*. 1978) and Cod (Plack *et. al.*, 1971).

The ovarian lipovitellin and phosvitin of most vertebrates are complexed together in the form of crystalline platelets. However, most teleost eggs contain a nongranular fluid, yolk and the yolk proteins are present in fluid filled spheres together with the other egg components.

The literature regarding the interpretation of the electrophoretically separated fractions of fish proteins in normal and under experimental conditions still needs attention of workers. However, very limited information is available. (Shakoori *et. al.*, 1976; Mukhopadhyay and Dehadrai, 1980. Bhattacharya and Alfred 1982; Freytag *et. al.*, 1984) on the plasma protein under changed physiological conditions of fishes. Pandey *et. al.*, (1988) studied that the electrophoretic variations in the blood protein fractions in Malathion treated fish *Channa punctatus*.

The purpose of the present study is to observe the yolk protein mobility of fish in the control and experimental pesticide stressed because none of the workers has so far studied this aspect.

MATERIAL AND METHODS

The live fishes *Lebistes reticulatus* were collected from the ponds situated in the vicinity of Ujjain city (India) and acclimatized for 15 days under laboratory conditions. Females were sorted out and maintained separately. Malathion 0,0-dimethyl-S-(1-2-dicarbethoxy ethyl dithiophosphate) 50 EC (Cynamide India, Bombay) was dissolved in 10 litres of water for preparing the 0.025 ppm, 0.020 ppm and 0.0125 ppm concentrations respectively. This was done after testing the LC 50, LC 100 values of malathion for *Lebistes reticulatus*. The control fish were maintained under similar conditions without the addition of insecticide. The fish were fed on alternate days and the water in the aquaria was not changed throughout the period of experiment (Chronic test). The duration of the experiment was seven days only. The fish taken for the experiment were approximately of same weight (0.28 ± .007 gm) size (2.4 ± 0.18 cm) and sex (female). During the experiment the physico-chemical properties of water employed in this experiment were analysed both in control and malathion added water (APHA 1975) Table I and II. After seventh days of the experiment, the fish was dissected under the binocular microscope and the ovary was taken out. The ovary was carefully focussed under the microscope and with the help of a fine syringe and the yolk was sucked and collected in a tube.

Table I. Physico-chemical analysis of normal and Malathion added water
(First day of experiment)

Physico-Chemical properties	*Normal water*	*Malathion Added water*		
		0.025 ppm	*0.020 ppm*	*0.0125 ppm*
pH	6.5	7.0	7.0	7.0
Dissolve Oxygen (mg/l)	7.5±0.1	7.5±0.1	7.5±0.1	7.5±0.1
Chloride (mg/l)	15.72±0.4	15.72±0.4	15.72±0.4	15.72±0.4
Hardness (mg/l)	161±3.03	161±3.03	161±3.03	161±3.03
Total Alkalinity (mg/l)	182.4±3.64	182.4±3.64	182.4±3.64	182.4±3.64
Calcium (mg/l)	26.8±1.92	26.8±1.92	26.8±1.92	26.8±1.92
Phosphate (mg/l)	0.01±0.001	0.011±0.001	0.011±0.001	0.01±0.001

Room temperature=23.7±1.32°C
Water temperature=20.6±1.14°C
All values are expressed in Mean±SD.

The yolk proteins are extracted from the yolk with a concentrated inorganic salt solution *i.e.*, 0.5 M NaCl and centrifuged at 3000 rpm. It facilitated to remove debris and fatty matterials. Alternatively, ovarian fluid and blood clots removed from the yolk by rinsing with 1% saline. The yolk proteins are then precipitated as lipovitellin phosvitin complex by dilution with a large volume (5-7 ml) of distilled water.

Polyacrylamide disc gel electrophoresis gave different fractions of yolk proteins. In the present work electrophoretic

Table II. Physico-chemical analysis of Normal and Malathion added water (After Seventh day of experiment)

Properties	*Normal water*	*Malathion added water 0.025 ppm*	*0.020 ppm*	*0.0125 ppm*
pH	6.7	7.1	7.1	7.1
Dissolve Oxygen (mg/l)	4.3±0.2	4.6±0.2	4.5±0.2	4.5±0.2
Chloride (mg/l)	12.5±0.5	12.5±0.5	12.5±0.5	12.5±0.5
Hardness (mg/l)	158.2±3.01	158.2±3.01	158.2±3.01	158.2±3.01
Total Alkalinity (mg/l)	190±3.24	196±3.26	196±3.26	195±3.26
Calcium (mg/l)	26.2±1.62	26.6±1.52	26.6±1.54	26.4±1.52
Phosphate (mg/l)	0.01±0.001	0.01±0.001	0.01±0.001	0.01±0.001

Room temperature=24.3±1.46°C
Water temperature=21.1±1.18°C
All values are expressed in Mean±SD.

studies were carried out using Tris glacial acetic acid buffer (pH 8.8—8.9) and polyacrylamide gel (7.5%) following the method of Smith (1976). Yolk protein sample were mixed with 15% of sucrose solution and tracking dye bromophenol blue was added to the top of the gel tubes along with the yolk protein sample. The instrument was ran 1-1½ hours with 5mA current per tube. After running for requisite period the tubes were removed and stained with Amidoblack stain. The gel tubes were destained till the over stain is removed and clear bands is observed.

RESULTS

The electrophoretic changes in the yolk proteins profiles of a teleost fish guppy, *Lebistes reticulatus*, in normal and after the treatment with malathion, an organophosphorus insecticide at 0.025 ppm, 0.020 ppm and 0.0125 ppm concentrations have been studied (Table III).

I. Control Group

The eggs of ovo-viviparous fish guppy, *Lebistes reticulatus*, contains a non-granular, semi-solid yolk and the yolk proteins are present in ball like spherule arrangement alongwith other components. In an electrophoretic run yolk proteins pattern revealed four distinct fractions which have the different Rf values (0.099±0.002, 0.188±0.002, 0.266±0.008. and 0.344±0.008 (respectively) were observed, Fraction IV has a maximum Rf 0.344±0.008 and fraction I has a minimum Rf 0.099±0.002.

II. Malathion Treated Group

The fishes were exposed to an organophosphorus very commonly used insecticide in 3 different sublithal concentrations, only for seven days, a short term asssessment.

(*a*) 0.025 ppm :

Malathion treated (0.025 ppm) altered the corresponding Rf values of different bands and some of the bands disappeared altogether. In this concentration only two fractions were appeared which have Rf value 0.033±0.0007 and 0.210±0.008 respectively.

(*b*) *0.020 ppm* :

In 0.020 ppm malathion treated groups, the Rf values of all the 4 fractions were decreased. The Rf values of all four fraction, in control group are 0.099±0.022, 0.116±0.002, 0.266±0.088 and 0.344±0.008 respectively while in the treated group yolk protein fraction has 0.049±0.001, 0.177±0.008, 0.244±0.008 and 0.333±0.014 respectively. In treated group first fraction showed a significant ($P<0.001$) change in Rf.

(*c*) *0.0125 ppm* :

In 0.0125 ppm malathion exposed fish revealed that of the all four fractions were present and Rf values were decreased

Table III : Rf values of the yolk protein fractions of teleost fish *Lebistes reticulatus*

Fraction No.	*Rf value (mm)*			
	Control	*Malathion treated*		
		0.025 ppm	*0.020 ppm*	*0.0125 ppm*
I	0.099±0.002	***0.033±0.0007	***0.049±0.001	***0.044±0.004
II	0.188±0.002	**0.210±0.008	**0.177±0.008	**0.116±0.007
III	0.266±0.008	—	‡0.244±0.008	‡0.244±0.008
IV	0.344±0.008	—	°0.333±0.014	*0.332±0.011

All values are expressed in SEM±

***	$P<0.001$	S
**	$P>0.05$	S
*	$P>0.5$	NS
‡	$P>0.1$	NS
o	$P<0.5$	NS

significantly. The Rf values of treated yolk proteins fraction are 0.044±0.004, 0.116±0.007, 0.244±0.008 and 0.332±0.011 respectively.

DISCUSSION

In the present electrophoretic study of the egg yolk proteins patterns of teleost fish guppy, *Lebistes reticulatus* exhibit four protein fraction in the egg yolk.

Chromatography of the atlantic salmon of yolk extract on TEAE—cellulose yields a single peak which is fractionated into three peaks on Bio-Gel with Tris-NaCl buffer. This middle peak is identified as vitellogenin by its immunoreactivity towards as antiserum raised against a fraction of ovarian yolk protein retarded on sephadex 4.200 (Idler *et. al.*, 1979). Wallace *et. al.*, (1986), observed that the multiple bands are revealed in egg yolk proteins of teleost *Fundulus heteroclitus* he demonstrated that a complex array of phosvitin like components are present within a single ovary of this fish. He postulated that one reason for the anomalous yolk proteins generally found therefore, in teleost eggs is that some of the proteins derived from vitellogenin during vitellogenesis undergo further proteolysis during oocytes maturation.

The similar number of fractions were observed in the malathion treated groups of 0.020 and 0.0125 ppm concentration respectively but in the 0.025 ppm exposure only two fractions were obtained, however, fraction displacement was noticed with marked decreased and increase in the mobility and Rf values of the malathion treated fishes was evident (Table III).

The egg yolk of malathion treated fish *Lebistes reticulatus* in the present study, suggests that a significant alteration in the some of the protein fraction is because of the some of the commercial organophosphorous insecticides contain thiophosphoryl (P=S) group. There are usually very weak inhibition of AchE in vitro, but are activated in vivo by mixed function oxidase (MFO) to the corresponding phosphates

(P=O). The organophosphorous compounds may show differential toxicity in animals. Malathion is rapidly activated to Malaoxon by oxidative disulphuration. Malasxon is a highly potent agent and is very toxic to animals. The compound which cause delayed effect seems to have a common property of binding to specific protein fractions of animals. The P=S compounds are poor inhibitors and must first be oxidized to the P=O analogue. The metabolic processes responsible for this change.

Insecticidal exposure under controlled conditions or in nature certainly bring changes in the gross structure (Shukla and Pandey, 1985 ; 1986) and in protein profiles of yolk in fishes. These changes may not be good for the developing embryos, perhaps it may lead to a reproductive or developmental failure. Therefore, needs caution against wide and frequent application of insecticides in agriculture.

REFERENCES

Ando, K. (1965), Vitra centrifugal analysis of yolk proteins in rainbow trout egg and their changes during development. *Can. J. Biochem. 43*, 373-379.

APHA (1975), *Standard methods for the examination of water* and *waste water*, 14th edition, American Public Health Association, New York.

Barman, T.E. ; Bai, N.K. and Thoai, N.V. (1969) ; Studies on a herring egg phospho protein. *Biochem. J. 90*, 555-558.

Bhattacharya, A. and Alfred J.R.B. (1982). Electrophoretic variation of soluble protein fractions in tissues of different size group of *Channa sterwartii* and *Danio daggilo*. *Matsya 8* : 23-32.

Campbell, C.M., and Idler, D.R. (1980). Characterization of an estradiol—induced protein from rainbow trout serum as vitellogenin by the composition and radio—immunological cross reactivity to ovarian yolk fractions. *Bid. Reprod. 22*, 605-617.

Freytag, V.G. and Spannhof, L. (1995). Electrophoretic patterns of blood plasmo proteins in *Hypophthalmichthys molitrix*, *Aristichthys nobilis* and *ctenopharyngodon idella*. *Zool. Anz.Jena. 215* ($1^1/_2$) ; 83-89.

Idler, D.R., Hwang, S.J. and Crim, L.W. (1979). Quantification of vitellogenin in Atlantic Salmon (*Salmo salar*) plasma by radioimmunoassary. *J. Fish Res, Board—Cana. 36* ; 574-579.

Idler, D.R. and Campbell, C.M. (1980). Gonodotropin stimulation of estrogen and yolk precursor synthesis in juvenile rainbow trout. *Gen. Comp. Endocrinol. 41*, 384-391.

Ito, Y., Fujii, T., Kanamori, M. Hattori, T. and Yoshioka, R. (1966). Phosvitin of the trout roe : *J. Biol. Chem. 60* : 726-728.

Markert, J.R. and Vanstone, W.E. (1968), Egg proteins of pacific salmon genus *Oncorhynchus* : Proteins in the soluble fraction. *Can. J. Biochem. 46* ; 1334-1337.

Mukhopadhyay and Dehadrai, P. (1960). Studies on Air-breathing cut fish *clarios batrachus* (Linn) under Sublethal malathion exposure. *India. J. Exp. Biol.* Vol. *18* April 1980.

Pandey, A.K. P K Joshi and R.K. Nagar (1988), Electrophoretic variation in the blood protein fractions in malathion treated fish *Channa punctatus* (Bloch). *Proc. Nat. Symp. Anim. Meta* and *Poll*. 147-149.

Plack, P.A. and Fraser, N.W. (1971). Effect of oestradiol—3—benzoate on the biosynthesis of egg proteins by Ced. *Biochem. J.* 118, 13P—14P (abstract).

Shakoori, A.R., Saleem A.Z. and Mohammed, S.A, (1976). Effect of malathion, dieldrin and endrin on blood serum proteins and free amino acids pool of *Channa punctatus* (Bloch). *Pak. J. Zool. 8* (2) : 129—139.

Shukla, Lata and Pandey, A.K. (1985). Effect of Endosulphan on the Hypothalmo Hypophysical complex and Fish reproductive physiology. *Bull. Environ. Contam. Toxicol. 36* : 122—131.

Smith, I (1975), *Chromatographic* and *Electrophoretic Technique*. Vol. *II* Pub. William Heinemann Medical Books Ltd.

Wallace, R.A., Jared, D.W. and Eisen, A.Z. (1966). A general method for Isolation and purification of phosvitin from vertebrate egg. *Can. J. Biochem. 44*. 1647-1655.

Wallace, R.A., Robin A. and Paul C. Begorac (1986). Phosvitins in *Fundulus heteroclitus* oocytes and eggs. Preliminary chromatographic and electrophoretic analysis together with biological considerations. *J. Biol. Chem. 260* (20) : 11268-11274.

19

ADVANCES IN HYDROBIOLOGY—MANAGEMENT AND EXPLOITATION OF WATER RESOURCES

Dr. P.V. Dehadrai
Deputy Director General (Fisheries)
Indian Council of Agricultural Research
Ministry of Agriculture
New Delhi-110 001

INTRODUCTION

The Hydrosphere constitutes about three fourth of the earth's surface and it was in water that life is said to have originated first. What led to its origin and how it gradually evolved into various forms leading to the advent of man over millions of years is now the scientific history. But the fact remains that water is one of the Pancha-tatva or five elements so essential for living beings even on land. No wonder the water shed attracted human beings from prehistoric times for habitation and they harnessed the water for food and sustainance and with the growth of civilisation for trade, transport and cultural cohabitation.

However, the world of water remained an alien medium for human due to physiological barriers and therefore, to understand its nature and behaviour, man had to be highly innovative in his approach and methods. Such indulgence of man has not been far less adventurous than his exploits in the space above.

SCIENCE OF HYDROBIOLOGY

In the process, the science of hydrobiology developed. Hydrobiology refers to the interactive science of aquatic organisms and their ambience. It concerns not only with individual species and their environmental relationships, but also with communities of organisms and their interrelationships. Hydrobiology is a multi-disciplinary science involving various aspects of geology, meteorology, physics, chemistry, life-sciences, biochemistry, microbiology and even mathematical statistics and economics of social relevance. Advances in hydrobiology thus have been the result of cumulative advances in overall science of nature and socio-economic demands.

UNSOLVED MYSTERIES

With the growth in scientific knowledge and development of sophisticated instruments for studying the physical, chemical and biological characteristics of waters in their entirety, man has been able to understand today a great deal about various types of aquatic ecosystems. But, as if to reaffirm that man is only a small part of the whole creation a valuable one though-is just a dwarf before the mighty nature and as such many secrets of the world of water still puzzle him. To cite a few examples :

The process of photosynthesis in aquatic medium and the links of bioenergetics in the entire food chain in water remains to be fully understood.

The changes in the oceanic currents such as *Al Nino* along the Peruvian coast, affecting the productivity of oceans and causing unpredictable changes in the climate are puzzling.

Oceanic influences on the behaviour of Monsoon are still unpredictable.

The rising level of the Sea and its association with ozone layer depletion and global warming effect are all uncertain correlations.

The universally constant ratio of chloride *via-a-vis* other salts in Sea water is one of the wonders of the world ; and

What becons Salmon or Hilsa from the Sea to the same river where they were borne ?

This should certainly deflate the ego of the Hydrobiologists ! !

APPLICATION OF HYDROBIOLOGY

Application of the hydrobiological knowledge helped man use the water and exploit it for his immense benefit. Environmental parameters as indexes of aquatic productivity has helped identifying highly fertile zones in oceans. Based on oxity, salinity, current directions and primary level productivity, the ever dynamic migrating shoals of Tuna could be tracked in the open oceans by the Purse Seiners supporting a multi million dollar Tuna fishery industry in the world. Hydrobiology studies in coastal areas unveiled the secrets of upwelling phenomena and imparted a quality of predictability to the data/information on resource behaviour and economics of productivity.

On the basis of the present state of our knowledge, management of riverine and flood plain fisheries, aimed at maximising sustainable yield, has to take congnizance of the hydrobiological parameters. The dynamics of fish populations as well as catch in rivers is a close positive correlates of hydrobiological regime *viz.* water level, discharge and flood behaviour. Stock assessment and catch predictions lean heavily on statistical correlations between hydrobiological parameters and the catch.

Knowing the hydrobiology of open and flowing waters in rivers, canals and in coastal protected bays in Seas, has helped in raising highly remunerative animal protein production through floating or submerged cages all over the world. In fact, 80% of the total aquaculture production in South East Asia Region comes from cage culture of fish.

The principles of hydrobiology have been most effectively utilized in raising the productivity of closed water systems through aquaculture. As such, the potential of productivity from a three dimensional water body per cubic meter volume is sky high against the productivity potential of a two dimensional land area per square meter.

Fish being cold blooded animal in water, with its positive hydrodynamics, low resistance in locomotion, and low energy budget for maintenance of routine metabolism reserves almost 80% of the energy from feed for growth. Whereas the terrestrial domesticated animals such as cattle, pigs, poultry birds etc. require more than 80% of the energy made available to them from feed for routine metabolism. Therefore, the profit from per unit investment efforts by rearing aquatic animals is much higher than from rearing *terrestrial* animals. Thus application of hydrobiological knowledge has transformed the art of extensive fish culture systems to semi-intensive, intensive and now to super-intensive or industrial aquaculture. Production rates as high as 600 tonnes/ha/yr. of *Tilapia* in Philippines, 155 tonnes/ha/yr. of Carps in Manheim in Germany, 200 tonnes/ha/yr. of Catfish in Thailand, 20 tonnes/ha/yr. of Shrimp in Taiwan, 55 tonnes/ha/5 months of Magur in Delhi were unthinkable even a few years ago.

Semi-treated domestic sewage tanks with very high BOD could be treated for oxidation through culture of green algae like *Spirulina*, which could be a food with over 20% protein. Brewery wastes are being utilised for *Tilapia* production of the order of 50 tonnes/ha/yr. near Madras.

Hydropunics or Aquapunics are recent innovations where cut flower plants or vegetable plants are grown on wire gauze suspended in trays filled with nutrient rich water drawn from treated domestic sewage as a high profit commercial activity in California and elsewhere.

Professor Hardy's (year) suggestion to construct a factory —submarine swallowing plankton from open oceanic waters to process edible protein thus imitating the *Whale*, is still a concept though. Use of large, transparent plastic balloons submerged in open Seas for controlled production of plankton is another innovation using hydrobiological principles.

DAMAGE TO WATER RESOURCES

It is, however, a matter of great concern that in the over-zealous utilisation and exploitation of water resources, rather indescriminately, man has already brought about situations when several aquatic ecosystems are at the verge of disruption. Cases of oil pallution in the Sea is of usual occurrence. Thailand overfished around its seas causing long term damage to its fisheries resource. India is not far behind by exerting thoughtless fishing pressure on its coastal shrimp resources.

Damage to water bodies all over the country through agricultural pesticide run-off is already a national concern. Industrial affluents have converted our estuaries into waste dumping grounds. Domestic sewage incursions have damaged productivity of rivers all over. Per unit effort Catch of fish from Ganga has decreased from about 1000 Kg in fiftees to about 400 Kg presently.

With man made engineering changes through construction of dams and barrages, the riverine ecosystems have changed adversely. Due to Farrakka barrage and industrial pallution in Ganga, the most valuable *Hilsa* fisheries worth millions of rupees have been lost for India. Thoughtlessly planned river valley projects on five fast flowing rivers in North Bihar has turned one of the most fertile lands in India into inundated saline tracts. Consequently, the ponds and lakes of North Bihar have become more or less swamps. While it took 17 years for British to bring about equitable natural balance in river Thames at enormous cost, the story of Ganga has been a repeatation with crores of rupees allocated to clean it. Similarly our national temples of prosperity the man made reservoirs-could end up shallowed, silted and unfunctional due to lack of monitoring and maintenance of the ecology of the catchment areas. Role of industrial and agricultural pallution flowing into the rural ponds and rivers could as well be a major cause of the Epizootic Ulcerative Syndrome (EUS)-the much dreaded fish disease which damaged the fish production and trade affecting the daily income of thousands of small fishermen in the eastern states of India during 1988-89.

To what use, we are putting our knowledge on hydrobiology ? Having assembled here in proximity of Ujjain, I am reminded of Mahakavi Kalidas. Before the wisdom dawned on him, he is reported to have been cutting the branch of the

Tree on which he sat. The analogy is too apt for the situation I believe.

ROLE OF HYDROBIOLOGIST

Therefore, I would urge the hydrobiologists that they must participate in strategy formulation based on their research and that they should ensure that the information generated in the field of hydrobiology must influence the policies and programmes of our country. The scientific community must assert with their considered opinion on the overall development of water resources in the country.

20

HYDROGRAPHIC AND HYDROBIOLOGICAL PATTERNS OF THE EAST COAST OF INDIA AND THEIR UTILITY IN COASTAL USE ZONATION

T. Rajyalakshmi

(Scientist (Retd.), Central Institute of Brackishwater Aquaculture (I.C.A.R.), Madras)

INTRODUCTION

The economic development of a nation or a region is directly dependant on the resources available to it and the way these resources are utilized by the people. Inadequate knowledge of these resources is likely to result in unplanned exploitation, some times over exploitation or even wasteful exploitation. This lecture is on a major resource, *viz.*, the coastal zone on the East Coast of India, with particular reference to the hydrobiological and hydrographic patterns and utilization of this zone, with emphasis on coastal aquaculture.

These hydrographic and hydrobiologic patterns have received considerable attention in selected regions of the East coast in early 1940s to 1950s. Mention can be made of studies conducted by La Fond and the team of scientists from the Andhra University, Waltair, off the Waltair coast extending upto the head of the Bay of Bengal. We have similar, if not such detailed studies off the Madras coast, off the Gulf of Mannar to Palk Bay. From 1970s, the National Institute of Oceanography has taken up detailed studies in different regions.

Of studies pertinent to coastal aquaculture, mention must be made of site-specific studies along Balasore coast and along Digha-Sagar coasts conducted by the IIT, Kharaghpur.

In this lecture, the emphasis is on the hydrographical and hydrobiological patterns which may help us ultimately in classification of zones of the coastal areas. These can act as tools as for example how and using what criteria ports are located in specific areas. Similar, if not more, detailed approach can be made to locate the coastal aquaculture farms or hatcheries. Such studies will dilineate the areas suitable for a purpose without conflict with another use of the coastal zone.

The coastline of nearly 2500 km along the four contiguous states on the East Coast *viz.*, West Bengal, Orissa, Andhra Pradesh and Tamil Nadu is governed by specific tidal, current, littoral drift and such factors. These are presented taking available data from the existing literature and giving necessary illustrations and interpretations as to their import on larval movements of prawns particularly and, on the broodstock of prawns. Since prawns are the one single group of animals that dominate the aquaculture industry today in India, this approach is used. Similar approaches can be made for other groups of animals.

In the following pages, descriptions are made on the state-wise coastal region basis to suit the understanding of coastal aquaculturists.

GENERAL TIDAL CHARACTERISTICS

In this context, it is necessary to mention about the method of determining the coastal characteristics, characteristics such as the tides, sea levels, winds and tidal forces.

Tidal levels

The tidal levels are generally recorded every hour over a period of 19 years to establish the starting level or sea level datum. These readings are then averaged to establish a mean datum known as Mean sea Level. This mean sea level (MSL) is a datum which fluctuates according to atmospheric pressure changes, winds, tidal forces etc. Variations of upto one metre have been noted (Buttler *et al.*, 1986).

Two more points are to be noted in regard to these datum. One is the depth and another is height. Normally, therefore, lowest Normal Water Level and in the intertidal zones, heights are measured upwards from the same level. The other datum is heights. This phenomenon termed as spring tides, which occurs twice a monthly peak of tidal range, is generally used to determine this datum. These are labelled as Higher Water Ordinary Spring Tides, Mean High Water Springs, Highest Astronomical Tide, Mean Higher High Water.

On the East Coast of India now we have coastal tidal heights of 7 m at the mouths of the Hooghly estuary, and less than 1 m at the Kaveri. In between the heights decline to 5 m northern Orissa, 2 m along southern Orissa 2-1 m along Andhra Pradesh. These are the highest high water levels. The lowest are at -0.1 m.

These features are of great importance when we consider the tidal dominance in the estuaries and backwaters, distances covered by these tides both ebb and low tides and consequently the enormous movements of soil, nutrients, fauna etc. involved here. This movement has the power to convey enormous loads of effluents from along the banks of rivers where the industries are located and move them miles away. This movement is also a interia for location of aquaculture farms.

WEST BENGAL

This coast line forms the head of the Bay of Bengal. Several rivers from the vast complex of Hooghly, Padma, Irrawady and others open into the Bay here.

Meteorological data

The humidity, wind speed, evaporation and rainfall data of two of the main coastal regions along this coast, Kakdwip and Sagar areas is shown in Table 1. The maximum temperature range is from 25.1°C (January) to 35.1°C (May). The

Table 1. Average metrological data of the coastal area

Month	*Jan.*	*Feb.*	*March*	*April*	*May*	*June*	*July.*	*Aug.*	*Sept.*	*Oct.*	*Nov.*	*Dec.*
Max. Temp. °C	25.1	29.4	33.5	34.9	35.1	33.1	31.3	31.0	31.4	30.7	28.8	26.1
Min Temp. °C	14.1	17.3	21.2	24.9	26.7	26.3	26.3	25.9	26.0	24.1	13.1	14.6
Relative humidity (I)	90	90	90	89	88	92	93	94	93	92	90	89
% (II)	46	43	44	56	63	80	83	85	81	69	52	47
Wind velocity km/hr	4.0	4.0	6.0	9.3	16.6	12.0	12.9	9.8	5.4	3.1	3.0	3.0
Rainfall (Cm)	1.5	2.2	3.5	6.1	14.2	30.0	38.3	37.2	30.0	30.0	2.4	1.2
												=196.6

Source : Report of the Department of Fishenes, Govt. of West Bengal, 1970.

minimum range is from 14.1°C (Jan) to 26.7 (May). Relative humidity is 88-92 ; wind velocity 3.0-16.6 km/hr and Rainfall 1.5 cm to 38.3 cm.

For the Contai-Digha coast the meteorological data is as follows :

Temperature	:	Varies from 22°C to 37°C
Humidity	:	upto 70%
Rainfall	:	2000 m.m.

Tidal pattern

Tidal characteristics and spring and neap tide pattern at Sagar shown in Table 2 is the highest for the East coast in general. Tidal high water elevation is 5.7 m on an average. This carries the tidal water for inland and even upto 190 km.

Coastal currents

The most pertinent studies relating to coastal currents were conducted in the Bay of Bengal by Andhra University, Waltair, starting during the year 1952. 52 oceanographic cruises were conducted and 700 stations were established for coverage (Ganapati and Rao, 1959). This study showed that drainage into the Bay of Bengal is maximum during the post monsoon months namely September-October. On an average, per annum, nearly 1300 million acre feet of fresh-waters are drained into the Bay from all the northern rivers from Ganges to Irrawady.

Warm, low saline waters in the surface region, cold saline waters at subsurface waters are present. This is not conducive for mixing.

There is a clockwise movement of waters during January-June, called as transparent currents and anti-clockwise movement southerly termed turbid current during July-December. These are termed as littoral drifts. The effects of these are felt 15 miles offshore.

These are discussed in greater detail later.

ORISSA

Temperature

The temperature of the coastal area from north to south (where sites for brackishwater farms are likely to be located) is shown in the Table 3 (Anon, 1984). This shows a minimum of 10°C to a maximum of 43°C.

The salinity and temperature of air and coastal water has been estimated for one site at Balasore coast in detail by studies conducted by IIT, Kharaghpur (Data by courtesy-State Department, Fisheries, West Bengal). This is as follows : During the monsoon months, the coastal water is diluted to some extent by the discharges of Dhamra-Mahanadi system of rivers at one end and by the River Gumei at the other. In summer, the rivers hardly carry any significant amount of freshwater and the salinity of the coastal water attains the maximum limit. Range of salinity variation of coastal water as obtained from the observation during the period of field investigation in the months of April-August shows 5 ppt-30 ppt in July-October, 25-35 ppt and 33-35 ppt in the remaining months. The salinity of coastal water attains a maximum of 33-35 ppt during the dry months of the year. With the onset of monsoon the salinity drop depends upon the quantum of freshwater discharges by the rivers. Since the rivers often bring flash discharges, the salinity of coastal water drops temporarily and rises sharply inbetween the intervening period of peak river discharges. Thus, the salinity of coastal water changes over a wide range during the monsoon months.

Considering the similarlity of coastal climate and proximate, the air temperature recorded at Balasore by IMD (Anon, 1985) show the monthly average of air temperature. Records of water temperature throughout the year are not available. However, temperature of sea water and air maintain a close relationship and based on the observations made and relationship obtained for the coastal water of neighbouring region, likely variation of temperature of the coastal water at the site is around 5°C.

Table 2. Monthly occurance of tidal high water at sagar

Tidal High Water elevation above Sagar Datum	*Number of exceedance*																	
	Jan.			*Feb.*			*Mar.*			*Apr.*			*May*			*June*		
in Meter	*Min.*	*Avg.*	*Max.*	*Min.*	*Avg.*	*Max.*	*Min.*	*Avg.*	*Max*	*Min.*	*Avg.*	*Max.*	*Min.*	*Avg.*	*Max.*	*Min.*	*Aug.*	*Max.*
4.90	4	7.2	11	4	7.2	10												
5.00	0	4.6	7	2	4.7	7	6	8.4	12	8	11.7	17	10	14.2	19	12	17.1	20
5.10	0	2.7	5	0	2.9	6	3	6.1	8	4	8.4	12	6	12.3	15	9	13.2	16
5.20	0	1.8	4	0	1.6	4	0	3.1	6	3	5.8	8	2	7.1	9	4	8.7	11
5.30	0	0.5	3	0	0.7	2	0	1.3	4	1	3.3	6	0	4.5	5	2	5.5	7
5.50	0	0	0	0	0	0	0	0.1	1	0	1.7	4	0	2.8	5	0	3.5	5
5.50							0	0	0	0	0.3	1	0	1.6	3	0	1.9	4
5.60										0	0	0	0	0.8	3	0	1.2	4
5.70													0	0		0	0.4	2

(Contd.)

Table 2. *(Contd.)*

Tidal High Water elevation above Sagar Datum	*Number of exceedance*																	
	Jul.			*Aug.*			*Sept.*			*Oct.*			*Nov.*			*Dec.*		
in Meter	*Min.*	*Avg.*	*Max.*	*Min.*	*Avg.*	*Max.*	*Min.*	*Avg.*	*Max.*	*Min.*	*Avg.*	*Max.*	*Min.*	*Avg.*	*Max.*	*Min.*	*Avg.*	*Max.*
4.90																		
5.00	19	20.9	24	25	27.5	31	26	28.6	32	20	27.6	32	12	1.9	24	5	9.5	15
5.10	14	16.5	19	18	22.4	26	23	25.3	27	178	23.9	29	10	15.2	18	0	6.9	11
5.20	10	12.6	15	15	18.3	23	19	21.2	24	13	19.8	26	6	12.6	17	0	4.2	8
5.30	5	8.1	12	10	13.5	16	13	17.3	21	10	15.1	22	4	8.6	18	0	3	6
5.40	4	4.8	6	7	9	12	9	11.3	13	5	11.7	17	4	5.9	11	0	1.2	4
5.50	3	3.8	5	1	6.1	9	4	8	10	3	7.9	12	2	3.9	7	0	0.7	3
5.60	0	2	3	0	4.1	7	0	6	8	2	5.4	9	0	2.5	4	0	0.1	1
5.70	0	0.4	2	0	2.4	5	0	4.8	8	0	3.6	6	0	1	0			

Source : Report of the Dept. of Fishence Govt. of West Bengal, 1990.

Rainfall in coastal areas

The pattern of rainfall for the coastal region along the four districts of Orissa for the years 1977-83 varied from 995 mm to 1914 mm (Anon, 1985).

Table 3. Temperature ranges by months and stations in coastal Orissa during 1984.

(in 0.0 Centigrade)

Month		*Station*			
		Balasore (Balasore district)	*Paradeep (Cuttack district)*	*Puri Puri district)*	*Gopalpur (Ganjam district)*
January	Max.	30	29	29	29
	Min.	10	13	15	16
February	Max.	33	29	30	31
	Min.	11	14	15	15
March	Max.	41	34	32	35
	Min.	13	16	18	15
April	Max.	43	33	36	36
	Min.	22	23	21	25
May	Max.	41	34	33	35
	Min.	19	23	19	24
June	Max.	39	37	38	39
	Min.	23	21	23	23
July	Max.	34	32	35	35
	Min.	25	25	25	25
August	Max.	35	33	34	33
	Min.	23	23	21	23
September	Max.	34	31	35	35
	Min.	22	24	24	24
October	Max.	34	34	35	35
	Min.	19	21	21	20
November	Max.	32	31	32	33
	Min.	13	14	15	13
December	Max.	33	33	33	33
	Min.	10	12	14	14

Source : Anon, Development of brackishwater Fishenes in Orissa States Report, 1984.

Specifically for northern coastline at Balasore coast, data collected by IIT pertaining to Chandhbali region for the period 1976-85 shows that maximum rainfall is associated with the period May-June to August-September *i.e.* south-west monsoon. Some rainfall occurs during the northeast monsoon between October-November. The period of minimum rainfall was December. Mean humidity is 61-80%.

Tides and tide patterns

The tidal volumes and velocities decline from northern coast of Orissa to southern coast at Ganjam. The average of 2 m occurs at the mid-coastal regions off Cuttack and Puri districts.

The tides are semi-diurnal in nature in this region and with considerable variation between spring and neap tides. At Balasore coast the spring and neap tide ranges are 2.90 m and 1.00 m respectively, spring tides occur on an average six times, in a month has been considered as "design tides" (Table 4) (Bose and Roy, 1985).

Table 4. Tidal high water reached or exceeded the levels indicated atleast six times in a month.

Months	*Spring tide high water reached or exceeded at least six times in a month at Shortt's Island*	*At Kaithkola (proposed site)*	*Remarks*
January	+ 1.40 m	+ 1.47 m	Elevation are referred to Mean Sea Level (GTS)
February	+ 1.40 m	+ 1.47 m	
March	+ 1.60 m	+ 1.68 m	
April	+ 1.60 m	+ 1.68 m	
May	+ 1.60 m	+ 1.68 m	
June	+ 1.60 m	+ 1.68 m	
July	+ 1.80 m	+ 1.89 m	
August	+ 1.80 m	+ 1.89 m	
September	+ 2.00 m	+ 2.10 m	
October	+ 2.00 m	+ 2.10 m	
November	+ 1.80 m	+ 1.89 m	
December	+ 1.60 m	+ 1.68 m	

Other meteorological data

from Bose and Roy, 1985)

Humidity

Humidity of air is important in estimating the evaporation loss from the aquaculture ponds. The loss not only necessitates replacement of pondwater but also increases the salinity which may require dilution if exceeded beyond the tolerable limit of the fish/prawn cultured. Although a site may be located on the coastline, the humidity of air in the coastal Orissa varies within the range 70-84% during different months of the year. During the winter, the humidity is lowest and it gradually increases with the onset of South-Westerly wind. Maximum humidity is observed during the monsoon.

Monthly averages of relative humidity of Chandbali is 70-84%.

Evaporation

Evaporation loss from the pond water surface has been estimated analytically on the basis of temperature, humidity, wind speed and sunshine hours of the locality. Penman's theory has been applied for the estimation. The estimated evaporation loss during different months of the year is 4.8% in December (being the minimum) and 8.0% (in April-May).

Wind and storm

The site located on the Balasore coast of the Bay of Bengal has the wind set-up at the place similar to that of the head of the Bay. There are mainly two dominant set up *viz.*, North-Easterly and South-Westerly wind. From December to February, the North-Wasterly wind is dominant and the period is characterized by low temperature and scanty rainfall. The South-Westerly wind dominates during the months from June to October and during this period a series of depression passes over the area accompanied by heavy precipitation. The relative humidity of air remains also high. During the pre-monsoon (March to May) and post monsoon (November to

December) months, the wind set up gradually changes from North-Easterly to South-Westerly and from South-Westerly to North-Easterly respectively. The nearest IMD station of the site is at Balasore where daily wind speed is recorded. Table 5 shows the monthly averages of wind speeds.

Table 5. Humidity at Chandbali, wind speed at Balasore and evaporration

Month	*Mean humidity* %	*Average wind velocity km/hr*	*Evaporation mm/day*
January	70	7.7	4.6
February	70	8.7	5.5
March	73	12.3	6.5
April	71	16.1	8.0
May	74	17.7	8.0
June	79	13.7	7.5
July	82	13.4	7.3
August	84	12.5	7.0
September	81	10.3	6.8
October	76	7.7	5.0
November	64	7.8	5.5
December	64	7.6	4.8

During the monsoon months, the normal wind set-up is frequently interrupted by the generation of depressions and storms over the bay. At this time cyclonic wind with much higher wind speeds than normal are experienced. A wind speed of 40-50 km/hr during depression and 70-120 km/hr during storm are common feature of the site. Examination of storm tracks over the head of the bay shows that majority of depressions and storms move towards the Balasore coast from the bay, cross the coastline at or near the site and enter into the country.

Frequency of storms and cyclones

Cyclones off Gopalpur Coast

The east coast is affected by cyclones. These cyclones normally prevail during the months May to November. The occurrence of cyclones is more frequent during South West monsoon than during North-East monsoon. During the months May to September, the storms originate in the northern half of the Bay of Bengal and these disturbances approach the Gopalpur coast from the east whereas during October to December, the storms originate much further south and curve towards the right running parallel to the coast. An examination of the frequency of occurrence of storms between 1891 to 1971 indicates that 77 storms had occurred in the vicinity of Gopalpur. This indicates that the sand drift would be in the region of 0.74 million m^3 which can also be used for Chilka Lake area, being very close to it (Anon, Report of the Committee constituted by the Government of India to examine all problems connected with the Chilka Lake and suggest Remedial measures, 1974).

Littoral drift

Littoral currents are induced due to breaking of waves at an angle to the shore-line. This littoral current is the primary factor causing sand movement along the coast. In the Bay of Bengal, during southwest monsoon, from April to September the waves coming from south southwest cause enormous sand movement northwards. Similarly, during northeast monsoon from October to December a predominance of northeastern waves causes a littoral drift from north to south. However, in view of the predominance of wave energy during southwest monsoon, the total drift from south to north along the coast is always higher than the drift from north to south with the result that the net littoral drift along Orissa coast is north-wards. For studies relating to Paradip, Visakhapatnam, Madras ports at the Central Water and Power research station, Poona estimates have been made from which it is seen that the net movement of sand from south to north is 0.6 million m^3 along the coast at Madras, 0.7 million m^3 along the coast at Visakhapatnam and 0.94 million m^3 at Paradip. The estimate of post monsoon (November to December) months the wind set up gradually changes from north-easterly to south-westerly and from south-westerly to North-easterly respectively.

At Balasore coast

During the monsoon months, the normal wind set-up is frequently interrupted by the generation of depressions and storms over the Bay. At this time cyclonic wind with much higher wind speeds than normal are experienced. A wind speed of 40-50 km/hr during depression and 70-120 km/hr during storm are common. The majority of storm tracts move towards the Balasore coast from the Bay. The average passages of depressions and storms over the Balasore area during a period of 70 years is given in the Table 6 which shows that frequency of severe storm in this area is of the order of once in 12 years. Sometimes depressions concentrate into heavy storm and lash this coastline with considerable damage. In recent past one such storm passed over the area in the year 1971 between 27.10.71 and 30.10.71.

Table 6. Average passages of the number of Depressions and storms in a month over the balasore coastline in Orissa (Period of study=70 years)

Months	*Depressions*	*Storms*	*Severe storms*	*Remarks*
July	54 nos	10 nos	3 nos	Frequencies :-
August	45 nos	2 nos	1 no	(*i*) Depression-2 nos per year approx.
September	25 nos	6 nos	1 no	(*ii*) Storms once in 4 years approx.
October	4 nos	×	1 no	(*iii*) Severe storm once in 12 yrs. approx.
Total	128 nos	18 nos	6 nos	

Storm-tides and waves at Balasore

Coast selected site

In the coastal location with the incidence of high wind during depressions and storms, a surface current is induced by the tangential stresses at the water surface between wind and water. The surface current is in the general direction of the wind movement and a strong offshore wind produces a piling up of water near the coastline increasing the water level above the normal tide level. The term storm-tide has been used to define that rise of normal water level due to the action of wind stress on the water surface during the storm as well as includes that rise in level due to atmospheric pressure reduction. Highest elevation of storm-tide level is attained when the surging effect occurs at the time of tidal high water. Since the occurrence of storm and depressions is a common feature in this area, the tidal high waters rise above the normal levels frequently during the monsoon months (Bose and Roy, 1985).

According to Bose and Roy (1985), one method of predicting storm surge in a coastal area is by the use of frequency charts obtained from a statistical analysis of water level data. This, they state, requires much data over many number of years and by appropriate analysis a relatively accurate empirical method of prediction can be developed. But these formulae are not applicable to places other than specific points where long period of observation is available and for which they are derived. For the place under consideration, such a statistical approach is not possible since long term observation for the place is not available. The authors mention that a good estimates may, however, be made by treating the problem analytically, considering the bottom topography of continental shelf, wind speed and direction during the storm, pressure reduction at the storm centre and normal astronomical tide. For the particular case, storm-surge level has been calculated for the storm of 1971. As per the analysis, maximum possible surge levels at the locality are about 3.21 m, 3.64 m, 4.28 m and 4.99 m above GTS for depressions and storms associated with maximum wind speeds of 75 km/hr, 100 km/hr, 125 km/hr and 150 km/hr respectively.

Currents in the bay of Bengal

Various occeanographic cruises werc conducted by the Andhra University during 1952-57 (La Fond 1957, 1958) and hydrographical data and plankton, surface and subsurface distribution of temperature and salinity in the western Bay of Bengal was obtained.

The area covered extends between parallels 13° 11.5′ latitude north (off Madras) and 21° 25.0′ latitude north (near swath of No Ground) and the meridians 80° 21.4′ longitude East and 59° 41.0′ longitude east.

The waters of the Bay of Bengal are subject to the influence of two important factors (Rao, 1958). Firstly, the major rivers of India and Burma empty themselves into the Bay of Bengal. To the East of it the Irrawady, to the North of it the Ganges and the Brahmaputra and on the west the Mahanadi, the Godavari, the Krishna and the Cauvery bring large volumes of silt-laden freshwater especially during the monsoons. Secondly, the north-east and the southwest monsoons set up, by their influence, four distinct patterns in the movements of the surface waters of the Bay during the various months of the year.

ANDHRA PRADESH

Environmental-climate and oceanographic data

The year can be divided into four clear seasons namely south-west monsoon (June-September), north-east monsoon (October-December), winter (January-February) and Summer (March-May).

Temperature of air and coastal water and salinity

The salinity and temperature fluctuations in the surface waters off Visakhapatnam are available for several years from 1954. These have been recorded by La Fond (1954) as factors affecting vertical temperature structure of the upper layer of the sea. These observations (Ganapati and Rao, 1954) are limited to the inshore and off shore waters extending between Latitudes 17° 03.′ 16 N and 17° 41.0′N and Longitudes 83° 18.2′E and 83° 57.6′E. This gives a profile of the coast about 50 miles off-shore and approximately 10 miles north and 10 miles south of Visakhapatnam harbour. Later observations in 1979 by National Institute of Oceanography, off Waltair coast also conform to this general pattern.

October is the transition month between the south-west and north-east monsoon. In the northern and central portions of the Bay the south-west monsoon diminishes rapidly The average salinity ranges from 18.26% to 28.59%o. The minimum salinity of the surface water is in the 5-10 mile zone. There is a steep rise in salinity in from the 20 mile line and at 50 miles offshore the salinity is 28.57%o. The temperature is also lowest in the 5 to 10 mile section and rises steeply in the 10 to 20 mile section after which it remains steady. There is a halocline developing with depth.

In December the north-east monsoon gets fully established over the Bay. Low salinity waters are restricted to a coastal belt extending to about 7 miles offshore and after this zone there is a steep rise to about 32% in waters 20 miles offshore. A halocline still persists. Surface temperature are also cooler from 75.9°F to 81°F due to waters discharged by the northern rivers at this season. There is as table stratification from October to January.

During February as the northeast monsoon retreats the currents are in a north-easterly direction. The salinity ranges from 32% to 33% within the 5-mile zone and upto 30 miles offshore. Surface temperature increases due to inflow of high saline waters from the mouth of the Bay and from the equatorial regions along the eastern coasts.

In March, the currents remain same but an upwelling occurs along the upper shelf upto 10 miles. There is a north-easterly current.

The peak salinity occurs in April and May and the minimum in November. The peaks in temperature are October, March, June and January, April, August respectively. The annual range in the mean being 4.4°C. The upwelling is more pronounced.

These studies were further affirmed by the investigations on hydrological conditions of the bottom water off Visakhapatnam during 1968-1972 (Mozumdar, 1979). Salinity showed lesser variation at bottom than at surfacc. Changes of salinity from maximum to minimum and vice versa were quite rapid with a minimum in October and a steady maintenance in the maximum for a long period from April to July. Temperature variation gave rise to double maxima and double minima in the annual trend.

The air temperatures during October are same as those of September. In November, the maximum temperature is around 28-30°C, while the minimum is around 20-22°C in the coast. In December, the pattern is similar.

Rainfall

Rainfall in Andhra Pradesh is influenced by both southwest and northeast monsoon. The normal annual rainfall of the state is 894 mm. The coast of Andhra Pradesh receives predominantly south-west monsoon in its middle to northern part (July-September) while predominant northeast monsoon prevails in the southern coastline (50-60% in Prakasam, Nellore districts ; thence in October-November). Coastal belt north of latitude 16°N receives 1000-1500 mm rain.

Tidal pattern

Tidal regime is generally low ranging between a maximum of 2 m in the northern part of Andhra Pradesh to 0.8 to 1 m in the southern part. At Kakinada Bay, highest high tides were recorded in May once and again November. These inundate the low-lying areas along with which incursions of postlarvae take place.

At Upputeru, Krishnapatnam areas (southern coastline of AP)

At Krishnapatnam (Nellore District), located on the left bank of Upputeru, about 1.5 km upstream of the Bay of Bengal. The tidal surge varies as per the littoral drift. In view of the large northerly drift along the eastern coast of India, tidal inlets along this coast are generally unstable and tend to get choked during the southwest monsoon when the quantity of drift is considerably high compared to the meagre tidal flow. However, the mouth of the Upputeru is an exception to this general rule. The tidal range here is very low (of the order of 0.8 m during a spring tide). However, the tidal rise in the ocean is transmitted through a distance of 21.5 km upstream from the mouth practically with no attenuation. As a result of this a total quantity of 8 million m^3 of water is exchanged between the river and the sea during a tidal cycle. This inflow and outflow through the mouth generate considerable velocities at the entrance so as to keep the entrance flushed against an estimated littoral drift of 0.6 million m^3 per year (Vidyaraman *et al.*, 1980). Thus, this is a self-maintaining inlet.

Kakinada bay (MID east coast, AP)

Kakinada Bay (or Coringa Bay as per old admirality charts) lies between 82° 14′ and 82° 22′E longitudes and 16°5′ and 17°N latitude, is a part of the Godavari estuarine system. It is an important brackishwater farm region in the mid-east coast of Andhra Pradesh and a zone of abundant seed availability.

The hydrography of this Bay has been described by Ramasarma and Ganapati (1967).

The water transparency in the 1-2 m depth zone nearest to shore is less than in the central part of the Bay (2-5m depth) and the mouth region (8-10 m depth) (Table 7).

Inflow of silt laden freshwaters into the Bay takes place during the southwest monsoon and post-monsoon also.

The Bay opens widely into the sea on the northern side. Freshwater inflow is from the south and from West. The hydrography of the bay is as follows :

Winds	*Months*	*Temp.*		*Salinity*	
		S	*B*	*S*	*B*
Northerly	March-May	26.2°C	34.0°C	28.03-34.42%	
South-west	June-Sept.	24°C	28.5°C	34.11-34.41%	
	Oct.-Nov.	26.6°C	28.8°C to 17.07	5.88-11.31 to 26.18%	
	Dec.-Feb.	23.0°C	26.5°C	26.76-31.02%	

Other meteorological data

Humidity and evapotranspiration

The humidity during the season from October to December ranges from 60-75%. The surface winds during October are very light and there are many days without wind. When the northeast monsoon is active, surface winds are generally from east or northeast with speeds of 5-10 km ph. During November-December, surface winds are from northeast/east with average of about 5 kmph.

The normal monthly and Annual potential evapotranspiration (mm) (Venkateswaralu, 1982) in coastal Andhra Pradesh is :

Jan	Feb	Mar	Apr	May	June	Jul	Aug	Sep	Oct	Nov	Dec	Annual
109	122	166	176	193		167	134	136	123	108	102	1655

Maximum temperature in January/February are 28-30°C and minimum 15-19°C. The surface winds are from north easterly/easterly direction in the morning become south easterly during afternoon. The speeds are 6-10 kmph.

Humidity is fairly high 60-70%. The driest part is between March to May. The relative humidities were highest during the early hours throughout most of the year.

Fequency of storms and cyclones

During the season of Northeast monsoon, October to December, depressions and storms take more southerly track and affect the coast of Andhra Pradesh (Venkateswarlu, 1982). During the season, 2 or 3 depressions occur from which one may attain severe storm intensity. These strike Andhra Pradesh in October, Tamil Nadu coast in October and South Tamil Nadu in December. These storms give widespread and heavy rainfall in coastal Andhra Pradesh. In addition these storms cause gales where wind speeds reach 100-200 kmph. In the case of severe storms, the gales may attain a speed of 150 kmph or more. Coastal Andhra Pradesh also encounters storms during the summer season particularly during May and June. On the basis of back data, it is estimated that a severe storm can strike once in 10 years.

Table 7. Net water movement in the Kakinada Bay in relation to the prevalling direction of coastal currents and wind

Month	*Predominant wind direction*		*Direction of the coastal current*	*Pattern of the residual or net movement of water in the Kakinada Bay*
	0830 hrs (I.S.T.)	*1730 hrs (I.S.T.)*		
Jan.	NE to N	E to SE	Northerly	Clockwise
Feb.	W to NW	S to SE	Northerly	Clockwise
Mar.	W to SE	S/SE to SW	Northerly	Clockwise
April	W to SW	W to SW	Northerly	Clockwise
May	W to SW	W to SW	Northerly	Clockwise
June	W to SW	W to SW	Northerly	Clockwise
July	W to SW	W to SW	Transition from Northerly to Southerly	Counter clockwise

(*Contd.*)

Aug.	W to SW	W to SW	Southerly	Counter-clockwise
Sept.	W to SW	W to SW	Southerly	Counter-clockwise
Oct.	NE to NW	E to SE	Southerly	Counter-clockwise
Nov.	N to NE		Southerly	Counter-clockwise
Dec.	N to NE		Transition from Southerly to Northerly	Clockwise

Source : Ramasarma and Ganpati 1968.

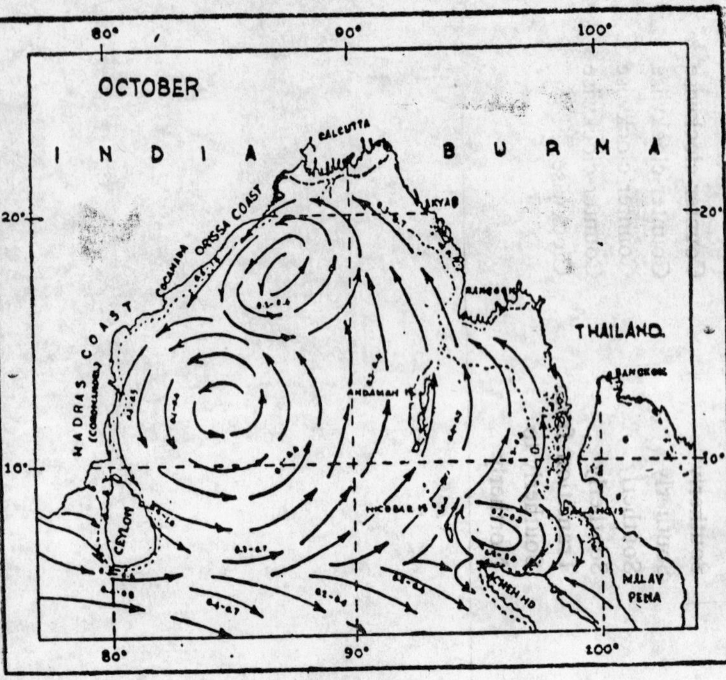

Fig. 1. Currents in the Bay of Bengal during October.

Source : Ganapati and Murthy, 1954.

Currents in the Bay of Bengal

Currents off the Visakhapatnam coast

The two opposite strong current skirting the coast, a south-westerly current from August to November and north-easterly current from mid-December till July. These two currents play the important role in distribution of salinity and temperature (Ganapati and Rao, 1954).

Figs. 1 & 2 shows the counter clockwise currents in October and December, one off the coast of Orissa and southern west Bengal and another off the Madras-Andhra Pradesh. The current off the Orissa coast moves in a southerly direction along the Visakhapatnam coast and merges with the other. It has a strong flow. It also brings large volume of freshwaters from the north from the Brahmaputra, Ganges and the Mahanadi.

By February and March the current is in a north easterly direction at a short distance from the land between Sacramento Shoal and Bheemunipatnam. From January to April both inclusive, it attains its greatest rate which is sometimes as much as 4 knots in the vicinity of Sacramento shoal and about 5 miles from the land. Between Kakinada and Bhimunipatnam it extends further offshore with a less rate ; inshore of this current there is generally slack water.

TAMIL NADU

Surface temperature and surface salinity in Palk Bay

Palk Bay on the south-east coast of Tamil Nadu is an important area where brackishwater farms are fast developing. The bay is shallow nowhere exceeding 15 m in depth. Murthy and Varma (1964) described some of the hydrographical features of this Bay during March 1963. This can be taken as representative of southern Tamil Nadu coastline in general.

Distribution of surface temperature

The surface waters are subjected to diurnal variations of temperature due to heating by day and cooling by night. The variations, it is mentioned, are much more appreciable in shallow basins due to the less thermal capacity of the basin waters. The observed surface temperatures are corrected from shore-station observations (*i.e.*, at Visakhapatnam, 17° 41′N, 83° 17′E). Surface temperature has been reported to be less in the southern region (27.4°C minimum) as compared to the coastal region of the west (28.8°C) and north-west of the Bay (29.6°C). In the neighbourhood of the Palk strait (north eastern region of the Bay) moderate values were reported (28.2°C).

Distribution of salinity

The salinity value depends not only on the origin of the water mass (Murthy and Varma, 1964) but also on the evaporation from the surface.

The incursion of the Bay of Bengal water into the Palk Bay through the strait splits the coastal water into two separate

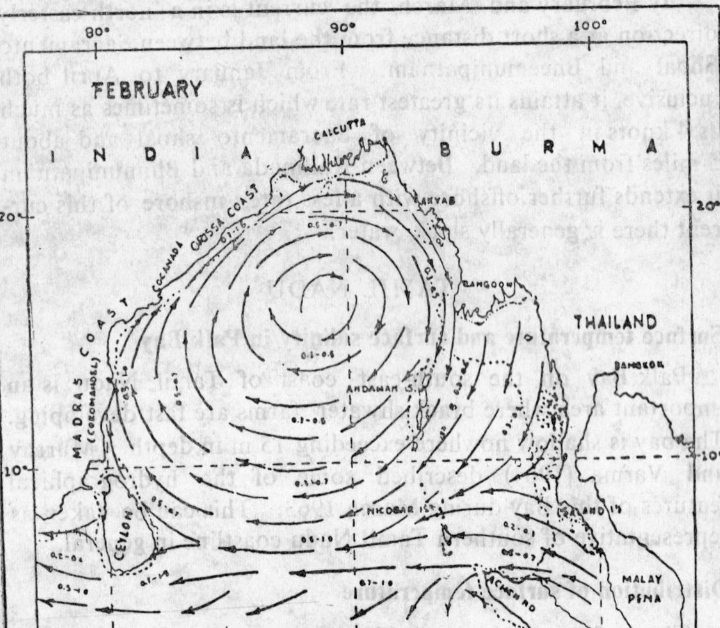

Fig. 2. Currents in the Bay of Bengal dur:ng February.
Source : Ganapati and Marthy, 1954.

parts. Except for a weak through (31% isohaline) at the centre of the Bay, the salinity decreases gradually along an axis in the south-west direction running from the strait. Highly saline water is pocketed in the south-south-western corner of the Bay. This was interpreted as due to Gulf water through the Pamban pass.

The circulation in the south-western region may be weak and the coastal configuration may be hydrodynamically protective constituting the causes for the entrapment of the high saline Gulf water without being mixed with the low saline coastal water in the south. The coastal water in the north is confined to a narrow region near the coast, whereas in the south the coastal water extends even to the middle of the Bay. The west south western region is having the lowest saline water.

Surface density

Density distribution shows coastal water is limited to a narrow region in the north-west and to a wider area in south-west.

Sharp gradients of density are present in the south-western region indicating that mixing of coastal water with gulf water is not prominent here. The influence of the Bay of Bengal water in the Palk Bay is much more extensive and intensive than the Gulf waters. The density decreases along an axis is the southwestern direction from the strait. The waters being more of sea origin at the beginning and more of coastal origin towards its further end. All the parameters are shown in Table-8.

Rainfall

The major rainfall is north-east monsoon confined to the months of November-December or October-November. The effect of south-west monsoon is felt to a little extent. The average of 6 years shows that south-west monsoon yielded only 196.4 to 398.8 mm (1980-85) whereas the north-east yielded 300.2 to 449.0 mm, both being above normal for this coast.

With particular reference to Palk Bay area the studies of Prasad (1954, 1957), Jayaraman (1954) give the distribution of hydrographical factors for the early years of 1950-54 and appear to be relevant for today also.

Tidal patterns

Tide patterns as recorded for the ports along the coastline of Tamil Nadu are as represented at Nagapattinam port is shown as an example :

Nagapattinam Port	*Standard level*	*Date*
MHWL (S)	0.48	August '88
MHWL (N)	0.46	
MLWL (N)	0.02	
MLWL (S)	0.04	
MLWL	0.00	

Table 8. Zonal Distribution of Physico-Chemical Properties of the Palk Bay

Zone	*Surface Temperature °C*		*Temperature °C*		*Salinity ‰*			*Surface Density σ_t*	*Dissolved Oxygen ml/1*			*Percentage saturation of Dissolved Oxygen*		
	Corrected for daily variations	*Uncorrected for daily variations*	*5m.*	*10m.*	*Surface*	*5m.*	*10m.*	*(t)*	*Surface*	*5m.*	*10m.*	*Surface*	*5m.*	*10m.*
North-West	29.1	29.0	28.8	..	31.85	31.86	..	19.71	4.7	4.7	..	106	105	..
North-East	28.6	28.8	28.7	..	32.29	32.24	..	20.19	4.7	4.6	..	106	103	..
South-East	28.4	28.5	28.7	28.4	31.10	31.02	31.23	19.50	4.6	4.8	4.6	102	105	102
South-West	29.0	28.6	28.6	..	30.26	30.24	..	18.70	4.6	4.5	..	100	100	..

Source : Murthy and Udaya Varma, 1964.

The currents and winds can raise this level to more than a metre in the Madras-Killai coast. The maximum distance the tidal impact is felt is upto 12 km with one or two exceptions in the southern part. This can be due to the flatness of the land.

Winds and coastal currents in Palk Bay

The speed and direction of the wind over the Palk Bay is computed from the daily weather reports of the shore stations of Nagapattinam and Pamban obtained from the Indian Meteorological Department.

The North-east winds are more frequent in March. The wind strength at Nagapattinam is about 8-10 knots on an average from the month and at Pamban only 2-4 knots. The winds over the Palk straight are considerable strong and they decrease along the axis in the south-western direction to attain maximum speed at the lower end of the axis (Murthy and Varma, 1964).

Dissoived oxygen

It is point of high interest that the waters everywhere in Palk Bay are almost saturated with dissolved oxygen. They sometimes ever exceed the saturation limit by a small percentage, especially in the northern region (Murthy and Varma, 1964).

GENERAL METEREOLOGICAL CHARACTERISTICS OF BAY OF BENGAL

Wave and wind characteristics

The seasonal wave and wind characteristics of both Arabian sea and Bay of Bengal have been studied by Srivastava *et al* (1968, 1970) and Vijayarajan *et al* (1976). Vijayarajan *et al* (1976) also made an attempt to compare the influences of cyclones and steady monsoon winds on the sea state.

Wave data given in the Indian daily weather Reports from 1960-64 were used.

From the IMD Atlas, the period from May to September, the direction of the wind in general, is south-west. The wind becomes south-westerly in the Bay of Bengal. This period from May to September has been classified as south-west monsoon period. In October, the direction of the wind is irregular. During November, to February the mean direction of the wind is North-east with variation of about 20 degree north or south of it, and hence this period is designated as north-east monsoon. The direction of wind in east coast is north-east. During March and April the wind direction in this area is highly variable and quite irregular. As this period represents the transition from north-east to south-west monsoon, it is defined as pre-monsoon or transition period.

A comparison of the wind and wave directions in this study showed that the direction of wave in all these seasons is almost the same as the corresponding direction of the wind in each season.

South-west monsoon (May-September)

During this season wave direction is mostly south-west. The sea in Bay of Bengal remains calm. The central part is comparatively rough. Except for a small rough region near Calcutta, the average (highest 10%) heights seldom exceed 2.3 m. As the wind, the direction of wave is south-west. Considering the maximum (highest 10%) height the area between Visakhapatnam and Calcutta is rough. The most pronounced period of the wave is 5 or less than 5 seconds.

Post-monsoon (October)

In October the direction of the wind is irregular and the direction of the wave follows the same pattern. The maximum (highest 10%) height is 8.9 m, slightly higher than that in monsoon. The average (highest 10%) sea state is very calm with the height 0.3 m.

North-east monsoon (November-February)

The direction of the wave is north-east in close conformity to prevailing wind. In some places like Visakhapatnam the maximum (highest 10%) height becomes 10 m. The most prominent wave period in this season is 5 or less than 5 seconds.

Pre-monsoon (March-April)

The direction of the wave, as the wind, changes very irregularly. Anti-cyclonic trend in circulation during this season can be observed. The same trend may be observed in the case of wave direction also. Average (highest 10%) heights are much calmer as compared to the two monsoon seasons and the heights reaching maximum of 1.7 m. The maximum heights (highest 10%) are also much lower, the maximum being 8.3 m than during south-west and north-east monsoon.

Effect of cyclones and depressions

From a study of cyclones and depression in different months and corresponding sea state and wind velocities Vijayarajan *et al* (1976) state that the cyclone/depression days are more in November while the maximum wave height is observed in June when the wind velocity is nearing the maximum. From this it was concluded that it is the steady and high monsoon which influences the sea state considerably and not, the cyclones and depressions, that strike infrequently. Wave height of each season shows direct relation to the corresponding velocity as shown below :

Bay of Bengal

	Average wind speed (knots)	*Average wave height (m)*	*Average of highest height (m)*
South-east monsoon	19.6	1.4	3.1
North-east monsoon	14.5	1.1	2.9
Pre-monsoon (March-April)	11.2	0.9	2.5
Post-monsoon (October)	11.0	1.0	2.6

Trends of larval and broodstock distribution

The trend of prawn larval and broodstock distribution off seem to be directly related to the current and drifts in the Bay. Similarly, the broodstock distribution appears to be related to the post-monsoon recovery in temperature and salinity in one major peak and lowering of temperature and salinity in pos-summer months coinciding with beginning of south-west monsoon. Transparency of waters also seem to play a part.

The distribution of *P. mondon* is related to northern part of Bay of Bengal, West Bengal and Orissa to Godavari estuarine mouths in Andhra Pradesh, while the tail end of the population moves down upto northern coast of Madras.

The reverse happens in *P. indicus* especially in the southern tip of Tamil Nadu where the temperatures do not go down below 25°C in November to October period.

Cyclonic storms on east coast 1970-1988

The occurrence of cyclonic storms in the East coast, frequency and place of crossing on the coast for the years 1970-1988 shows that severe cyclonic storms occurred along the entire east coast but the intensity was more along the coast of Andhra Pradesh followed by Orissa, Tamil Nadu and Bengal. The average occurrence was once in an year.

CONCLUSIONS

The Date given above which indicates the patterns of salinity, temperature, rainfall, humidity, evaporation and, tidal characteristics of each region are useful points of basic dassification of zones. From this the zonation can be made as highly suitable, medium and less suitable for purposes of coastal aquaculture.

For example, on the basis of tide characteristics, the very high tides of West Bengal are not optimally suitable for culture. Then the culture areas can be taken for inland to reduce the impact. The very low tidal volumes makes it imperative to locate farms very close to shore or coastal zone in Tamil Nadu and parts of southern Andhra Pradesh.

Secondly, type of classification can help in reducing multiple-use pressure on the coastal zone by dileanating specific areas for specific purposes.

Thirdly, the above study shows only broad descriptions of a lengthy coastal areas. For specifics, detailed studies of these features in a 20-50 km stretch both along the coast as well as extending to the upper tidal limit of rivers might be useful as shown in two studies of the IIT.

The zonation itself can take two forms: One is on the basis of suitability of location of the aquaculture farm (*i.e.*, the characteristics or parameters discussed in the earlier pages as pertaining to the four coastal States); the second is on the basis of multiple uses.

(*i*) The first category shows that from the viewpoint of all characteristics, the coast along the Orissa State, a mid-point between the two littoral drifts in the year, is the most favourable for aquaculture. This is particularly so for the most eminently culturable species, *Penaeus monodon.* For the same species, the coast along the West Bengal State and Andhra Pradesh are medium and northern Tamilnadu, is low. The last one is because of the closure of barmouths, increasing evapotranspiration, very lowtidal volumes and velocities requiring farm location very near tide lines.

—The same characters reverse the order of suitability in the case of the second important species, *P. indicus* and lesser-used species such as *P. semisulcatus*, *P. canaliculatus* and *P. japonicus.*

—Using similar criteria more location-specific zonation can be made.

—Unnecesary expansion of aquaculture in less suitable areas can thus be avoided reducing pressure on the coastal zone.

(*ii*) On the basis of multiple uses, the same hydrographic patterns may be useful for location of ports, location of harbours and so on. These two and aquaculture are to some extent mutually exclusive in the sense that most dumpings, wastages and pollutants emanating from the ports and harbours etc., are harmful to aquaculture industry.

In the interests of protection and rational utilization of coastal zone of India, I suggest that biologists, hydrographers, engineers and others conduct detailed studies with this multiple utilization in view and develop pragmatic utilization principles.

ACKNOWLEDGEMENTS

I take this opportunity to thank Dr. K.S. Rao, Director of the 'National Seminar on Recent Advances in Hydrobiology' for having given me this excellent opportunity to present this lecture.

REFERENCES

Anon, 1985. Development of brackishwater prawn farming in Orissa. Status Report. Contribution No. 2 BFDA, Puri, Orissa.

Bose, A.N., S.C. Roy, 1987. Microlevel survey of brackishwater fish farm complex near Basudevapur in Orissa. Vol. I. Aquaculture Engineering Section, IIT, Kharaghpur (Mimeo).

Butler, M J.A. et al. 1986 Marine resource mapping : An infraluctory Manual. F.A.O. Fisheries Technical paper, 274-257 p.

Ganapati, P.N. and V.S.R. Murthy, 1954. Salinity and temperature variations of the surface waters off the Visakhapatnam coast. Andhra Univ. Ser. No. 42 *Mem in Oceanogr*. Vol. 1 : 125-142.

Ganapati, P.N. and V.S.R. Murthy, 1955. Preliminary observations on the hydrography and inshore plankton in the Bay of Bengal off Visakhapatnam coast *Indian J. Fish*. 2. 84-95.

Ganapati, P.N. and T.S.S. Rao, 1959. Some remarks on the hydrography and biology of the Bay of Bengal. *J. Mar. Biol. Ass.* India 224-227.

Jayaraman, R. 1954. Seasonal variations in salinity, dissolved oxygen and nutrient salts in the inshore waters of the Gulf of Mannar and Palk Bay near Mandapam (S. India). *Indian J. Fish* , 1 : 345-364.

La Fond, E.C., 1954. On upwelling and sinking off the east-coast of India. *Memoirs in Oceanography*. Andhra Univ., Waltair, 1 (49) : 117-121.

La Fond, E.C. 1951. Proceeding oceanographic data M.O. Pub. No. 614. U.S. Navy Hydrographic Office, Washington, D.C., 114 pp.

La Fond, E.C., 1957. Oceanographic studies in the Bay of Bengal. Proc. Ind. Acad. Sci. 46 : 1-46.

La Fond, E.C., 1958. On the circulation of the layers off the east coast of India. Andhra Univ. Mem. in Oceanogr. 2 : 1-11.

Mojumder, P. 1979. Studies on the oceanographic conditions of the surface and bottom waters of the Bay of Bengal off Visakhapatnam dnring 1968-1972. *J. Mar. Biol. Ass., India* 21 (1 & 2) : 119-124.

Murthy, A.V.S. and P. Udaya Varma, 1964. The hydrographical features of the waters of Palk Bay during March, 1963. *J. Mar. biol. Ass.* India VI. (2) : 207-216.

Prasad, R.R., 1957. Seasonal variation in the surface temperature of sea at Mandapam from January 1950 to December 1954. Indian J. Fish., 4 : 20-31.

Prasad, R.R., 1954. The characteristics of marine plankton at an inshore station in the Gulf of Mannar near Mandapam 1 : 1-36.

Report of the Govt. of West Bengal, Director of Fisheries, 1990.

Roy, S.C. 1988. Feasibility of brackishwater fish farm construction along the Balasore coast, Orissa. *In* : Seminar on status and prospects of brackishwater aquaculture in Orissa. 14th October, 1988. Puri. Special Publication, Central Institute of Brackishwater Aquaculture, Madras. pp. 21-38.

Rao, T.S.S., 1958. Studies on the penetration of light in the Bay of Bengal Part I-Transparency of the waters on the east coast of India and its significance. *Proc. Nat. Inst. Sci.* India. 23B (5-6) : 165-190.

Ramasarma, D.V. and P.N. Ganapati. 1968. Hydrography of the Kakinada Bay. *Bull. Nat. Inst. Sci.* India. 38 : 49-79.

Vaidyaraman, P.P., P. Srinivasan, L.K. Ghosh and A.P. Dange. 1980. Studies on the improvement of bar levels at the river mouth by training works for Navigation to Krishnapatnam Port, Andhra Pradesh. 48th Research Session, 11-14 March, Hyderabad, Central Board of Irrigation and Power Proc. GOI II-Hydraulics pp. 147-159.

Venkateswarlu, V. 1982. "Climate". *In* : "Agriculture in Andhra Pradesh. Vol. I. Resources. The Farmer's Welfare Trust and the Society of Scientists for Advancement of Agriculture. Hyderabad, pp. 37-49.

Vijayarajan, P.K., M.J. Varkey and M.X. Joseph, 1979. Seasonal wave characteristics of the Arabian sea and the Bay of Bengal. *J. Mar. Biol. Ass.* India. 18 (1) : 99-109.

21

A TEXT OF A LECTURE TITLED, "MANAGEMENT OF AQUATIC ECOSYSTEMS"

Dr. G.A. Shirgur
Research Officer (Prof.) and Head, Taraporevala
Marine Biological Research Station, Bombay, 400 002

INTRODUCTION

Under the organised aquaculture practices, using fresh-water, brackishwater or seawater, the ideal fish/prawn farms comprise a series of excavated ponds and a laboratory space for breeding activities. The hatchlings either of prawns of fishes are to be properly produced using appropriate hatchery systems. Then, when the hatchlings start feeding, the live food organisms need to be stocked in the well provided nursery ponds. Once the fry/prawn juveniles are produced, these are to be further reared into larger size *e.g.* the fingerlings in case of fishes, using the rearing ponds. In the third stage, larger fish/prawn seed are to be stocked in larger sized fish culture ponds including the stock ponds.

For management of above farms, under extensive or semi-intensive systems, it is necessary to provide live feed organisms, as well as the organic substances used as direct feeds. In this context, the technique of generating rapid zooplankton had not been standardised until 1971. Unlike the natural food chain in the aquatic ecosystem and technique of phased fertilisation in this context has effectively accelerated the production process of zooplankton in required quantity and quality in relation to the measured input of the organic substances used as phased fertilisers. In the natural ecosystems, the zooplankton in very negligible quantities is being generated due to the primary productivity under very slow process. In contrast to this natural system, the phased fertilisation technique involves measured application of organic substances in liquid form of water media. Due to this, a different short and rapid food chain comprising the liquid organic mannure leads to formation of extensive planktonic bacterial biomass which in turn is consumed by the third link in the food chain comprising the filter-feeding zooplankters. Thus, it is possible to generate the zooplankton which is proteinaceous diet at all the levels of aquaculture, regulated in proportion to the level of organic inputs and the expected fish/prawn output. Therefore, phased fertilisation technique has become vary essential for production of fish/prawn seed and for their further growth by providing complementary diets comprising the organic substances by directly feeding in the baskets or by provision of pellets.

Phased fertilisation therefore, has opened a new dimension for organised aquaculture activities (Shirgur 1989 c & d).

It is thus possible to systematically plan the fish farm management activities in advance. For example, in respect of major carp culture forms, it is possible to accurately assess the number of free swimming larvae (spawn) available, by operating the running water hatcheries. In the follow up action, on the basis of the stocking density of one crore fish larvae/ha of nursery space, the nursery ponds can be kept ready. In these ponds, by adjusting the pH to alkaline side around 8.0—8.2, a schedule of phased fertilisation and direct feeding for 12 days, cycle can be implemented to produce the fish fry. In the further stage of management, the fish fry can be grown into fingerlings by applying a 28 days' schedule, involving phased fertilisation and direct feed methods.

To illustrate above type of systematic management, four Tables are displayed, describing spawn to fry and fry to fingerling production in respect of silver carp. These experiments for the first time were carried out at Walwhan Fish Farm of Tata Electric Company, Lonavala, Maharashtra. Table-I is self-explanatory giving all the details of methodological norms adopted for the experiments. Under Table-II and III, nursery pond management and rearing pond management schedules of

Table I. Various methodological norms adopted for conducting the experiments on spawn to fry and fry to fingerling rearing in respect of Silver carp at Walwhan Fish Farm, Lonavala, Maharashtra.

Abbreviations :
DOCG — De-oiled cake of groundnut
RB — Rice bran
RCD (Wet) — Raw cattle dung ;
PF — Phased fertilisation
DF — Direct feed provision.

Sr. No.	*Particulars*	*Pond-1* (*Spawn to fry rearing*)		*Pond-2* (*Fry to fingerling rearing*)	
		Rate/ha	*Pond-1*	*Rate/ha*	*Pond-2*
(*1*)	(*2*)	(*3*)	(*4*)	(*5*)	(*6*)
1.	Date of fish seed stocking	—	87.90	—	19-7-90
2.	Initial weight range (mg)	—	4.0—10.0	—	12.0—45.0
3.	Average weight (mg)	—	8.0	—	23.22
4.	Initial length range (mm)	—	5.0—9.0	—	11.0—17.0
5.	Average length (mm)	—	7.0	—	13.28
6.	Culture schedules (days) (Ref. Shirgur 1989c)	13.0 (Modified)	13.0	28.0 As per carp schedule)	28.0

(*Contd.*)

7. Surface area of pond (m^2)	$100 \times 100 \times 1.0$	$214 \times 17.5 \times 0.8$	$100 \times 100 \times 1$	$21.0 \times 17.3 \times 0.8$
8. Volume of water in the pond (m^3)	10,000	300	10,000	290
9. No. of seed stocked	20,00,000 (Spawn)	60,000 (Spawn)	10,00,000 (Fry)	29,000 (Fry)
10. Quantity of single super phosphate (SSP) used (kg)	22.0	0.6	60.0	1.75
11. Frequency of application of SSP	One application on only one day before spawn stocking		Four applications *i.e.* on 1st, 8th, 15th and 22nd days.	
12. Mode of application of SSP	By phased fertilisation		By phased fertilisation	
13. Total quantity of organic substances used under respective schedules (See Table II & III)				
(*a*) DOCG (kg)	985.80	29.57	2010.10	58.29
(*b*) RB (kg)	540.80	16.22	1310.40	38.00
(*c*) RCD (Wet) (kg)	1111.50	33.35	296.10	8.88

(*Contd.*)

Table 1. (*Contd.*)

(*1*)	(*2*)	(*3*)	(*4*)	(*5*)	(*6*)
14.	Scheme of application of the organic substances and their percentage proportions under PF & DF				
	(*i*) *PF* (%) (*a*) DOCG	38.5	38.5	50	50
	(*b*) RB	Nil	Nil	Nil	Nil
	(*c*) RCD	100.0	100.0	100.0	100.0
	(*ii*) *DF* (%) (*a*) DOCG	61.5	61-5	50	50
	(*b*) RB	100.0	100.0	100.0	100.0
	(*c*) RCD	Nil	Nil	Nil	Nil
15.	Total quantitative distribution of organic substances as per above scheme (kg) (as per Col. No. 15)				
	(*i*) *PF* (*a*) DOCG	379.5	11.38	1005.05	29.15
	(*b*) RB	166.66	5.00	Nil	Nil
	(*c*) RCD (wet)	1111.5	33.345	296.10	8.88

(*Contd.*)

(*ii*) *DF*	(*a*) DOCG	606.3	18.18	1005.05	29.15
	(*b*) RB	374.14	11.22	1310.40	88.0
	(*c*) RCD (Wet)	Nil	Nil	Nil	Nil
16. Scheme of sub-dosages of the organic substances/day under PF and DF (kg)					
(*i*) *PF*		RCD applied on 1st day only. DOCG was applied on 1st, 3rd, 5th and 7th days (See Table-II)		DOCG and RCD (wet) were used on every day for 28 days in decreasing order (See Table-III)	
(*ii*) *DF*		DOCG was applied on 9th to 12th days. RB was applied on 3rd, 5th, 7th and 9th to 12th days (see Table-II)		DOCG and RB were used in direct proportion to collective body weight (CBW) of fish fry in the increasing order daily for 28 days (Table-II)	
17. Assumed average weight and percentage recovery of the fingerlings at the end of 28 days' culture as per the schedule for carps (Ref. Shirgur 1989 c).					

(Contd.)

Table 1. (*Contd.*)

(1)	(2)	(3)	(4)	(5)	(6)
	(*a*) Average weight (g)	Not applicable		6.227	6.227
	(*b*) Percentage survival expected	—	—	85.0	85.0
18.	Total collective body weight (CBW) of the fingerlings (Col. 17 above) harvestible at the end of 28 days (See table-III)	Not applicable		5292.95	153.5
19.	Cummulative collective body (CCBW) of fish by summation of daily linear progressive weight wise growth values (Table-III)	Not applicable		76746.00	2224.92
20.	Percentage values of the organic substances in relation to CBW/day of the				

fingerlings by summation of linear progressive weight-wise growth values for 28 days (Ref. Col. No. 19, Table-III)					
(*i*) *PF values* :	(*a*) DOCG	Not applicable		1.31	1.31
	(*b*) RB	—	—	Nil	Nil
	(*c*) RCD (Wet)	—	—	0.395	0.395
(*ii*) *DF values* :	(*a*) DOCG	—	—	1.31	1.31
	(*b*) RB	—	—	1.707	1.707
	(*c*) RCD (wet)	—	—	Nil	Nil

N.B. : Daily dose of organic substances in Pond-2 under PF were applied in the morning and under DF in the afternoon (Table III).

Table II. 13 days' schedule of fertiliser application for spawn to fry rearing in respect of silver carp in pond-1 at Walwhan Fish Seed Farm, Lonavala, (Kceping over-all quantities constant a few modifications of fertiliser component done as against the standard schedule (Shirgur 1989 C)

Sr. No.	Day-wise schedule for 13 days	Mode of application	Quantities of Fertilisers used											
			DOCG			RB			RCD (Wet)			SSP		
			ppm soluble part	Quantity per hectare	in Kg pond wise	ppm soluble part	Quantity per hectare	in Kg pond wise	ppm soluble part	Quantity per hectare	in Kg pond wise	ppm gross part	Quantity per hectare	in Kg pond wise
1.	1 day before spawn stocking	PF	8.76	333.33	10.0	4.01	166.67	5.0	39.00	1066.67	33.345	2.00	20.0	0 6
		DF			—		—	—		—	—	—	—	—
2.	1st day		6000 Nos. of spawn of silver carp stocked in the pond. (@20,000,00/ha).											
3.	2nd day		Phased fertilisation and direct feeding not done in the pond.											
4.	3rd day	PF	1.74	65.00	1.95	0.9			—	—	—	—	—	—
		DF		—	—		37.5	1.125						
5.	4th day		Phased fertilisation and direct feeding not done in the pond.											
6.	5th day	PF	1.74	65.00	1.95	0.9	—	—	—	—	—	—	—	—
		DF		—	—		37.5	1.125						
7.	6th day		Phased fertilisation and direct feeding not done in the pond.											
8.	7th day	PF	3.42	130.0	3.90	1.80	—	—	—	—	—	—	—	—
		DF	—	—	—		75.0	2.25						
9	8th day		Phased fertilisation and direct feeding not done in the pond.											

(Contd.)

10.	9th day	PF	3.42	—	—	1.80	—	—	—	—	—	—	—	—
		DF		130.0	3.90		75.0	2.25						
11.	10th day	PF	1.74	—	—	0.9	—	—	—	—	—	—	—	—
		DP		65.00	1.95		37.5	1.25						
12.	11th day	PF	3.42	—	—	1.80	—	—	—	—	—	—	—	—
		DF		130.6	3.90		75.0	2.25						
13.	12th day	PF	1.74	—	—	0.9	—	—	—	—	—	—	—	—
		DF		65.00	1.95		37.5	1.125						
14.	13th day		Harvesting day.											
	Total		26.0	528.33	29.50	13.00	279.17	16.25	39.00	1066.67	33.345	2.00	20.00	0.6

Abbreviations : DOCG-De-oiled cake of groundnut ; RB—Rice bran ; RCD (wet)—Raw cattle dung (1 kg wet= 0.2 kg dry) S.S.P.—Single super phosphate.

N.B. :—1) Estimated solubility.

DOCG —25.0%
RB —24.8% and
RCD (Wet) —35.0%

(2) Phased fertilisation method :—Under this pattern all the measured quantities of fertilisers are mixed together with the water and the liquid is spread in the pond water [broadcasting into the pond].

(3) DF—Direct feeding method :—Under this pattern, measured quantity of fertilisers are mixed together and directly introduced in the pond in the form of dry powder.

(4) PF+DF—Phased fertilisation+Direct feeding ; Under this pattern DF and PF methods are to be followed as per schedule.

Table III. A Chart showing distribution of organic substances for each day during total of 28 days of culture of silver carp fry into fingerlings in Pond-2 at Walwhan Fish Seed Farm, Lonavala, near Bombay (Please See Table—I)

General features—(1) Volume of the pond —290.0 m³ (2) PF (Phased fertilisation—In decreasing order.
(3) DF (Direct feeds —In increasing order
(4) SSP (Single Super Phosphate)—4 times each on 1st, 8th, 15th and 22nd days by phased fertilisation method.

Date of application	*Phased fertilisation applied*									*Organic fertilisers applied as direct feeds*					
	DOCG			*RCD (Wet)*			*SSP*			*DOCG*			*RB*		
	% Appli-cation	*Application (kg) Per hectare*	*Pond wise*	*% Appli-cation*	*Application (kg) Per hectare*	*Pond wise*	*% Appli-cation*	*Application (kg) Per hectare*	*pond wise*	*% Applica-tion*	*Application (kg) Per hectare*	*Pond wise*	*% Applica-tion*	*Application (kg) Per hectare*	*Pond wise*
1	*2*	*3*	*4*	*5*	*6*	*7*	*8*	*9*	*10*	*11*	*12*	*13*	*14*	*15*	*16*
20.7.90	6.87	70.69	2.050	50.00	148.06	4.294	22.86	13.79	0.400	0.86	8.62	0.250	0.66	8.62	0.250
21.7.60	6.01	60.34	1.750	—	—	—	—	—	—	0.86	8.62	0.250	0.66	8.62	0.250
22.7.90	6.01	60.34	1.750	—	—	—	—	—	—	0.86	8.62	0.250	0.66	8.62	0.250
23.7.90	6.01	60.34	1.750	—	—	—	—	—	—	0.86	8.62	0.250	1.05	13.79	0.400
24.7.90	6.01	60.34	1.750	—	—	—	—	—	—	0.86	8.62	0.250	1.05	13.79	0.400
25.7.90	6.01	60.34	1.750	—	—	—	—	—	—	1.72	17.24	0.500	1.31	17.24	0.500
26.7.90	5.15	51.72	1.500	—	—	—	—	—	—	1.72	17.24	0.500	1.31	17.24	0.500
27.7.90	5.15	51.72	1.500	—	—	—	25.71	15.52	0.450	1.72	17.24	0.500	1.31	17.24	0.500
28.7.90	5.15	51.72	1.500	—	—	—	—	—	—	1.72	17.24	0.500	1.31	17.24	0.500
29.7.90	5.15	51.72	1.500	—	—	—	—	—	—	2.42	24.14	0.700	1.97	25.86	0.750

30.7.90	4 29	43.10	1.250	—	—	—	—	—	—	2.42	24.14	0.700	1.97	25.86	—
31.7.90	4.29	43.10	1.250	—	—	—	—	—	—	2.42	24.14	0.700	3.16	41.38	—
1.8.90	4.29	43.10	1.250	—	—	—	—	—	—	3.44	34.48	1.000	3.29	43.10	1.250
2.8.90	4.29	43.10	1.250	—	—	—	—	—	—	3.44	34 38	1.000	3.29	43.10	1.250
3.8.90	3.44	34.48	1.000	50.00	148.06	4.294	25.71	15.52	0.450	4.29	43.10	1.250	3.95	51.72	1.500
4.8.90	3.44	34.48	1,000	—	—	—	—	—	—	4.29	43.10	1.250	3.95	51.72	1.500
5.8.90	2.42	24.14	0,700	—	—	—	—	—	—	4.29	43.10	1.250	4.60	60.34	1.750
6.8.90	2.42	24.14	0.700	—	—	—	—	—	—	4.29	43.10	1.250	4.60	60.34	1.750
7.8.90	2.42	24.14	0.700	—	—	—	—	—	—	5.15	51.72	1.500	5.26	68 96	2.000
8.8.90	1.72	17.24	0.500	—	—	—	—	—	—	5.15	51.72	1.500	5.26	68.96	2.000
9.8 90	1.72	17.24	0.500	—	—	—	—	—	—	5.15	51.72	1 500	5.26	68.96	2.000
10.8.90	1.72	17.24	0.500	—	—	—	25.71	15.52	0.450	5.15	51.72	1.500	5 26	68.96	2.000
11.8.90	1.72	17.24	0.500	—	—	—	—	—	—	6.01	60 34	1.750	5.26	68.96	2.000
12.8.90	0.86	8.62	0.250	—	—	—	—	—	—	6.01	60.34	1.750	5.92	77.59	2.000
13.8.90	0.86	8.62	0.250	—	—	—	—	—	—	6.01	60.34	1.750	6 58	86.21	2.000
14.8.90	0.86	8.62	0.250	—	—	—	—	—	—	6.01	60.34	1.750	6.58	86.21	2.500
15.8.90	0.86	8.62	0.250	—	—	—	—	—	—	6.01	60.34	1.750	7.24	94.83	2.750
16.8.90	0.86	8.62	0.250	—	—	—	—	—	—	6.87	70.69	2 050	7.24	94.83	2.750
Total	100.00	1005.09	29.150	100.00	295.12	8.586	99.99	60 35	1.750	100.00	1005.09	29.150	99.98	10.39	38.000

application of organic substances as phased fertilisers and direct feeds has been described. In Table-IV the actual results are shown (Shirgur *et. al.* 1990c).

In this manner, all the aquaculture activities can be properly planned in advance and the aquatic ecosystems of freshwater, brackishwater and seawater can be suitably managed for optimum productivity.

Table IV. Results of harvesting of silver carp fry and fingerlings at the end of respective 13 days and 28 days' rearing schedule from the ponds at Walwhan Fish Seed Farm, Lonavala, near Bombay (M.S.).

Sr. No.	*Particulars*	*Pond No. 1*	*Pond 2*
1.	Pond volume (m^3)	300.0	290.0
2.	Total numbers harvested and date of harvesting	29,500.0 (on 19-7-90)	16,515 (on 17-8-90)
3.	Survival percentage	49.16	56.9
4.	Weight range of harvested silver carp seed	12.0–45.0 (mg)	0.275—2.0 (g)
5.	Average weight	23.22 (mg)	0.85 (g)
6.	Length range of harvested fingerlings	11.0–17.0 (mm)	2.9–5.5 (cm)
7.	Average length	13.28 (mm)	4.4 (cm)
8.	Hydrobiological parameters		
	(*a*) Days of observations	On 8-7-90	On 19-7-90, 5-8-90 and 16-8-90
	(*b*) Water temperature range (°C)	24.0	22.5–23.5
	(*c*) pH	8.0	7.8–8.0
	(*d*) Oxygen (mg/lit)	—	8.4—8.8
	(*e*) Zooplankton biomass (g/m^3)	—	4.9–6.5

REFERENCES

SHIRGUR, G.A. (1984a). Technique of phased fertilisation—A new Dimension in Aquaculture. Workshop on food from the sea to feed a billion by 2000 A.D. 28-29 January 1984. Article No. VI : 1-19, organised by Institute for Coastal and Offshore Research, Vishakhapatnam—530003 (A.P.).

SHIRGUR, G.A. 1971. Observation on rapid production of zooplankton in fish nurseries by intensive phased manuring *J. Indi. Fish. Assoc. I* (2) : 25-50.

Shirgur, G.A., (1989c). Presenting the formulated rearing schedule for optimum production of carp fry and fingerlings using same organic substances as direct feeds and phased fertilisers. *Natioual Seminar on forty years of Freshwater Aquaculture in India.* 7-9. Nov., 1989. CIFA, Kausalyaganga, Bhubneshwar, Orissa.

Shirgur, G.A., (1989d). On evolving a model of time bound feed management schedule for optimum pond culture of carps, *National Seminar on forty years of Freshwater Aquaculture in India* 7-9 November 1989. CIFA, Kausalyaganga, Bhubaneshwar, Orissa.

Shirgur, G.A., Ogale, S.N. and Indulkar, S.T. (1990c). On rearing the fry and fingerlings of Silver carp (*Hypopthalmicthys mollitrix*) at Walwhan Fish Farm, Lonavala (Maharashtra) using phased fertilisation and direct feed techniques. "The National Seminar on 'Recent Advances in Hydrobiology", Indore (M.P.), 1-3 October., 1990. organised by Vikram University, Ujjain (M.P.).

22

CONSERVATION IN RELATION TO NUTRIENT ACCUMULATION IN A HIMALAYAN ECOSYSTEM

A. Wanganeo
and
R. Wanganeo
Department of Limnology,
Barkatullah University,
Bhopal (M.P.)

Abstract—The chemical characteristics of an aquatic ecosystem play a major role in defining its trophic status. Its accumulation and availability has a profound impact on the metabolic activity that is going on in the system. As such, their regulation can help in maintaining the balanced metabolism. Accumulation beyond the limits warrents their removal from that fraction of the system which, otherwise, would result in undesirable consequences.

INTRODUCTION

While conserving an aquatic ecosystem it is important to work out each component before fruitful strategy can be adopted. In the present investigation an attempt has been made to evaluate various chemical parameters in three segments of the anchar aquatic ecosystem. The lake is situated at an altitude of 1584 m.a.s.l. within lower Himalayan range.

The lake is heavily infested with macrophytes on account of its shallow depth. Soil type and nutrient composition of water seem to have limited value in controlling macrophytic distribution. Water depth and turbidity are important factors in the colonization of submerged macrophytes. Maximum area of Anchar lake is under submerged flora, though emergents contribute more than 50% of the total macrophytic biomass in the Anchar lake (Kaul *et. al.* 1978).

MATERIAL AND METHODS

Chemical characteristics of water were analysed following the methods as given in APHA (1976), trace element analysis of macrophytes was carried out on Atomic absorption spectrophotometer.

RESULTS AND DISCUSSION

Morphometric and physical features of the lake are given in Table 1.

Table 1. Selected Morphometric and Physical Features of Anchar lake.

Altitude	m	1584
Area	ha	700
Max. depth	m	2.75
Volume	$m^3 \times 10^6$	3.90
Secchi	m	0.25–1.75
Max. Surface water temperature	°C	28
Flushing rate	ty^{-1}	5

Nutrients in Water

The usual cation content at different sites is Ca>Mg>Na>K. Concentration of divalent cations is very high while bicarbonate is the most abundant form among inorganic anionic components. The waters are 'calcium rich'. As per alkalinity values the water of Anchar lake is of 'hard water type'. The concentration of free carbon dioxide is generally low, the dissolved inorganic carbon necessary for photosynthesis being obtained from bicarbonates which presumably serve as an essential source of carbon for autotrophs, particularly submerged plants.

Despite large supplies of nutrients from the catchment areas low values of dissolved inorganic phosphate phosphorus (DIPP) seem to prevail. The depletion of DIPP from these waters is being associated with photosynthetically induced precipitation, besides, its utilization by the autotrophs. The high pH values, associated with high temperature, and high alkalinity, are favourable conditions for precipitation of $CaCO_3$ due to rapid carbon assimilation from dissolved bicarbonates. Under such conditions phosphate ion is reported to co-precipitate with carbonate (Otsuki and Wetzel, 1972). The co-precipitation of phosphate ion with $CaCO_3$ being the main mechanism of phosphate-phosphorus removal from the waters is borne out by the fact that relatively higher amounts of unreactive phosphate phosphorus occur at the time of maximum $CaCO_3$ precipitation. This unreactive phospate-phosphorus probably represents the co-precipitated portion of phosphorus. NO_2—N and NH_3—N are present in low quantities in comparison to NO_3—N (Table 2).

Nutrients in sediments

The sediment analysis reveals silt and clay to be the major constituents of sediment system. The relative magnitude of the coarse sand is comparatively lesser (0.06—1.7%). On the basis of granulometric proportion of the sediments of Anchar lake can be classified within silt-loam type (Table 3).

Sediment pH remained almost constant (6.0-7.2) throughout the study. Sediments, in general, are rich in nutrients and are essentially organic with appreciable quantities of calcarious deposits. On an average basis organic carbon content of sediments was 10.55%.

The seasonal values of total P, N and K show a decrease towards summer. Their concentration, however, starts increasing in late summer and early fall when macrophytic growth enters the stationary and declination phases. A consistent decrease is evidently due to the active growth of macrophytes. This decreasing type of tendency was not observed in regard to calcium and magnesium. Sodium concentration remained

Table 2. Average chemical parameters of Anchar Lake water.

Parameters	*Units*	*Values*
Cond.	μ moh cm^{-1}	500
pH		8.3
DO	mgl^{-1}	8.2
Tot. hardness ($CaCO_3$)	mgl^{-1}	111
Ca	mgl^{-1}	35
Mg	mgl^{-1}	9.5
Na	mgl^{-1}	6.2
K	mgl^{-1}	1.34
Cl	mgl^{-1}	30.00
DIPP	mgl^{-1}	0.03
TOPP	mgl^{-1}	0.248
TPP	mgl^{-1}	0.290
NO_2—N	mgl^{-1}	ND
NH_4—N	mgl^{-1}	0.11
NO_3—N	mgl^{-1}	0.30
Fe	mgl^{-1}	0.15
SiO_2	mgl^{-1}	4.00
Cu	mgl^{-1}	0.28
Zn	mgl^{-1}	0.005
Md	mgl^{-1}	0.009
CO	mgl^{-1}	0.10
Pb	mgl^{-1}	ND
Hg	mgl^{-1}	ND
Cd	mgl^{-1}	ND

Table 3. Granulometric composition of bottom sediments of Anchar lake.

(Kaul *et. al.* 1978)

	Percent codtribution
Coarse sand	2.50
Clay	27.50
Silt	62.50
Fine sand	7.50

almost constant throughout the study period with no appreciable losses or gains. Table 4, summarises the average elemental composition of Anchar lake sediment.

Table 4. Average elemental composition of Anchar lake sediment.

Ca	mg/g	7
Mg	,,	8
P	,,	0.5
Fe	,,	3.13
Na	,,	5
K	,,	39
Mn	μg/g	651
Zn	,,	168
CO	,,	6
Cu	,,	9
Ni	,,	11

Fe and Mn in the inflow and the sediment samples has been given in Table 5, while Table 6 gives the particle size dependence of iron and manganese in the inflow channel. It is observed from Table 6 that the distribution of iron and manganese was mainly in the grain size spectrum of the lake

Table 5. Iron and Manganese Concentration of sediments in Anchar lake.

(Kango *et. al.* 1987)

	Fe	*mg/g*	*Mn*	
	Range	*Mean*	*Range*	*Mean*
Inflow	3.18—3.47	3.32	0.65—0.84	0.71
Sediments	2.79—3.69	3.32	0.48—0.96	0.71

Table 6. Particle size dependence of iron and manganese in inflow of Anchar lake.

(Kango *et. al.* 1987)

	Fe	*mg/g*	*Mn*
Sand	2.11		0.46
Silt	3.33		0.68
Clay	4.07		0.72

having smaller sized groups with a higher metal content (Kango *et. al.* 1987).

On computing the accumulation of iron and manganese per gram of sediment samples in different fractions it was observed that more than 70% of the metals were concentrated in the silt portion of the sediments. Humic material of the lake showed iron in the range of 0.45 to 1.89 g/l and manganese in the range of 0.12 to 0.62 g/l. Iron and manganese are the two important mobile metals (Kempt *et. al.* 1976) in an aquatic environment. These metals play an important role in the binding of nutrients and toxic elements by virtue of their high specific area in oxide form. Kango *et. al.* (1987) reported that significant quantities of iron and manganese were present as reducible phases and more than 60% of these were leached out. Less than 25% of iron was present in an exchangeable form. Fe showed a higher degree of reactivity than manganese as it forms an integral part of the clay mineral (Davison and Dickson 1984).

Macrophytes

In the Anchar lake, stands of Typha-phragmites occur all along the shoreline of the lake forming large aggregations over floating islands.

The most distinguishing feature of this lake is the existance of extensively large stands of *Potamogeton natans*, occupying 10.67 ha, and encroaching fast into the deeper regions, leaving only a central area of the lake uncovered. The other prominent feature of the lake is the occurrence of small stands of *Trapa natans* at a number of places, occupying a total area of 1.02 ha. *Nymphoides peltata* is widely distributed and occupies an area of 9.62 ha.

Submersed types are dominated by *Ceratophyllum demursum*, *Potamogeton lucens* and *Myriophyllum spicatum*, covering a total area of 16.52 ha against 0.57-1.77 ha under the colonization of other submergeds like *Potamogeton pectinatus*, *P. crispus*, *P. pusillus*, *Hydrilla verticillata* and *Najas graminea*.

Primary production on account of macrophytes

The primary productivity values reveal emergents to be highly productive followed by rooted floating leaf types. The submerged macrophytes despite convering major portions of the lake area contribute only 14% of the total macrophytic biomass (Table 7). *Hydrilla verticillata* among submergeds and *Nelumbo nucifera* and *Trapa natans* among rooted floatings contained maximum percentage of crude protein.

Table 7. Primary production and its crude protein content in Anchar lake, during maximum growth period.

	*gm² day**	*Percent crude protein*	*Percent* biomass*
Emergents			
Typha angustata	10.13	14	
Phragmites communis	5.44	10	70.00
Sparganium erectum	4.25	18	
Myriophyllum verticillatum	3.21	8	
Rooted floating leaf types			
Potamogeton natans	1.83	16	
Nelumbium nucifera	5.48	20	16.00
Trapa natans	2.96	20	
Nymphoides peltata	0.82	18	
Hydrocharis morsurs ranae	1.47	12	
Submersed			
Ceratophyllum demursum	0.94	16	
Potamogeton lucens	1.09	10	
P. pectinatus	1.16	9	
P. crispus	1.09	11	14.00
P. pucillus	1.18	10	
Hydrilla verticillata	0.74	22	
Najas graminea	0.94	15	
Myriophyllum spicatum	0.86	15	

*after Kaul *et. al.*, 1978.

Nutrients in macrophytes

Table 8 sums up the contribution of various nutrients in

Table 8. Chemical characteristics of macrophytes in Anchar lake

	Mg	*Na*	*K*	*Cu*	*Mn* *mg/g dry wt.*	*Fe*	*Zn*	*Ni*	*CO*
Emergents									
Typha angustifolia	2.177	5.203	13.27	0.005	0.146	1.327	0.029	0.0065	0.0026
Phragmites communis	1.240	1.292	8.99	0.064	0.201	0.105	0.024	0.0041	0.003
Sparganium erectum	1.286	5.716	24.00	0.006	0.375	0.214	0.031	0.0068	0.005
Rooted floating									
Potamogeton natans	3.684	7.148	11.05	0.013	0.092	1.842	0.032	0.021	0.010
Nymphoides peltata	1.222	2.531	12.22	0.012	0.218	0.218	0.086	0.0083	0.003
Trapa natans	0.864	1.178	12.57	0.0068	0.079	0.126	0.052	0.0059	0.003
Submerged									
Ceratophyllum demursum	1.230	4.472	11.17	0.0064	0.138	0.089	0.074	0.0053	0.0035
Potamogeton crispus	1.178	1.766	11.18	0.0034	0.041	0.165	0.047	0.006	0.003
Hydrilla verticillata	3.595	6.03	13.34	0.0066	0.263	1.392	0.046	0.022	0.008
Myriophyllum spicatum	1.618	5.933	16.18	0.0108	0.216	0.054	0.055	0.0072	0.0037
Potamogeton lucens	3.802	8.449	14.29	0.0053	0.132	0.634	0.058	0.012	0.0063
Potamogeton pucillus	1.392	3.595	13.92	0.0099	0.174	0.093	0.038	0.0078	0.003

different plant species belonging to different types *viz.*, emergents, rooted floating and submergeds.

Among the various nutrient constituents in Anchar lake the maximum amount is due to nitrogen, followed by K, Ca, Na, Mg and P (Table 8 and 9).

Potassium is the most abundant element present within the tissue of macrophytes Table 8. Li, Cd and Pb were below detection level and as such have not been included.

Table 9. Mineral nutrient accumulation of various macrophytes in Anchar lake.

(Kaul and Trisal, 1985)

	N	*P*	*K*	*Na*	*Ca*	*Mg*
Emergents	7.88	0.42	3.27	0.98	5.48	2.53
Rooted floating	1.49	0.04	0.86	0.19	1.48	0.65
Submergeds	1.33	0.04	1.02	0.11	1.46	0.64
Total nutrients m tons	10.70	0.50	5.15	1.28	8.42	3.64

Table 10 compares the nutrients in three segments of a water body *viz.* water sediments and macrophytes. On the whole it is seen that sediments are rich in nutrients compared to water and macrophytes.

In order to maintain the balanced trophic status of the water body it is suggested that their should be periodical removal of nutrients by way of dredging the sediments near the mud water interface and using them as fertilizers. This will rapidly help in reversing the nutrient enriched condition of an aquatic ecosystem. Further their entry from both the point and diffused sources should be checked in order to increase the life span of the water body.

Removal of macrophytes by both manual and mechanical methods can result in further compounding (complicating) the conservation strategy programme as their removal/cutting leads to unwanted algal blooms, because of the fact that a

Table 10. Average values of various chemical parameters in three segements of the lake.

Parameters	*Water* *mg/l*	*Sediments* *mg/g*	*Macrophytes* *mg/g*
Ca	38	7	
Mg	8	8	1.94
Na	5	5.00	4.50
K	0.72	39	13.45
Cl	29		
PO_4-P	0.03		
P		0.5	
NO_3-N	0.32		
NH_4-N	0.21		
SiO_2	3.33		
Fe	0.15	3.13	0.522
Cu	0.28	0.009	0.012
Zn	0.005	0.168	0.048
Mn	0.009	0.651	0.173
CO	0.10	0.006	0.0045
Pb	—	—	ND
HG	—	ND	ND
Cd	—		ND
Ni	—	0.011	0.0094
Li	—	—	ND

prominent competitor with algac regarding nutrients is being removed. Under such circumstances where, the system has attained a stage where conventional methods fail in conserving the system, the sediment removal method can only held in restoring back the ecosystem to its normal condition.

ACKNOWLEDGEMENTS

The senior author would like to thank BARC, Trombay, Bombay for extending help in the macrophytic analysis.

REFERENCES

Davison, W. and Dickson, D.P.E. (1984) ; Mossbauer spectroscopic and chemical studies of particulate iron material from a seasonally anoxic lake. *Chem. Geol. 42* : 177-187.

Kango, R.A. ; Dubey, K.P. and Zutshi, D.P. (1987). Environmental geochemistry of Iron and Manganese in six Himalayan lakes. *Acta Hydrochim. Hydrobiol. 15* (1) : 57-64.

Kemp, A.L.W., Thomas, R.J., Dell, C.I. and Jaquet, U.M. (1976). Cultural impact on the Geochemistry of sediments in lake Erie. *J. Fish Res. Bd. Can. 33* : 440-462.

Otsuki, A. and R.G. Wetzel, 1972. Coprecipitation of phosphate with carbonates in a marl lake. *Limnol. Oceanogr. 17* : 763-767.

Kaul, V., Trisal, C.L. and Handoo, J.K. (1978). Distribution and production of macrophytes in some water bodies of Kashmir. *In Glimpses of Ecology* (J.S. Singh and B. Gopal, Eds). Jaipur, India : International Scientific Publications.

Kaul, V. and Trisal, C.L. (1985). Ecology and conservation of the fresh water lakes of Kashmir. *Proc. Nat. Sympos. Evalu. Environ.* (spl. vol. Geobios). 164-170.

23

PAST, PRESENT AND FUTURE OF UDAIPUR LAKES AS A SOURCE OF DRINKING WATER

V.S. Durve
Department of Limnology and Fisheries, Rajasthan Agricultural University, Udaipur

INTRODUCTION

Udaipur, the city of lakes and sometimes called the Kashmir of Rajasthan had its origin at the hands of Maharana Udaisingh in 1559 A.D. situated in a picturesque valley of Aravalli hills – geographically called the Ahar basin, the terrain of Udaipur is studded with land depressions, high hills and seasonal and semi-perennial streams.

Founders of Udaipur city had far sighted vision and a tremendous planning efficiency. From a rugged terrain, they created one of the most beautiful cities of India adored with a necklace of a chain of lakes. Besides being impregnable in relation to the military science as existing in 16th century, the city was self-sufficient in food and water.

Out of the six lakes situated within the city of Udaipur, one was in existance even prior to the establishment of the city. The six lakes are Pichhola, Rangasagar, Swaroopsagar, Fatehsagar, Badi or Jiyansagar and Udaisagar. The lake Pichhola is the oldest one having been built in 13th century by a Banjara chieftain. The lake was named Pichhola after the village Piccholia existing then. This village is now a small locality even now called Pichholia in the city of Udaipur. The early settlers of Udaipur had their first glass of drinking water from the lake Pichhola. The portion of the lake Pichhola now called Rangasagar was improved in its embankments and shores a little later by Shri Amar Singh Badava, the Prime Minister of Maharana Hari Singh in sixteenth century. Thus in the early days of Udaipur city the lakes Pichhola and Rangasagar remained the principal source of drinking, domestic use and recreational waters of Udaipur. This situation continues even to-day. The lake Swaroopsagar was added a little later.

For the convenience of the public, the rulers of Udaipur constructed most tasteful bathing ghats—some of which are of marble-and temples on the boundaries of these lakes. A couple of bridges were constructed by early rulers to cross the lakes. Two additional bridges have come up only recently.

The lake Fatehsagar was built in 1678 A.D. as an irrigation reservoir and continued to be so till recently. There was no habitation around this lake and even to-day except for a few 3 and 5 star hotels, the lake is relatively free of human interference. The lake is a tourist attraction and is thus the most beautiful lake of Udaipur.

The lake Jiyansagar of Badi is situated a few kilometers away from the city. This lake was tapped for city water supply only two years back.

The lake Udaisagar is about 15 km away from Udaipur city and was built by the end of sixteenth century. Till two decades back, this was an irrigation reservoir but is now being used exclusively by Hindusthan zinc Ltd., a public sector undertaking mining and smelting zinc and related metals.

The primary source of water is the rainwater runoff from the catchment areas of the lakes and the seasonal rivers Sisarma and Ahar. The six lakes are so built that the excess water of each lake automatically flows into the other thereby ensuring no loss of water at any stage. The river Sisarma pours its water in the lake Pichhola. Water from Pichhola flows to Rangasagar and then to Swaroopsagar and Fatehsagar.

The lake Fatehsagar also receives water from the lake 'Madar' situated about 15 km up in Udaipur basin or Ahar basin. Overflow from the lake Fatehsagar falls in the river Ahar and reaches Udaisagar through Ahad river. The overflow of Udaisagar reaches the lake Vallabhnagar about 30 km away through Berach river.

The lake Jiyansagar or Badi also opens in the lake Fatehsagar though there is no much water flow from Badi to Fatehsagar. There is a small water area called Dudhtalai adjuscent to the lake Pichhola.

As stated earlier, from drinking water point of view, the lakes Pichhola and Rangasagar are the most important ones followed by the lake Fatehsagar. As such, in the present talk only these three lakes are dealt with.

About 50 years back, Udaipur lakes used to be full for the whole year except during the years of draught. Even then, the population of the city being low, the water was abundant. Though the domestic pollution existed, the water was reported to be relatively clean. This was probably because the self-purification processes of the lake balanced the incoming domestic and sewage wastes. There were no algal blooms, rooted and floating vegetation. With the availability of adequate natural food, alkaline waters, moderately high temperatures, the lakes supported good fish fauna. The religious status of the lakes declared by the rulers, forbiddened any fishing and this helped considerably in keeping the lakes clean.

The physical and chemical features of the lake water show a higher transparency of the lake Fatehsagar and the lowest of the lake Rangasagar with Pichhola showing an intermediate value of transparency. Other pollution indicative parameters are phosphates, nitrates, chlorides, electrical conductivity and BOD. The values of total and faecal coliform are significantly higher in Rangasagar and Pichhola.

High quantum of Cyanobacteria especially the bloom forming marker species *Anacystis aeruginosa* and *Anacystis flosaquae* are again indicative of biological pollution and eutrophication. Cyanobacteria are dominant in summers and decline in monsoon showing minimum density in winter. While there is a wider assemblage of different planktonic algae, *Anacystis*, spp, *Fragilaria, Closteridium, Synedra* and *Oscillatoria* are perennial or semi-perennial in their occurrence with high counts, so is the case with *Euglena* and *Chlorella*. The pollution tolerant species have established themselves permanently in the lake Rangasagar and to a greater extent even in the lake Pichhola.

Table I. Water chemistry of the drinking water lakes of Udaipur

		Pichhola	*Rangasagar*	*Fatehsagar*
pH		8.5	7.8	8.13
Depth of visibility	cm	62	48	73.27
Dissolved Oxygen	ppm	5.8	4.3	6.79
Free CO_2	ppm	3.2	7.5	1.97
Carbonates	ppm	5.30	1.66	3.08
Bicarbonates	ppm	140	168	136.92
Calcium	ppm	17.2	22.5	18.55
Magnesium	ppm	7.2	9.2	7.72
Silicates	ppm	12.6	7.2	5.11
Orthophosphates	ppm	0.095	0.38	0.091
Nitrates-N	ppm	0.008	0.742	0.08
Chlorides	ppm	17.5	83.73	18.13
E.C.	μmhos/cm	404.0	516.0	426.56

Total zooplankton count of 680.00 Ind/ml indicates the unsuitability of water for drinking purposes. Out of this, the copepodes form the dominant plankton with a contribution of 84.58% of the total zooplankton. Rotifers are second in dominance. The species *Branchionus falcatus* and *B. forficula* and *B. calyciflorus*, and species of *Keratella* and *Filinia* formed the bulk of Rotifer population. These are tolerant of organic pollution. The protozoans common in the water body are *Diflugia, Phacus* and *Actinosphaerium*. The sensitive form like *Arcella* are low being between 10-20 per ml of water.

Table II. Water chemistry of the drinking water lakes of Udaipur

Parameter	*Pichhola*	*Rangasagar*	*Fatehsagar*
pH	8.5	7.8	8.13
Depth of visibility cm.	62	48	73.27
Dissolved Oxygen ppm	5.8	4.3	6.79
Free CO_2 ppm	3.2	7.5	1.97
Carbonates ppm	5.30	1.66	3.08
Bicarbonates ppm	140	168	136.92
Calcium ppm	17.2	22.5	18.55
Magnesium ppm	7.2	9.2	7.72
Silicates ppm	12.6	7.2	5.11
Ortho Phosphates ppm	0.095	0.38	0.091
Nitrates—N ppm	0.008	0.742	0.08
Chlorides ppm	17.5	83.73	18.13
E.C. μmhos/cm	404.0	516.0	426.56
B.O.D.$_5$ ppm	9.70	15.40	3.47
MPN/100 ml total coliform	540	2400	221
MPN/100 Faecal coliform	63	170	7

The bacterial counts of total and faecal coliform are substantially higher. All these confirm the unsuitability of water for drinking purposes.

The lakes especially Rangasagar, Pichhola and Swaroopsagar occasionally get a heavy cover of floating *Eichhornia.* This is manually removed by the voluntary organisations in Udaipur. The mat of rooted hydrophytes on the lake bottom has been estimated to be 4170 g/m² at the densest colonisation. They heavy occurrence of rooted and floating vegetation itself is a sign of excessive nutrient loading in the lakes, thereby making the water improper for consumption. The drinking water source should preferably be from oligotrophic water bodies.

The silt deposition over the past several hundred years as around 8 meters and a part of this is removed occasionally by the voluntary organisations through 'Shramdan' during summers when the water level falls. However, bottom areas permanently inundated receive the silt loading almost incessantly in varying quantities. This silt loading has progressively increased during the last around 50 years owing to the excessive deforestation of the catchment area during this period. This has reduced the water-holding capacity of the lake basins.

The lake Pichhola is a principal source of drinking water for the city of Udaipur. Out of 7.5 million gallon of tap water supplied daily to the city of Udaipur, 4.5 million comes from the lake Pichhola. This is besides the direct use of both Pichhola and Rangasagar.

The part of the city where water from Pichhola is supplied, has higher incidence of water-borne diseases. 52% of the population of these areas has chronic diarrohoea and 70 to 80% children have running stomachs. Outbreak of jaundice and incidence of amoebic dysentry is a common feature in the locations where water from Pichhola is supplied.

Table III : Limnobiological features of drinking water lakes of Udaipur

	Pichhola	*Rangasagar*	*Fatehsagar*
Total phytoplankton cell/ml	5780	13869	3715
Blue-greens	1520	5204	965
Greens	1230	5197	1230
Diatoms	1140	3368	1520
Total Zooplankton No./ml	385	680	391
Rotifers	95	61	149
Copepods	160	152	126
Cladocerans	80	334	69
Misc.	50	133	47
Primary productivity (GPP) $gC/m^2/day$	3.53	8.56	1.03

Table IV : Income based distribution of households around the lakes Pichhola and Rangasagar (1982)

Income limit	*Percentage*
Below Rs. 400	41.60
Rs. 400—600	31.84
Rs. 600—800	13.12
Rs. 800—1000	08.10
above 1000	05.30

Table V : Bacteriological features of the drinking water lakes of Udaipur

	Pichhola	*Rangasagar*	*Fatehsagar*
Total Coliform MPN/100 ml	540	2400	221
Faecal Coliform MPN/100 ml	63	170	7

Pollution indicative phytoplankters in the lake Rangasagar

	Cells ml^{-1}
Fragilaria brevistriata	2252.2
Synedra ulna	1223.0
Anacystis aeruginosa	5301.8 (Bloom)
Anacystis flosaquae	5434.6 (Bloom)
Oscillatoria sp.	1036.4

Pollution indicative zooplankters in the lake Rangasagar

	Ind ml^{-1}
Brachionus falcatus	141.2
Brachionus forficula	400.8
Brachionus calyciflorus	311.8
Keratella tropica	94.0
Filinia longiseta	162.4
Arcella discoides	20.0 Highly sensitive
Arcella mitrata	10.6 Highly sensitive

The population of the city of Udaipur is around 3.5 lakhs. Out of this, approximately 14% inhabit the marginal areas of the lake Pichhola and Rangasagar. The region of the city being the oldest and comparatively unhygienic, is favoured by economically weaker section of the society. The majority of the inhabitants of the area are around poverty line. These people have unhygienic habits of using the lake-shores for relieving themselves. Everyday, average 675 persons use the lake shores for this purpose. The washing of clothes and bathing is a common feature.

Thus 62.56% population of the area uses the lake directly, rest 37 43% uses tap-water which though treated is drawn from the same lake. The water-closets in the residential areas around the lake are 33.79% flush-type, 43.83% of traditional baskets type and 22.34% open fields. In the houses, where the basket-type water closets are provided, the watery wastes are drained to the lakes. The people in the area have the most unhygienic habits of making their children relieve nature on the open drains which empty into the lake. This is the major cause of all alimentary disorders prevailing in the area. Besides this, the other domestic washings incessantly flow into the lake from the surrounding habitation.

Thus from the population of 49,000 about 4.9 MT of organic matter of O_2 and at least 98 kg of Phosphate flows daily into the waters of the lake Pichhola and Rangasagar (Golterman—Physiological Limnology). The yearly figures would be 1764 MT of organic matter and 34.5 MT of Phosphate. The amount of liquid wastes getting into the lake every year is 3-6 million cubic meters. At present, around 30.5 million tonnes of BOD_5 is being discharged every year in the lakes Pichhola and Rangasagar.

Probable Corrective Measures

It is unlikely that the inhabitants of the area will ever convert their water-closets to flush type and connect them to the underground sewer. There has to be some incentive in terms of subsidy or financial help for the conversion of basket type closets to flush-type ones.

Mass education and bringing awareness amongst the people for the adoption of cleaner habits should be resorted to on priority basis. The area is the most conjested one in Udaipur.

The narrow roads with open drains as per the system in olden days has further aggrevated the problem. It may be possible to shift at least a part of this population to other newly developed areas or elsewhere by launching special housing schemes for them.

The provision of bathing ghats was understandable in the early days. Their importance is now over in view of the modern hygienic requirements and the easy availability of tap water. The use of the bathing ghats should now be prevented and these may be preserved as historical or cultural monuments.

Chain of hotels is being established around Udaipur lakes and proper check on their discharges has to be maintained.

Afforestation of the surrounding hills should be done on war footing so as to check the silt loading of the reservoirs.

The most appropriate step would be to develop city-water supply from a source far away from the city of Udaipur and keep the city lakes full with water as tourist attraction. This would develop tourism thereby bringing considerable money simultaneously keeping the city free of the epidemics of gastric disorders and other water borne diseases.

24

"STATUS OF WATER QUALITY DURING PRE AND POST TREATMENT STAGES IN A PROTECTED WATER WORKS"

Praveen Tamot, Pradeep Shrivastava*
and
Sadhna Tamot**
Post-Graduate Department of Zoology,
Govt. Science & Commerce College (Benazeer),
Bhopal (M.P.), 462001, India

Abstract—Limnological studies of the Upper lake as a raw water source for 5 M.G.D. water treatment plant (PHED) Bhopal at intake point and water quality of treated water which is supplied to the city of Bhopal, were carried out from January to December 1988. Reduction percentage in physico-chemical and biological parameters during the course of water treatment were found out with regard to temperature, turbidity, pH, free carbon di-oxide, phenolphthalein alkalinity, total alkalinity, chloride, dissolved oxygen, total hardness, calcium hardness, magnesium hardness, nitrate, total phosphorus, B.O.D., phytoplankton, and zooplankton. Different biological indices were also applied so as to assess the quality of raw and treated water. Planktonic organisms causing number of problems in relation to water treatment practices were also found out.

INTRODUCTION

Bhopal, the picturesque capital of the State of Madhya Pradesh, is also known as the "City of lakes", on account of

*Department of Limnology, Barkatullah University, Bhopal.
**Department of Zoology, S.V. College, Bairagarh, Bhopal.

a large number of water bodies present in and around Bhopal. The upper lake, the most magnificent of all these lakes and reservoirs has been the lone source of drinking water supply for the entire population of the capital city. The water is drawn from various parts of the lake and treated in fifteen water works units controlled by different authorities. The lake is badly managed as still large number of houses and huts of slums are located on the northern bank of the lake. Number of activities such as religious (idol and tazia immersion), agriculture, colonisation etc. are going on in the catchment area. As a result of which variety of materials including residual fertilizers, pesticides, waste water, sewage, slit and various possible disease/parasites organisms are finding their way into the lake. These factors are responsible for eutrophication, deterioration of water quality, degraded environmental status of the water body and enhancement in the cost of water treatment at the same time.

The available knowledge on the Upper lake, related to water works limnology still excludes a large number of parameters which are important from the point of view of drinking water quality. In the present study, an attempt has been made to assess the raw water quality of Upper lake and quality of treated water which is supplied for drinking purposes to the city. Percentage reduction in different parameters during the course of treatment were also found out. The information generated on the basis of this study is useful to assess the quality of 5, M.G.D. water treatment plant (PHED) Bhopal and line of treatment to be recommended in order to get wholesome, economical and aesthetically clear potable water supplied at a low cost of treatment.

MATERIAL AND METHODS

All the methods used for the analysis of physico-chemical and biological parameters were followed as given in "Standard Methods for the Examination of Water and Waste Water (A.P.H.A. & A.W.W.A., 1985). Samples were collected once in a month from raw water of Upper lake at intake point of 5,

M.G.D. water treatment plant (PHED), Kamla Park Pumping Station and treated water (filtered water).

RESULTS AND DISCUSSION

Studies on the limnological features of the raw and treated water were carried out from January to December 1988. Variations in the following limnological parameters were studied and found to be within the permissible limits as prescribed by W.H.O. for drinking water quality. Average values of physico-chemical and biological parameters are shown in Table 1.

Table 1. Average values of physico-chemical and biological parameters during pre and post water treatment (All values in mg/litre except pH or unless specified)

S. No.	*Parameter*	*Raw water*	*Treated water*
1.	Water temperature °C	24.2	24.1
2.	Turbidity	13.5	4.9
3.	pH	8.3	7.5
4.	Free carbon di-oxide	2.5	2.2
5.	'P' alkalinity	3.6	3.0
6.	Total Alkalinity	83.4	76.6
7.	Chloride	19.9	22.9
8.	Dissolved oxygen	9.5	8.4
9.	Total hardness	76.1	68.2
10.	Calcium hardness	43.4	37.7
11.	Magnesium hardness	35.4	30.9
12.	Nitrate	0.07	0.05
13.	Total phosphorus	0.37	0.36
14.	B.O.D.	4.4	2.0
15.	Phytoplankton units/lit	7411	61
16.	Zooplankton org./lit.	2700	17

Physico-chemical parameters

Temperature is one of the most important factors in an aquatic environment. Changes in air temperature naturally affect the water temperature. It has universal influence and is a limiting factor for the growth and distribution of flora and fauna in an aquatic body (Birge & Juday, 1982 ; Hutchinson, 1967). In the present study seasonal variation in water temperature was noted. No significant variation in water temperature was recorded in raw and treated water. Turbidity of the water was recorded higher in the monsoon period, due to heavy winds the entire water gets churned up and bottom material comes into suspension and suspended particles, chemicals, fertilizers, domestic waste, animal dung and other materials join the lake water along with surface run off from the catchment area. In raw water average value of turbidity was recorded as 13.5 mg/lit. During the course of treatment 63.7% turbidity of raw water was reduced in treated water. Even after treatment considerable amount of turbidity still remained in treated water. The average value of turbidity in treated water was recorded as 4.5 mg/lit. As stated by Jayangoudar (1967) turbidity has great importance in the occurrence of viral diseases *viz.* jaundice and polio etc.

The water of Upper lake is slightly alkaline and pH is regulated by the carbon di-oxide, bicarbonate and carbonate system. Average of pH in raw water was noted as 8.3, over 9.6% decline in pH values from raw water to treated water was owing to the removal of algal matter during the course of treatment. According to Spence (1967), the pH of a eutrophic lake ranges from 7.7 to 9.6. The present findings are in support of the observations made by Spence (1967) that Upper lake is eutrophic on the basis of its pH range. Oxygen and carbon-di-oxide are reciprocal directly influencing the ecology of water stock. It is assumed that these are interrelated in maintaining their balance in any living sphere. In the present observation free carbon di-oxide shows an inverse relationship with dissolved oxygen. Only 12% reduction was noted in free carbon di-oxide content during the course of treatment. Carbonate or phenolphthalein alkalinity was recorded as 3.6 mg/lit and 16.6% reduction was noted in treated water. Juday *et. al.* found that the surface water were alkaline to phenolphthalein in all the "drainage" lakes investigated by them. Upper lake is also a "drainage" type reservoir and water is always alkaline to phenolphthalein. Total alkalinity in raw water was always higher mainly due to bicarbonates. In raw water its average value was noted as 83.4 mg/lit and nearly 8.0% reduction was noted during the course of treatment.

Chloride status in natural water is indicative of its degree of pollution especially of animal origin, sewage input, and urination in the water bodies (Tamot, 1988, 1989). In the present study, average value of chloride in raw water was noted as 19.9 mg/lit, slightly higher values were noted in treated water due to the addition of chlorine in water during the process of pre and post chlorination. Dissolved oxygen has been attributed great significance as an indicator of water quality, especially the magnitude of eutrophication. Dissolved oxygen content in raw water was noted as 9.5 mg/lit and depletion was noted in treated water. This has to be attributed to very little biological activities in filter media when the water passes through filter units and treated water is not exposed directly to atmosphere. Higher values of dissolved oxygen content in water increase the corrosion in raw water and distribution pipe system. The total hardness of raw water was recorded as 76.1 mg/lit. On the basis of total hardness value and Taylor (1949), we can consider the Upper lake water under the moderately soft category. During the course of treatment 10.38% of total hardness is reduced. In Upper lake water, main contribution of the hardness is from calcium which exceeds magnesium hardness. Its average value in raw water was recorded as 43.4 mg/lit whereas in treated water it was 37.7 mg/lit. Reduction in calcium hardness value was noted to be 13.13% from raw water to treated water. Magnesium is an essential mineral for the body and is relatively non-toxic to man in concentration normally encountered in nature (Aboo, 1967). In the present study, the average value of magnesium hardness in raw water of Upper lake was noted as 35.4 mg/lit and 12.71%, of average reduction was recorded during the course of treatment. The major plant nutrients, nitrates and phosphates cause eutrophication of the inland waters. In raw water of Upper lake average value of nitrate was recorded as 0.07 mg/lit. It has been found that the nitrate concentration declines by 28.57% in the treated water, which may be attributed to the elimination of planktonic biomass and organic matter by flocculation and filtration. However, the processes of treatment is not meant for removing nitrates from the water. Presence of nitrate in water indicates the final stage of the mineralisation of nitrogenous organic matter. (Nema *et. al.*, 1984). Knowledge of the amount of phosphorus in fresh water from a particular environment would be of great value in evaluating the nutrient level and naturally occurring blooms of problems causing algae and bacteria. Phosphorus has origin in surface run off and domestic sewage entering the lake. Phosphorus is an important plant nutrient and is readily taken up by phytoplankton. Increase in phosphate contents is an indication of eutrophication (Wetzel, 1975). In raw water its average value was noted as 0.37 mg/lit but no significant variation was observed during the course of treatment. Bio-chemical oxygen demand (B.O.D.) in streams or lakes is due to the bacterial activity on organic matter and a direct method for measuring of organic pollution. The higher values of B.O.D. produce obnoxious condition in the environment (Bewtra and Radha Charan, 1968). In the present study average value of B.O.D. in raw water was noted as 4.4 mg/lit and in treated water 54.54% reduction was observed. The reduction in B.O.D. load in treated water may be due to the removal of organic matter by the process of flocculation, sedimentation and filtration.

Planktonological study

The biota in an aquatic system positively reflects the conditions existing in the environment and information can be utilized for biological monitoring of water pollution levels. Planktonic study is a very useful tool for the assessment of water quality in any type of water body and also contributes to an understanding of the basic nature and general economy of the lake.

Study of plankton in water works limnology is very important because they produce aromatic or fishy odours in water and also cause clogging of filter units and screen nets. The biological management of a water treatment plant is as important as the engineering or physico-chemical water quality aspect.

Investigations on phytoplankton population in fresh water is one of the most important steps in the complex analysis of aquatic ecosystem, because phytoplankton being the primary producers forms the basic food source of the lake which supports the aquatic animals. During the study period, 33 genera of phytoplankton belonging to three major group were identified. They are as follows :

Cyanophyceae
Anabaena
Arthrospira
Anacystis
Agmenellum
Nostoc
Oscillatoria
Spirulina
Chlorophyceae
Chlorella
Chlorococcum
Cosmarium
Closterium
Cladophora
Mougeotia
Oedogonium
Pediastrum
Scenedesmus
Spirogyra
Staurastrum
Tetraedron
Trochicia
Ulothrix
Volvox

Bacillariophyceae

Amphora

Cymbella

Fragillaria

Frustulia

Gomphonema

Melosira

Navicula

Nitzschia

Pinnularia

Rhopalodia

Synedra

During the period under study, average number of phytoplankton in raw water was recorded as 7411 units/lit and 99% of phytoplankton were eliminated in treated water during the course of treatment, about 1% of phytoplankton passed through filter beds and still recorded in treated water. During the study period, some problems creating algae such as : *Anacystis, Anabaena, Arthrospira, Chlorella, Closterium, Cymbella, Fragillaria, Melosira, Mougeotia, Navicula, Nitzschia, Oscillatoria, Spirogyra, Spirulina, Synedra, Ulothrix* were noted in the water sample collected. They create severe problems in water treatment procedures like taste and odour, filter and screen clogging, colouration in water, corrosion of concrete structures, persistence in distribution system, interfering with coagulation. Their presence is strictly objectionable as they burden the water works authorities with extra expenditure for their removal. Some important forms of algae were also identified which characteristics of organically enriched area *i.e.*, *Anabaena, Arthrospira, Anacystis, Oscillatoria, Spirulina, Chlorella, Closterium, Pediastrum, Scenedesmus, Melosira, Navicula, Nitzschia, Synedra*, (Palmer, 1980) indicating eutrophication in the lake. Even after treatment, some forms of algae were identified in filtered water *e.g. Anacystis, Cosmarium, Pediastrum, Nitzschia* which penetrate through rapid sand filters and are responsible for turbidity in treated water.

In fresh water system zooplanktonic organisms form an important group, as most of them feed upon and incorporate the primary producers into their bodies and make themselves available to higher organisms in food chain (Michael, 1973). Being heterotrophic in nature they play a key role in cycling of organic matter in aquatic ecosystem and also play a major role in the energy transfer at secondary level. They contribute significantly to biological productivity of fresh water ecosystem. Thus, they can be used as "bio-indicators".

In the present study twenty two genera of zooplankton were identified, represented by three major group *viz.*, Protozoa, Rotifera, and Crustacea.

Protozoa
Amoeba
Ceratium
Euglena
Paramecium
Rotifera
Asplanchna
Brachionus
Filinia
Keratella
Notholca
Trichocerca
Crustacea
Alona
Bosmina
Ceriodaphnia
Cypris
Cyclops
Daphnia
Diaphanosoma
Eubranchipus
Moina

Mesocyclops
Nauplius
Sida

During the study period average number of zooplankton was noted as 2700 organisms/lit and 99.37% was eliminated in treated water during the course of treatment. Presence of *Brachionus, Ceriodaphnia, Daphnia, Diaptomus, Filinia, Keratella, Moina* are indicative of eutrophication of the lake. Some genera of Crustacea and Protozoa group they create problems in water treatment procedure such as *Bosmina, Ceriodaphnia, Daphnia, Moina, Mesocyclops, Nauplius, Amoeba, Euglena* may even penetrate through filter media and were observed in treated water. Presence of *Cyclops* and *Amoeba* in treated water is dangerous from the health point of view for being the carriers of larvae of *Dracunculus medinensis* (guinea worm) and the latter causing amoebiasis, dysentry etc.

On the basis of the present study it can be assumed that the system is under process of eutrophication and limnological parameters of treated water which is supplied for drinking point of view were found to be within the limits prescribed by W.H.O. for drinking water quality.

REFERENCES

Aboo, K.M. and Manuel, A.C. (1967). Preliminary observations on the Upper lake at Bhopal. *J. Environ Hlth.*, 9 : 22-23.

American Public Health Association (1985). *Standard Methods for the Examination of water and Waste Water.* APHA-AWWA-WPCF, Washington.

Bewtra, J.K. and Radha Charan (1968). Further observations on progression of Bio-chemical oxygen demand. *Ind. J. Environ. Hlth.*, 10 : 51-60.

Birge, E.A. and Juday, C. (1922). The inland lakes of Wisconsin. The plankton I Its quality and chemical composition *Wisc. Geol. Nat. Hist. Surv. Bull*, 64 *Sci. Ser.* No. 13 : 1-222.

Hutchinson, G.E. (1967). *A treaties on limnology.* II. *Introduction to lake biology and the limnoplankton.* John Wiley & Sons Inc., N.Y. London, Sydney.

International Standards for Drinking Water (1963). World Health Organisation, Geneva.

Jayangoudar, J. (1967). *Hydrobiological studies on the Ajwa reservoir and Nimeta water works at Baroda*. Ph. D. Thesis, M.S. Univ., Baroda.

Nema, P., Rajgopalan, S. and Mehta, C.G. (1984). Quality and treatment of Sabarmati river, Ahmedabad. *J.I.W.W.A.*, 16(1) : 99-107.

Palmer, C.M. (1980). *Algae and water pollution*. Castle House Pub. Ltd.

Spence, D.H.N. (1967). Factors controlling the distribution of fresh water macrophytes with particular reference to the Lochs of Scotland. *J. Ecol.* 55 : 147-170.

Tamot, P. and Bhatnagar, G.P. (1988). Limnological studies of Upper lake, Bhopal. *Proc. Nat. Symp.* P.P. & F. of Bhopal lakes, 37-40.

Tamot, P. and Bhatnagar, G.P. (1989). Studies of raw water quality of Upper lake and its change during various stages of treatment at five M.C.D. water treatment plant (P.H.E.D.), Bhopal (M.P.). *J. Hydrobiol.*, 5(1) : 35-38.

Tayler, E.W. (1949). *The examination of water and water supplies*. Blakiston Son & Co., London.

Wetzel, R.G. (1975). *Limnology*. W.B. Saunders Co., Philadelphia.